Maruyama · Kajiya · Hoffman · Spaan (Eds.)

Recent Advances in
Coronary Circulation

With 100 Figures

Springer-Verlag
Tokyo Berlin Heidelberg
New York London Paris
Hong Kong Barcelona
Budapest

Yukio Maruyama M.D., Ph.D.
Professor and Chairman, First Department of Internal Medicine, Fukushima Medical
College, Hikarigaoka 1, Fukushima, 960-12 Japan

Fumihiko Kajiya M.D., Ph.D.
Professor, Department of Medical Engineering and Systems Cardiology, Kawasaki
Medical School, 577 Matsushima, Kurashiki, 701-01 Japan

Julien I.E. Hoffman M.D.
Professor of Pediatrics, Senior Staff Member, Cardiovascular Research Institute
University of California, San Francisco, CA 94143-0130, USA

Jos A.E. Spaan Ph.D.
Professor of Department of Medical Physics, Faculty of Medicine, University of
Amsterdam, 1105 AZ Amsterdam, The Netherlands

ISBN-13:978-4-431-68251-6

Printed on acid-free paper.

Library of Congress Cataloging-in-Publication Data
Recent advances in coronary circulation / Maruyama . . . [et al.], (eds.). p.   cm. Includes biblio-
graphical references and index. ISBN-13:978-4-431-68251-6   e-ISBN-13:978-4-431-68249-3
DOI: 10.1007/978-4-431-68249-3   1. Coronary circulation—Congresses.   2. Coronary
circulation—Measurement—Congresses.   3.   Coronary   heart   disease—Pathophysiology—Congresses.   I. Maruyama, Yukio, 1941–   .   II. Title: Coronary circulation. [DNLM: 1. Coronary
Circulation—congresses.   2. Coronary Disease—congresses.   WG 300 R295 1993] QP108.R43
1993 612.1'7—dc20 DNLM/DLC for Library of Congress 93-33408

© Springer-Verlag Tokyo 1993
Softcover reprint of the hardcover 1st edition 1993

Preface

Coronary circulation research is advancing at a rapid rate. Not only are innovative techniques for studying myocardial perfusion being developed, new mechanisms related to coronary blood flow control and mechanics are also being discovered. The progress in this field justifies an update in the form of this new monograph.

The book is divided into the following sections:

"Measurement of Coronary Blood Flow and Assessment of Myocardial Perfusion" discusses advances in perfusion measurements in humans as well as nonradioactive microsphere methods.

"Coronary Flow Dynamics" elucidates the effect of heart contraction on coronary flow, perfusion, and reserve distribution as well as systolic-diastolic interaction.

Models—a frequent topic of debate—are used to quantify hypotheses. "Models of Coronary Circulation" attempts to elucidate the concept of tissue pressure.

"Regulatory Mechanisms of Coronary circulation and its Clinical Relevance": Numerous mechanisms affecting coronary flow have been defined and studied at the level of isolated vessels and whole organs. The chapters in this section provide an in-depth analysis of a selection of these mechanisms and their interactions. "Pathophysiology of Coronary Circulation in Ischemic Heart Disease" considers important aspects of factors which restrict perfusion of the myocardium in ischemic heart disease. An understanding these factors is of crucial importance in the management of patients.

"Small Vessel Disorder in Coronary Circulation" describes circulatory flow and how it can be influenced by drugs.

In all, the chapters cover a wide range of research. The scope of this book has been augmented even further by the addition of several abstracts.

Thanks to our distinguished colleagues from Japan, basic scientists and clinicians worldwide were able to meet in satellite symposia related to the World Congress on Medical Physics and Biomedical Engineering (1991) as well as the Cardiovascular Systems Dynamics Society (1992). Both symposia were organized in Fukushima under the direction of Prof. Yukio Maruyama. The papers presented in this book are to a large extent the result of presentations and discussions that took place during these conferences. The initiator of the book was Prof. Yukio Maruyama. The outline of the book was further discussed with

V

the other editors, Prof. Julien Hoffman, Prof. Fumihiko Kajiya, and myself.

This monograph can be seen as a successor to an earlier publication "Coronary Circulation: Basic Mechanism and Clinical Relevance", also published by Springer-Verlag. Once again, Springer-Verlag has provided proficient advice and language editing. As a result of all these efforts, the end product is a high-quality book which will be helpful in further developing the research field of coronary circulation.

Jos A.E. Spaan
Yukio Maruyama
Fumihiko Kajiya
Julien I.E. Hoffman

Contents

A. Measurement of Coronary Blood Flow and Assessment of Myocardial Perfusion

Abstract

B. Coronary Flow Dynamics

Abstract

C.   Models of Coronary Circulation

D.   Regulatory Mechanism of Coronary Circulation and Its Clinical Relevance

E. Pathophysiology of Coronary Circulation in Ischemic Heart Disease

Abstracts

F.   Small Vessel Disorder in Coronary Circulation

Abstracts

List of Contributors

Miura, M. 267
Miyazaki, Y. 261
Moncada, S. 171
Mori, H. 17
Morinobu, S. 201
Morioka, T. 235
Morozuki, T. 315
Muehrcke, D.D. 123
Mullani, N.A. 27
Nakagomi, A. 267
Nakamura, Y. 200
Nakayama, H. 262
Natsume, T. 318
Ogasawara, K. 69
Ogasawara, Y. 283
Ohmori, M. 264
Ohtsuka, S. 265
Onodera, K. 307
Owada, T. 260
Palmer, R.M.J. 171
Poitevin, P. 83
Revenaugh, J.R. 2
Saito, T. 114, 261, 267
Saitoh, S. 114

Sakauchi, Y. 261, 320
Sammuel, J.L. 83
Sasayama, S. 251
Sato, F. 319
Sato, H. 206
Sato, M. 114, 320
Satoh, S. 124
Sekiguchi, H. 318
Sekiguchi, N. 313
Shiraishi, A. 317
Sideman, S. 135, 263
Sipkema, P. 54
Spaan, J.A.E. 60, 182, 203
Stork, M.M. 123
Sugishita, Y. 265
Sunagawa, K. 91
Suzuki, A. 261
Takahashi, K. 267
Takashima, S. 206, 235
Takijiri, C. 315
Takishima, T. 124, 202, 313, 319
Tani, A. 315

Thornton, J.D. 223
Todaka, K. 91
Toshima, H. 262, 317
Tsujioka, K. 69, 283
Tujimura, E. 264
Ueno, T. 262, 317
Uno, K. 200
Van Winkle, D.M. 223, 266
Van Bavel, E. 182, 203
Watabiki, H. 318
Watanabe, H. 265
Watanabe, J. 124
Watanabe, N. 320
Westerhof, N. 54
Yaginuma, T. 318
Yamada, T. 264
Yamaga, A. 317
Yamagami, H. 315
Yoh, M. 317
Yokoyama, M. 189, 204
Yokoyama, M.M. 262
Yoshiyama, H. 317
Zinemanas, D. 135

A. Measurement of Coronary Blood Flow and Assessment of Myocardial Perfusion

Evaluation of Coronary Perfusion by Positron Emission Tomography

James B. Bassingthwaighte[1], James R. Revenaugh[1], Andreas Deussen[2], Michael M. Graham[1], Thomas K. Lewellen[1], Jeanne M. Link[1], and Kenneth A. Krohn[1]

Summary. The methods used in positron emission tomography (PET) studies for the estimation of myocardial blood flow are based upon the principle of mass conservation; what goes in, must come out. Whether or not there are barriers, any nonmetabolized tracer has a mean transit time through each region that is exactly its volume of distribution divided by the flow. When retention is long, the tracer is almost an analog of a mechanically deposited microsphere. When the tracer is flow-limited in its exchange, the measurement of its mean transit time is most secure when the volume of distribution and mean transit time are large. Complicating factors, such as slow permeation across capillary and cell barriers and the metabolism of the tracer, require more detailed models for analysis. The movements of the heart and the fact that the spatial resolution of PET images is almost of the same magnitude as the thickness of the wall of the ventricle, complicate the form of the signal. Nevertheless, with the improved temporal and spatial resolution of newer tomographs, there are good prospects for the evolution of protocols that will give local metabolism and flow with usable accuracy.

Introduction

The interest in myocardial perfusion imaging stems from more than just the question of whether or not there is flow to a region, or whether or not obstruction to flow can be relieved, but extends to more basic levels. Does local demand for nutrients exceed the capacity for flow to deliver them? Has the state of the tissue changed, in response perhaps to the ambient conditions, so that its demands have diminished to an unnatural extent and that, while viable or "hibernating", it is not working well? Or does the tissue, even while being perfused, have no capacity for work, or for recovery of contractile cells to a useful functional state?

The desired measure of blood flow is F $(ml\,g^{-1}\,min^{-1})$. The most direct measure of myocardial energy consumption is oxygen consumption, MRO_2 (μl $O_2\,min^{-1}\,g^{-1}$). Indirect measures of substrate utilization are its rate of clearance into tissue, which is governed by a combination of membrane barrier limitation, the PS or permeability-surface area product $(ml\,g^{-1}\,min^{-1})$, and the consuming

[1] Center for Bioengineering, Departments of Medicine and Radiology, University of Washington, Seattle, WA 98195, USA
[2] Zentrum für Physiologie und Klinische Physiologie, Universität Düsseldorf, Moorenstraat 5, 4 Düsseldorf 1, Germany

reaction limitation, the gulosity G ($mlg^{-1}min^{-1}$). There are many sources of complexity in the anatomic arrangements within tissues. The heterogeneities of the several processes involved, diffusional factors, and non-linearities in reactions and in binding, can be summarized by the relationships between F, PS, G, the local concentrations C, and the volumes of distribution V_d.

In this review we concentrate on the use of simple principles and the potential for their application in practical methods for "perfusion imaging" and "metabolic imaging". The terms in quotations are a reflection of our instinctive desire to have the image of an answer; an image of the physiological status or of recoverability of the tissue, or of the ratio of local metabolic demand to substrate and oxygen supply.

Principles Underlying Experimental Methods

Conservation of Mass. It is this principle on which indicator dilution, external detection image interpretation, and metabolic modeling are based. The first statement is that the tracer content of an organ is the integral of the difference between influx and efflux. Since influx and efflux are carried by flow, it follows that the mean transit time, $\bar{t}$, from inflow to outflow for an unconsumed tracer is

$$\bar{t} = V_d/F. \tag{1}$$

The Volume of Distribution, V_d, is the volume of plasma equivalent fluid that would contain an amount of the solute equal to the actual amount present. For example, the concentration of K^+ in the plasma is $5\,mM$, while that in cells is $150\,mM$. The volume of distribution is composed of plasma, interstitial fluid, and cells, each weighted according to its plasma equivalent concentration. So V_d ($mlg^{-1}min^{-1}$) is $V_{pl} + V_{isf} \times 5/5 + V_{cell} \times 150/5 = V_{pl} + V_{isf} \times 30V_{cell} = 0.07\,mlg^{-1} + 0.18\,mlg^{-1} + 30\,(0.54\,mlg^{-1}) = 16.45\,mlg^{-1}$. This value for V_d is just over 20 times the water space of the heart, which is $0.78\,mlg^{-1}$ in normal hearts (data in dogs from Yipintsoi et al. [1]), and acts as a storage reservoir for tracer potassium. Whether or not there are permeability barriers, the mean transit time for tracer potassium would be about 20 times longer than for tracer water: the actual ratio would be $V_d(K^+)/V_d(water)$. This is the prime basis for the use of K^+, Rb^+, Tl^+, and Cu PTSM being retained for long enough times to serve as nearly flow-proportional markers.

The Classes of Markers commonly used to estimate regional flows are

Flowlimited tracers: (a) vascular markers, (b) markers which enter cells because of high permeability, and (c) markers which bind easily and firmly to receptor sites.
Partially barrier-limited markers, which are nearly completely extracted.
Mechanically deposited markers, which are completely extracted.

We shall discuss each of these, giving examples, and discuss the advantages and disadvantages from the point of view of the above principles and the natural compromises that occur because of low signal to noise ratios and other such practicalities. At the end we shall extrapolate from the current state of the art to suggest more advanced methods which depend on mathematical models for the interpretation of data, but which in the tomograph suite may be reasonably simple to perform.

The Evaluation of Methods requires that there be a "gold standard" against which the unproven method is compared. For flows the best standard that we have is the "molecular microsphere" *iododesmethylimipramine* (IDMI). It has a molecular weight of 396 Da, is highly lipophilic, and binds avidly to serotonin receptor sites and to the sites of uptake of norepinephrine on neuronal endings. It is nearly 100% extracted on single passage through the capillary bed of the heart [2, 3]. Simultaneous injections of 15 μm diameter polystyrene microspheres along with IDMI showed there was little systematic difference between them, *but* a small systematic error in the microsphere technique was revealed by the test, as suggested by Fig. 1. The error is that, because of their large size, microspheres have a tendency at branch point in the small arterioles to enter preferentially those branches with more flow; the overall result is to exaggerate the values of the flows to high flow regions and to slightly underestimate the flows in low flow regions.

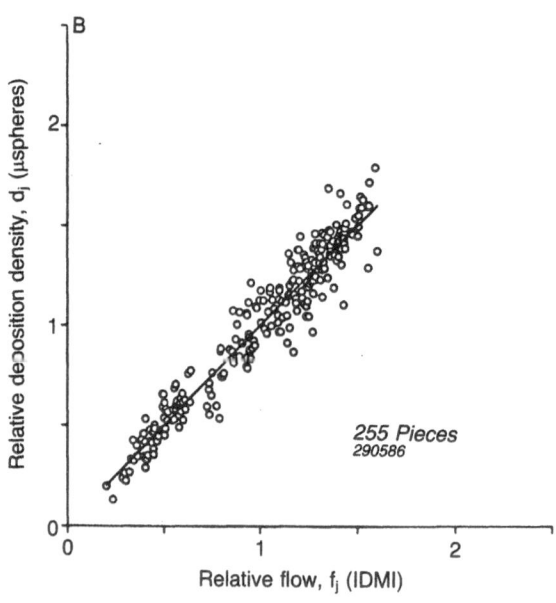

Fig. 1. Microsphere deposition densities (d_j) versus IDMI deposition densities (f_j) which are used as the measure of relative regional flow. Each d_j is the average of two observations in each piece, as is each f_j. The wide range of flows gives an estimate of the heterogeneity of flows. (From [3], with permission)

This is reassuring since microspheres have served as the standard for a long time and, for practical reasons, will not be easily replaced by "molecular microspheres" which cannot be labelled with a large number of different tracers. Moreover, non-tracer fluorescent microspheres are becoming the new standard since they are not radioactive [4]. Our conclusion is that the fluorescent sphere technique should replace the radioactive microsphere technique, for reasons of both safety and radioactive disposal, even though the technique is more labor intensive than the standard microsphere technique. Further, though all 15 µm diameter spheres will suffer the systematic error shown in Fig. 1, the error is small enough not to be regarded as serious.

Strictly Flow-Limited Markers are Uninfluenced by any Barrier-Limitation. A purely intravascular marker, such as ^{131}I-albumin or indocyanine green or an intra-vascular contrast marker for nuclear magnetic resonance spectroscopy (NMR), or (almost) [11]CO-hemoglobin for PET, can be regarded as flow-limited within its volume of distribution. As long as a marker does not have to cross any membranes or other diffusional barriers which impede its continuous moment-to-moment equilibration between plasma and any other space which it enters, that marker will be flow-limited in its rate of traversal of the organ. Albumin, remaining within the vascular space, qualifies. Iodoantipyrine, butanol, and other highly lipophilic tracers, which cross membranes easily and have high diffusion coefficients within the tissue, are also flow-limited; this latter group has volumes of distribution equal to the water space of the organ. A first estimate of the ratio of butanol transit time to that of albumin is the organ water space $(0.78\,\mathrm{ml\,g^{-1}})$ divided by the plasma space within the region where there is blood–tissue exchange $(0.07\,\mathrm{ml\,g^{-1}})$. The actual ratio, Eq. 2, will be modified from this by the dwell times for both of these tracers within the large non-exchanging vessels within the organ, about an additional $0.05\,\mathrm{ml\,g^{-1}}$:

$$\frac{\bar{t}_{\mathrm{but}}}{\bar{t}_{\mathrm{alb}}} = \frac{V_{\mathrm{vasc}} + V_{\mathrm{pl}} + V_{\mathrm{isf}} + V_{\mathrm{cell}}}{V_{\mathrm{vasc}} + V_{\mathrm{pl}}}. \tag{2}$$

These points are made to emphasize that when attempting to measure flow per gram of tissue, myocardial perfusion, the only available measure is transit time. When volumes of distribution are known, perhaps from prior data like the water space per gram of tissue, then estimates of F $(\mathrm{ml\,g^{-1}\,min^{-1}})$ can be made.

Estimating Vascular Space. There is less accuracy in estimating V_{d} for vascular markers: in PET studies [11]CO is commonly used as a vascular marker because it binds tightly to hemoglobin. (Actually Deussen and Bassingthwaighte [5] have found that CO, like oxygen, diffuses rapidly into the tissue, binding also to myoglobin, and therefore has a volume of distribution larger than the blood space.) In principle, PET image reconstruction can give signals adequate for the estimation of vascular space. One requires however the recording of two signals, both with good temporal resolution: an input function on a region-of-interest (ROI) containing only blood, and a tissue ROI. Any spreading of the apparent source of the signal, such as scatter and partial volume effects, compromises this method, so the technique is more accurately applied to X-ray computed tomographic reconstructions and NMR than to PET.

Defining terms, we use $q(t)$ as the quantity of tracer within the ROI, and subscript the term blood or tissue (tiss). If one obtains measures of apparent concentration from a specified volume by external detection, one must use the estimated tissue specific gravity, ρ, to obtain an estimate of the volume of distribution V_d ($ml\,g^{-1}$). For the left ventricle ρ is $1.06\,g\,ml^{-1}$ and ρ is only slightly less for the right ventricle. The volume of distribution within the tissue region is found from the area of the tissue ROI time-activity curve divided by that of the blood ROI time activity curve, as outlined by Bassingthwaighte et al. [6]:

$$V_{tiss} = \frac{1}{\rho} \cdot \frac{\int_0^\infty q_{tiss}\,dt}{\int_0^\infty q_{blood}\,dt}. \tag{3}$$

Use of the ratio of the peaks of the two time-activity curves has poorer accuracy, as described in [6], but can be used when the volume of distribution is small and the input curve broad.

Estimating the Flow F ($ml\,g^{-1}\,min^{-1}$) is trickier and less accurate because the requirements are for a large, rather than a small, volume of distribution. The most useful model-free formula is

$$F = \frac{[q_{tiss}]_{peak}}{\rho \int_0^\infty q_{blood}\,dt}. \tag{4}$$

The rationale for this expression, derived by Bassingthwaighte et al. [6], is that $[q_{tiss}]_{peak}$ should represent the activity of the *whole* quantity of tracer that passes through the tissue, that is, the bolus must be narrowly enough dispersed that all of the tracer passing through the ROI lies within it at one moment, and that the peak of $q(t)$ represents this amount with the same sensitivity as was obtained for the input curve q_{blood}. The difficulty with this is seen in Fig. 2.

It is evident that this simple model-free expression is quite restricted in its application to intravascular indicators: their V_d is usually far too small to allow the use of intravenous injections, and nearly too small for even left atrial injections. Restricting the use of Eq. 4 to left ventricular injections of short duration makes sense, but then one must be cautious because it is difficult to place an ROI for the input function that is insensitive to streaming or poor mixing.

However, it is much easier to apply Eq. 4 to flow-limited indicators such as ^{11}C-butanol, which is often used in brain studies because its volume of distribution is the tissue water space, so much larger than the vascular space, therefore allowing retention in the tissue ROI of a much more spread input function. The other side of the coin is that the input bolus should still not be too spread,

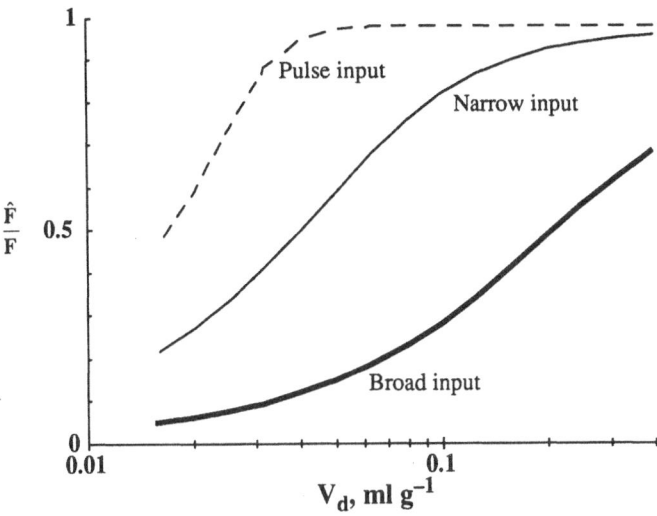

Fig. 2. Estimates of local myocardial blood flow F $(\mathrm{ml\,g^{-1}\,min^{-1}})$. Responses to three forms of input are used for the calculation to obtain the curves, a 2-s pulse, a narrow input with $\bar{t} = 6\,\mathrm{s}$ and SD $= 1.8\,\mathrm{s}$, and a broadly dispersed input with $\bar{t} = 30\,\mathrm{s}$ and the same relative dispersion of 30%, namely SD $= 9\,\mathrm{s}$. The calculation of Eq. 4 uses the peak height of the tissue ROI curve and the area of the curve in the left ventricular cavity ROI. The estimates are good only if the volume of distribution V_d is large. $\hat{F} = q_p/(\rho \times$ area of $q_b \times t)$. (From [6], with permission)

because if it is the recirculation will mask the tail and tend to overestimate the area of q_{blood}. This source of error is minimized by using exponential extrapolation of the downslope to project the shape of the first passage tracer and eliminate the recirculation, as originally described by Hamilton et al. [7] for outflow dilution curves, but which also works well for signals from intracavity ROIs that are not overly dispersed.

An Approach to Model-Dependent Methods

Model-Dependent Methods of Analysis are the next step. The goal is the same: to obtain a good estimate of mean transit time and then to interpret it via one's best estimate of the volume of distribution as in Eq. 1. To be clear, the underlying premise and (often unstated) goal of the modeling is to determine the mean transit time, V_d/F. There may be other goals for which models are used, such as permeability surface area products or consumption rates, but these are usually dependent on first knowing the flow. The difficulties with modeling analysis are exactly those discussed above, namely the problems in defining the input function and the time-activity curve in the ROI with sufficient accuracy to make the interpretation. What the models may have to offer is an adequate estimation of mean transit time even if there is substantial recirculation and partial volume effect degrading the signal. Recirculation can be accounted for by using the continuously-recorded input function as the input to the model.

With respect to partial volume effects, the contamination of the ROI with activity from nearby regions, appropriate modeling can theoretically account for the time courses of activity, the $q(t)$'s, in several regions simultaneously, and use several such simultaneous curves to make corrections by estimating the degrees of spillover from one region into its neighbors. The current state of the art is to try to account for the spillover from the left ventricular cavity into the myocardium and vice versa, a 2-by-2 matrix inversion, but accounting for multiple ROIs simultaneously has not been attempted, so far as we know. A particular situation that illustrates how big the spillover effects are is seen with intravenous or right atrial injection of $^{15}OH_2$-labelled water for the estimation of flow. As the bolus passes through the right ventricle strong signals of activity are obtained from not only the left ventricular (LV) septum but also the ventricular cavity, vitiating the use of the LV cavity signal as a good representation of the input function. For such reasons, it is now common to rely more on the left atrial (LA) signal as being freer of contamination. This is only relative, for there is no doubt that the LA signal includes spillover from the lung and pulmonary veins, even though their influence may not always be so obvious.

Because regional myocardial flows are so heterogeneous and show further heterogeneity within each piece no matter how small the heart is divided, it is important to take into account this flow heterogeneity in modeling [3, 8]. Residue curves, or tracer washout curves in general, show multiexponential washout. When using the correct input function, fitting such curves well with any model will give the correct mean transit time. This does not justify believing that an oversimplified model represents the physiology of the organ, but may justify its use for getting the answer to how to estimate flow. A general principle which links the residue and outflow dilution data is that one model should be able to fit both simultaneously, as illustrated for sodium exchange in the heart by Guller et al. [9] and demonstrated in a detailed myocardial exchange model by Bassingthwaighte et al. [10]. Taking this point of view tends to make one more cautious about believing in particularly simple models.

An example of the complex nature of the shape of washout curves is shown in Fig. 3. These are from an isolated blood-perfused dog heart study [11] in which the tracer xenon, Xe^{133}, was injected into a single coronary artery supplying a small region of the heart. Of the six washout curves recorded at different flows, none is monoexponential, even though the injection duration was about 0.3 s and all the tracer injected was in the field of the detector. They showed that the model-free height over area technique and multiexponential analysis both yielded good estimates of the flows. The basic height/area method, from Zierler [12] is

$$F = \frac{V_d(q_{tiss})_{peak}}{\displaystyle\int_0^\infty q_{tiss}\, dt}. \tag{5}$$

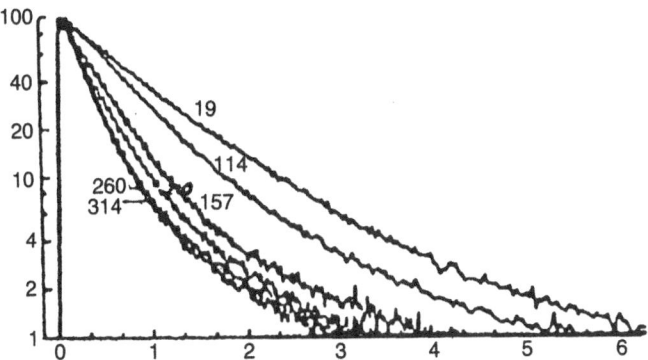

Fig. 3. Semilogarithmic plots of the time course of washout of ^{133}Xe injected into the coronary artery of a dog. Tracer residue curves recorded by external detection using NaI crystal detectors. Heart Segment-32g. (From [11], with permission)

That the two- and three-compartmental analyses gave nearly as good results is not a tribute to the model, but attests to the adequacy of any model that gives a reasonably correct estimate of mean transit time. Multicompartmental models, having more shaping parameters, can fit the data more precisely than can single compartmental models, but unless a model truly represents the anatomic and physiologic situation, its solution tends to misfit the data in small ways and therefore errs to some extent. So the model-free techniques, like Eq. 5, tend to be better since there are no inbuilt biases. Model-free Eq. 5 does have the problem that the integral goes to infinity, and assumes the absence of recirculation, so it too requires correction; there are different tricks to terminating the integral, as Bassingthwaighte et al. [11] describe.

One- and Two-Compartmental Models are Inadequate for Myocardial Kinetics. The simple compartmental models are wrong not just in the sense that all models are wrong since they are either incomplete or inexact in their features, but wrong in the sense that they are usually based on demonstrably incorrect presumptions. This may lead users of compartmental models to misunderstand the processes of flow, transmembrane exchange and diffusion in tissues. We pick as an example the description of such a model used for the estimation of flow in an article by two strong and innovative investigators, Hutchins and Schwaiger [13]. In their nice review of the use of inert diffusible markers, they describe the model of Kety [14], which has been used by many investigators, including Iida et al. [15]. They wrote the equation (with different symbols)

$$\frac{dC_{tiss}}{dt} = F\left(1 - e^{-\frac{PS}{F}}\right)C_{blood} - F\frac{V_{blood}}{V_{tiss}}C_{tiss}, \qquad (6)$$

where C_{tiss} and C_{blood} are concentrations in tissue and blood of tracer activity in cpm/ml, F is flow in $ml\,g^{-1}\,min^{-1}$, PS is permeability-surface area product in

$mlg^{-1}min^{-1}$, and V_{blood} and V_{tiss} are the volumes of tracer distribution in blood and tissue in mlg^{-1}, and the ratio V_{tiss}/V_{blood} is the tissue-to-blood partition coefficient, dimensionless. The input function is the recorded curve C_{blood} as a function of time. They say this equation describes a two-compartmental model, but since there is only one equation it resembles a one-compartmental model.

Some assumptions are implicit in the model, and we will bring these out, and thereby ferret out inconsistencies in the approach. The first assumption implied is that C_{blood} is the arterial blood concentration and that it remains unchanged in its passage through the tissue; this can never be exactly correct because the equation states that the tissue gains material from the blood, the first phrase on the right of the equality, so the secondary implication is that the flow is infinitely fast so that the material lost to the tissue is negligible. This isn't what the authors wanted to assume, because their goal is to estimate the flow, not assume that it is infinite. So this is contradiction number one. The equation for a single mixing chamber, wholly instantaneously mixed between blood and tissue within the ROI, could be written as a single differential equation of form similar to Eq. 6, omitting for the moment any consideration of barrier-limitation:

$$V_{tiss}\frac{dC_{tiss}}{dt} = FC_{blood} - FC_{vein},\qquad(7)$$

where C_{vein} is the concentration in the effluent blood. Eq. 7 is dimensionally balanced; each term has the units of quantity of tracer per unit time. If this is rearranged to be closer to the form of Eq. 6, the differences are more apparent:

$$\frac{dC_{tiss}}{dt} = \frac{F}{V_{tiss}}C_{blood} - \frac{F}{V_{tiss}}C_{vein}.\qquad(8)$$

Each phrase now has the dimensions of concentration divided by time. The phrases in Eq. 6 differ from those in Eq. 8 in several respects, one of which is dimensional balance. Eq. 6 can have dimensional balance only if the units of F are defined to be inverse time, min^{-1}. This would mean the F of Eq. 6 must be redefined to be F/V_{tiss}, with the units $mlg^{-1}min^{-1}$ divided by mlg^{-1}.

Comparison of the last phrases in Eq. 6 and Eq. 8 makes it clear that it is assumed that $C_{vein} = C_{tiss}$; this is the second implicit assumption, now made explicit, and we must examine its validity. The idea is completely consistent with the assumption that blood and tissue exchange is infinitely rapid, so this not a new assumption, only a restatement of it. The question is whether or not it is consistent with the rest of the expression.

The answer to this is no, for there is a second contradiction. The inconsistency is between the second and third phrases of Eq. 6, not in terms of dimensionality, but in terms of the definition of the model itself and of the concept of the uniformly mixed blood–tissue exchange region. The problematic term is the modifier in the second phrase, namely the term for the fractional extraction, $E =$

$1 - e^{-(PS)/F}$, which introduces the concept of a diffusional or membrane barrier between the blood and the tissue. Kety [16] introduced this anomaly into the literature because he saw that blood–tissue equilibration was not always complete. But using it as a modifier in the one-compartmental model equations was wrong then and is still wrong; one cannot assume equilibration in the third phrase while assuming a barrier limitation in the second phrase! The existence of a barrier limitation means that the concentrations in blood and tissue are *not* in equilibrium, violating the basic assumption of the mixing chamber model.

The resolution is to recognize the differences between intracapillary blood and extravascular tissue and to use a pair of equations so that proper definitions and proper mass balance are obtained. Writing equations for the blood component of the organ and for the extravascular part of the organ separately, we have

$$V_{blood}\frac{dC_{blood}}{dt} = F(C_{art} - C_{blood}) - PS(C_{blood} - C_{tiss}), \qquad (9)$$

where it is assumed that C_{vein} is represented in the first phrase on the right by C_{blood}, which is the mixing chamber assumption, and

$$V_{EVT}\frac{dC_{EVT}}{dt} = PS(C_{blood} - C_{EVT}). \qquad (10)$$

Unfortunately, here we have introduced a new term for extravascular tissue (EVT) distinguishing it from the blood within the tissue. Now it is clear that not only does EVT differ from blood, but the previous definition of V_{tiss} may not have been clear. In fact it is clear, since Eq. 7 can only be correct if V_{tiss} includes both extravascular tissue and blood, and if C_{blood} really means C_{art} and not blood somewhere between the artery and vein. Actually no new parameters or concepts are introduced in these last two equations. The kinetic parameters are F and PS and the two volumes; although the two volumes in Eq. 6 appeared as a ratio, and one might have thought that only the ratio, the partition coefficient, was necessary, the partition coefficient is not a physicochemical constant but a ratio of volumes and therefore the "partition coefficient" depends on the ambient state of the tissue and on the actual volumes.

A "conservation equation" combining these two equations can be formulated by substituting for the phrase including the PS, because it appears in both equations. The expression, which is another version of the mass balance statement that says what goes in comes out, is

$$\frac{V_{blood}dC_{blood}}{dt} + \frac{V_{EVT}dC_{EVT}}{dt} + FC_{blood} = FC_{art}. \qquad (11)$$

In the special case when PS is infinite, then C_{EVT} is equilibrated continuously with C_{blood}. The equation can be rewritten after dividing through by V_{blood}, using $C_{EVT} = C_{blood}$ and gathering like terms together:

$$\frac{dC_{blood}}{dt}\left(1 + \frac{V_{EVT}}{V_{blood}}\right) + \frac{F}{V_{blood}}(C_{blood} - C_{art}) = 0. \tag{12}$$

Tissue activity is the equilibrated average in the ROI, $C_{ROI} = C_{blood}(1 + V_{EVT}/V_{blood})$, so that the externally detected concentration is

$$\frac{dC_{blood}}{dt} + \frac{F}{(V_{blood} + V_{EVT})}(C_{blood} - C_{art}) = 0, \tag{13}$$

which shows by comparison with Eq. 8 that one must identify C_{vein} as identical to C_{blood}. This in turn alerts us to the corollary statement that there must be a discontinuity in the concentration at the inflow to the blood–tissue exchange region, namely the concentration instantaneously jumps from C_{art} to C_{blood} at the entrance. This is a physical impossibility but is not a mathematical problem.

Such anomalous situations can be avoided by taking spatial distances along the length of the vessels into account, as has been done in the indicator dilution field since the papers of Goresky [17] for flow-limited tracers and Crone [18] for barrier-limited tracers. Preceding both of these was the pioneering paper of Sangren and Sheppard [19] providing a mathematical solution that described both of these cases by means of a model which accounted for gradual gradients along the length of the capillary, and the specific but incomplete model of Renkin [20] which accounted for unidirectional escape of tracer from blood to tissue but did not account for the return flux from tissue to blood. The Sangren-Sheppard model has exactly the same parameters as Eqs. 9 and 10 and so is just as easy to use. It has the great advantage that it is more realistic and can be fitted to high resolution curves which the compartmental model fails to fit. Our conclusion from this discussion is that the application of the distributed models is straight-forward and can provide good fits to the data, as is exemplified by the many papers of Goresky and colleagues, as well as those of our own group, e.g., that of Kuikka et al. [21], and Deussen and Bassingthwaighte [5].

Application of Distributed Models to Barrier-Limited Uptake in the Myocardium

The non-metabolized tracer ^{86}Rb and the metabolized substrate IPPA (iodo-phenylpentadecanoic acid) are partially barrier-limited in their uptake into the myocardium. Both have large intracellular volumes of distribution inside the myocardial cells; rubidium is close enough to potassium that it uses the same ionic channels and ionic pumps as potassium; IPPA, like other fatty acids, is taken up and trapped as acyl-CoA and di-and triglyceride prior to being metabolized. Thus for the first moments after an injection these substances are strongly taken up by myocytes and very little is released. If they were completely taken up during a single transcapillary passage and none were released then they would be excellent markers for estimating regional blood flow. But the extraction of neither is complete.

Rubidium extraction is dependent on the flow, but ranges from 50% at very high flows to 85% at low flows. An example of the relationship between Rb deposition in 150 mg pieces of tissue and the regional flow, as measured by the molecular microsphere IDMI, is shown for a sheep heart in Fig. 4.

Two model solutions are shown in Fig. 4, both for a three-region axially distributed model accounting for both capillary and cell membrane barriers to Rb exchange. The volume of distribution in the myocytes was taken to be the same as for potassium, i.e., the cellular concentration of the ion was taken to be 30 times the plasma concentration at equilibrium. Two different forms of the model were tested against the data. (This is in accord with the principle put forward by Platt [22] that one should not only have an explicit hypothesis, but that one should have an alternative hypothesis, and that the experiment should be designed to distinguish between the two hypotheses. When successful, the experimental data rule out one hypothesis and so advance the science, which can only proceed by disproof of an hypothesis, for it can never be proven.) The model is a multicapillary one, using 20 paths in parallel, in order to account for the broad range of flows seen in both Fig. 1 and Fig. 4. The two hypotheses were (1) the capillary permeability-surface area products, PS, are the same in all regions in spite of the differences in flows, and (2) the PSs are proportional to the regional flows, $PS \propto F$, as if high flow regions were more fully recruited. The

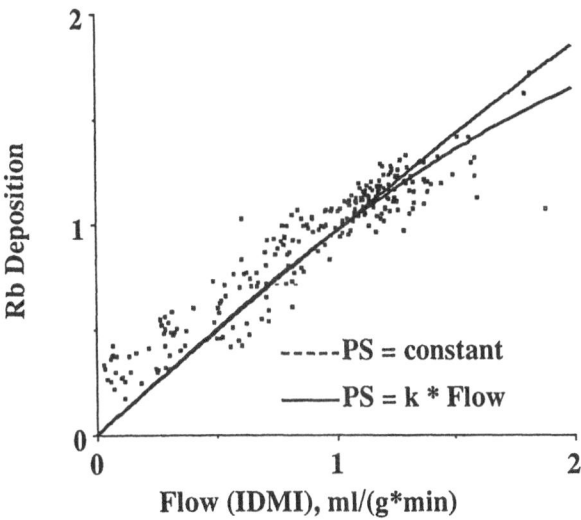

Fig. 4. Deposition of ^{86}Rb in a sheep heart plotted versus the deposition of IDMI, a molecular tracer which is extracted virtually 100%. The ordinate and abscissa values are both normalized to 1.0. Tracers were injected simultaneously into the left atrium of an anesthetized sheep and the animal sacrificed after 90 s. The excess deposition of ^{86}Rb compared to IDMI, most evident in the low flow regions, shows that Rb enters tissue regions which IDMI does not reach, and also shows that in these regions either IDMI deposition underestimates the local flow or ^{86}Rb deposition overestimates it

result is inconclusive: the myocardial blood flows in this study were low, and the models differ only in the higher flow regions; there are not enough data in the region of the difference to distinguish between the two hypotheses. In this sense the experiment is a failure.

However, as happens so often in a failed study, something interesting can be seen in the data at very low flows. The deposition of Rb is much higher than that of the IDMI, which at first seems impossible since IDMI is 98% extracted and Rb only 80% extracted on the average. But consider the observations in the light of an axially distributed capillary from which no IDMI emerges: this means that at the downstream end of the capillary there must be no IDMI; since none emerges practically all of it must have been sequestered by the tissue farther upstream, nearer the capillary entrance or even in small arterioles. In contrast, the Rb escapes from the venous end at a significant concentration, so at the down stream end of the capillary-tissue unit there is Rb present which can be taken up in this segment of the capillary. If the downstream ends of large numbers of capillaries are close together, then there will be sizable regions with higher Rb concentration than IDMI. (One of the events that contributes to this result is that Rb is taken up by erythrocytes, just as is potassium, and so there is an additional barrier involved, the red blood cell membrane, which helps to keep the Rb within the vascular space for a longer time.) Although there is as yet no good way of providing the explanation in quantitative detail, it appears to be due to the axially distributed nature of the blood–tissue exchange region, and argues against the concept of a mixing chamber analogy.

Closing Comment

For thirty years workers using high resolution indicator dilution methods have placed reliance on carefully anatomically structured, kinetically descriptive models for blood–tissue exchange. See Bassingthwaighte and Goresky [23] and Goresky [24] for an overview. Such models were needed to fit the data obtained from multi-tracer indicator dilution experiments which provide 1–2 s resolution and the concentrations range over three to four orders of magnitude. The functioning of these models in the most exacting situations shows them to be useful examples of "working hypotheses" or "state of the art" descriptions of the physiological events. Zierler [25] has pointed out that the use of inappropriately simplified models is risky. Such oversimplified models can give inexact or biased estimates of parameters, and even perhaps directionally wrong answers. Without extensive testing we cannot say what the errors are in applying compartmental analysis in PET applications, but we do say if we use such models for the sake of their simplicity we must determine the level of error by evaluating the compartmental models against high resolution data and against the distributed models of demonstrated reliability.

References

1. Yipintsoi T, Scanlon PD, Bassingthwaighte JB (1972) Density and water content of dog ventricular myocardium. Proc Soc Exp Biol Med 141:1032–1035
2. Little SE, Link JM, Krohn KA, Bassingthwaighte JB (1986) Myocardial extraction and retention of 2-iododesmethylimipramine: A novel flow marker. Am J Physiol 250 (Heart Circ Physiol 19):H1060–H1070
3. Bassingthwaighte JB, Malone MA, Moffett TC, King RB, Chan IS, Link JM, Krohn KA (1990) Molecular and particulate depositions for regional myocardial flows in sheep. Circ Res 66:1328–1344
4. Glenny RW, Bernard S, Brinkley M (1993) Validation of fluorescent-labeled microspheres for measurement of regional organ perfusion. J Appl Physiol 74:2585–2597
5. Deussen A, Bassingthwaighte JB (1993) Blood-tissue oxygen exchange and metabolism using axially distributed convection-permeation-diffusion models. Am J Physiol (in press)
6. Bassingthwaighte JB, Raymond GR, Chan JIS (1993) Principles of tracer kinetics. In: Zaret BL, Beller GA (eds) Nuclear Cardiology: State of the Art and Future Directions. Mosby-Year Book Inc., St Louis, pp 3–23
7. Hamilton WF, Moore JW, Kinsman JM, Spurling RG (1932) Studies on the circulation. IV. Further analysis of the injection method, and of changes in hemodynamics under physiological and pathological conditions. Am J Physiol 99:534–551
8. Bassingthwaighte JB, King RB, Roger SA (1989) Fractal nature of regional myocardial blood flow heterogeneity. Circ Res 65:578–590
9. Guller B, Yipintsoi T, Orvis AL, Bassingthwaighte JB (1975) Myocardial sodium extraction at varied coronary flows in the dog: Estimation of capillary permeability by residue and outflow detection. Circ Res 37:359–378
10. Bassingthwaighte JB, Chan IS, Wang CY (1992) Computationally efficient algorithms for capillary convection-permeation-diffusion models for blood-tissue exchange. Ann Biomed Eng 20:687–725
11. Bassingthwaighte JB, Strandell T, Donald DE (1968) Estimation of coronary blood flow by washout of diffusible indicators. Circ Res 23:259–278
12. Zierler KL (1965) Equations for measuring blood flow by external monitoring of radioisotopes. Circ Res 16:309–321
13. Hutchins GD, Schwaiger M (1993) Quantifying myocardial blood flow with PET. In: Zaret BL, Beller GA (eds) Nuclear Cardiology: State of the Art and Future Directions. Mosby-Year Book Inc., St Louis, pp 305–313
14. Kety SS (1960) Blood-tissue exchange methods: Theory of blood-tissue exchange and its application to measurement of blood flow. In: Bruner HD (ed) Methods in Medical Research, vol 8. YearBook Medical Publishers Inc., Chicago, pp 223–228
15. Iida H, Kanno I, Takahashi A, Miura S, Murakami M, Takahashi K, Ono Y, Shishido F, Inugami A, Tomura N, Higano S, Fujita H, Sasaki H, Nakamichi H, Mizusawa S, Kondo Y, Uemura K (1988) Measurement of absolute myocardial blood flow with $H_2{}^{15}O$ and dynamic positron-emission tomography: Strategy for quantification in relation to the partial-volume effect. Circulation 78:104–115
16. Kety SS (1949) Measurement of regional circulation by the local clearance of radioactive sodium. Am Heart J 38:321–328
17. Goresky CA (1963) A linear method for determining liver sinusoidal and extravascular volumes. Am J Physiol 204:626–640

18. Crone C (1963) The permeability of capillaries in various organs as determined by the use of the "indicator diffusion" method. Acta Physiol Scand 58:292–305
19. Sangren WC, Sheppard CW (1953) A mathematical derivation of the exchange of a labeled substance between a liquid flowing in a vessel and an external compartment. Bull Math Biophys 15:387–394
20. Renkin EM (1959) Transport of potassium-42 from blood to tissue in isolated mammalian skeletal muscles. Am J Physiol 197:1205–1210
21. Kuikka J, Levin M, Bassingthwaighte JB (1986) Multiple tracer dilution estimates of D- and 2-deoxy-D-glucose uptake by the heart. Am J Physiol 250 (Heart Circ Physiol 19):H29–H42
22. Platt JR (1964) Strong inference. Science 146:347–353
23. Bassingthwaighte JB, Goresky CA (1984) Modeling in the analysis of solute and water exchange in the microvasculature. In: Renkin EM, Michel CC (eds) Handbook of Physiology, Sect 2, The Cardiovascular System, vol IV, The Microcirculation, Am Physiol Soc, Bethesda, pp 549–626
24. Goresky, CA (1985) Biological barriers: Their effects on cellular entry and metabolism in vivo. Microvasc Res 29:1–17
25. Zierler KL (1981) A critique of compartmental analysis. Annu Rev Biophys Bioeng 10:531–562

Regional Blood Flow Measurement with Non-Radioactive Microspheres by X-ray Fluorescence Spectrometry

Hidezo Mori[1] and Julien I.E. Hoffman[2]

Summary. X-ray fluorescence spectrometry has now become available for measuring regional blood flow with microspheres loaded with heavy nonradioactive elements. Drawbacks in previous studies were leaching of elements from the nonradioactive microspheres, and insufficient sensitivity for detecting small amounts of elements. We developed 9 sets of new nonradioactive microspheres (Ti, Br, Y, Zr, Nb, In, I, Ba, Ce) with a mean diameter of about 15 µm, and 1 set of Br-microspheres with a diameter of 60 µm which did not leach tracer elements even in strongly alkaline solution. We used a wavelength dispersive X-ray fluorescence spectrometer, which is characterized by great sensitivity and is commercially available (PW 1480 Philips Almelo, The Netherlands). We validated the method by comparing duplicate flows measured with radioactive and nonradioactive microspheres in acute dog experiments, expanded the application of the method into long term experiments in dogs up to 6–12 months, and into systemic and pulmonary blood flow distributions in rats. By applying a monochromatic synchrotron radiation as a primary X-ray source, we increased the signal-to-background ratios of X-ray fluorescence spectra 50 dB or more above those with the wavelength dispersive system. This system allowed us to measure accurately blood flow distributions in contiguous small regions of 10–20 mg or less.

Key words: Nonradioactive microspheres—Regional blood flow—X-ray fluorescence

Introduction

Radioactive microspheres are widely used to measure regional blood flow in various physiologic experiments [1]. However, the method has disadvantages in terms of biohazards and the tight restrictions in handling and disposal of radioactive materials and exposed animals. Recently, several alternatives for radioactive microspheres have been developed [2–5]. One is the method described here, in which heavy element-loaded non-radioactive microspheres are analyzed by X-ray fluorescence spectrometry [6–10].

---

[1] Department of Physiology, Tokai University School of Medicine, Bohseidai, Isehara, Kanagawa, 259-11 Japan
[2] Department of Pediatrics and Cardiovascular Research Institute, University of California, San Francisco, CA, 94143, USA

17

Principle of X-ray Fluorescence Spectrometry and a Wavelength Dispersive Spectrometer

When the target elements are exposed to primary X-rays, inner orbital electrons are ejected and certain outer electrons move into the inner orbits while radiating X-ray fluorescence. This fluorescence is characteristic for the atomic number of the relevant element. Therefore, X-ray fluorescence spectrometry allows the separation and quantitation of multiple elements with an atomic number of 5 or more [6] even in mixtures. Kaufmann et al. [7, 8] initially tried to adapt X-ray fluorescence spectrometry to non-radioactive tracer analysis in medicine, and Morita et al. [9] applied X-ray fluorescence spectrometry for regional blood flow measurement with heavy element-loaded nonradioactive microspheres. Mori et al. [10] established the present method as a practical tool to measure regional

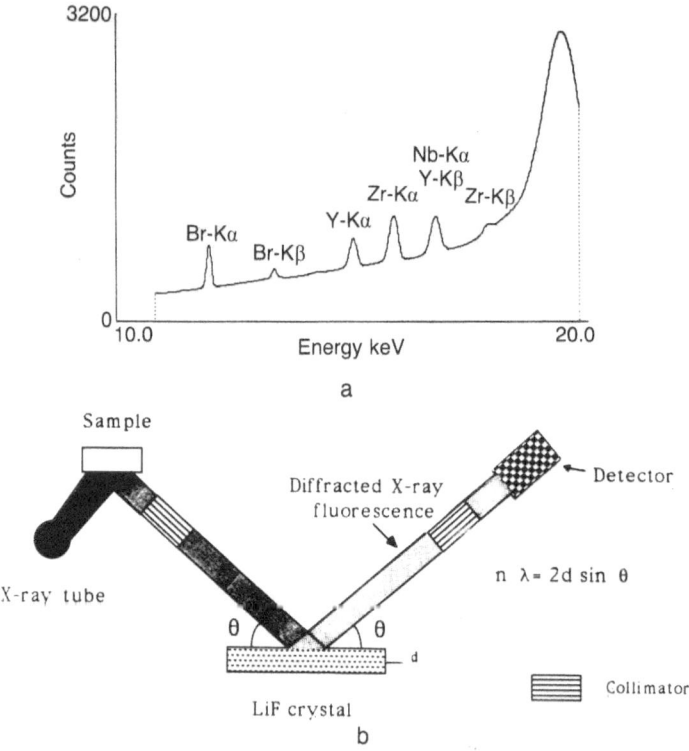

Fig. 1. A schematic representation of the wavelength dispersive spectrometer. A representative K-spectra derived from a sample containing ^{35}Br-, ^{39}Y-, ^{40}Zr- and ^{41}Nb-loaded microspheres are shown. Abbreviations: n, integer; λ, wavelength of fluorescent X-ray; d, distance between the lattice planes; θ, Bragg angle; Br, bromine; Y, yttrium; Zr, zirconium; Nb, niobium; K_α, X-ray fluorescence during the transition of electrons from the L- to the K-shell; K_β, X-ray fluorescence during the transition of electrons from the M- to the K-shell. (From [10] with permission)

blood flow in animal experiments by developing new nonradioactive microspheres and by using a more sensitive X-ray fluorescence spectrometer. The principles used in the present method are almost the same as those used in the radioactive microsphere technique described by Heymann et al. [1].

We used a wavelength dispersive spectrometer obtained from Philips Co., Ltd. (PW1480, Fig. 1a,b) to measure regional blood flow in pieces of tissue about 1 g or more in mass. This spectrometer has several advantages. First, there is no exposure to X-rays outside the machine, permitting manipulation of the measurement system without any particular precautions against high-energy X-rays. Second, the wavelength spectrometer has a high energy resolution, which can even separate the X-ray fluorescence spectra of multiple elements with sequential atomic numbers (Fig. 1a). In this system, fluorescent X-rays are diffracted by LiF crystals (Fig. 1b), and the radiation at specific wavelengths, determined by Bragg's law ($n\lambda = 2d \sin\theta$; where n is an integer, λ is the wavelength of the electromagnetic wave, d is the distance between lattice planes and θ is the Bragg angle), is detected by a combination of gas-flow and scintillation detectors. Third, many sets of elements can be measured by the spectrometer. The combination of a programmable generator with a maximum power of 3 KW and interchangeable tubes (Au-, W-, Rh-, Mo-, Cr-, Sc-anodes) provides several options regarding the energy level of the primary X-ray beam. This is important, because X-rays with an energy peak just above the K-absorption edge of the target element are the most effective in exciting the relevant atoms [6]. We usually use a ^{45}Rh-anode with a generator voltage of 75–100 kV to evoke K-fluorescence in elements from ^{35}Br to ^{41}Nb, and a ^{24}Cr-anode with a generator voltage of 40–60 KV to evoke L-fluorescence of the elements from ^{49}In to ^{58}Ce and K-fluorescence of ^{22}Ti. By applying one of the 6 anodes, we can quantitate any element with an atomic number ranging from 5 to 92.

Nonradioactive Microspheres Loaded by Stable Heavy Elements

Sekisui Plastic Company, Tokyo, Japan, developed 9 sets of nonradioactive microspheres with a nominal diameter of 15 μm, which were labeled with one of the following stable elements (Table 1): titanium (^{22}Ti), bromine (^{35}Br), yttrium (^{39}Y) zirconium (^{40}Zr), niobium (^{41}Nb), indium (^{49}In), iodine (^{53}I), barium (^{56}Ba), and cerium (^{58}Ce). We recently developed a Br-loaded microsphere with a mean diameter of 60 μm. Seven of the 9 sets of nonradioactive microspheres with a nominal diameter of 15 μm were manufactured by polymerizing styrene monomer in which fine particles of heavy elements, with a diameter of approximately 0.1 to 0.5 μm, had been homogeneously distributed. On the other hand, nonradioactive microspheres containing I and Br were manufactured by polymerizing iodized-phenolmethacrylate and by copolymerizing methylmethacrylate and bromized-phenolmethacrylate. The diameters of all these microspheres were distributed normally about the mean diameter. The mean diameters for the

Table 1. Descriptions of nonradioactive microspheres

Element	Atomic no.	Fluorescence peak (KeV) Kα	Lα	Specific gravity	Diameter (μm)	number of MS per 1 g	Polymer
Ti	22	4.5		1.34	15.0 ± 1.5	4.27 × 10⁸	polystyrene
Br	35	11.9		1.34	14.9 ± 1.5	4.31 × 10⁸	methacrylates
Y	39	14.9		1.29	14.8 ± 1.5	4.57 × 10⁸	polystyrene
Zr	40	15.7		1.36	15.3 ± 2.2	3.92 × 10⁸	polystyrene
Nb	41	16.6		1.32	15.2 ± 1.7	4.12 × 10⁸	polystyrene
In	49		3.3	1.20	14.7 ± 1.9	5.01 × 10⁸	polystyrene
I	53		3.9	1.60	14.7 ± 1.9	3.76 × 10⁸	methacrylates
Ba	56		4.5	1.61	15.7 ± 2.3	3.07 × 10⁸	polystyrene
Ce	58		4.8	1.29	15.9 ± 2.0	3.69 × 10⁸	polystyrene
Br (large)	35	11.9		1.17	59.7 ± 4.1	7.66 × 10⁶	methacrylates

Abbreviations: *Ti*, titanium; *Br*, bromine; *Y*, yttrium; *Zr*, zirconium; *Nb*, niobium; *In*, indium; *I*, iodine; *Ba*, barium; *Ce*, cerium; *MS*, microspheres. (From [10] with permission)

smaller microspheres were 14.7 μm per set for the smallest (In) and 15.9 μm for the largest (Ce), and their standard deviations were equal to or less than 2.3 μm (Table 1), and those for the large Br-loaded microspheres were 59.7 ± 4.1 μm.

Because we had to extract microspheres for X-ray fluorescence spectrometry the tissue was dissolved in a 2N-KOH solution, so we tested the possibility of elements leaching from the microspheres in a 2N-KOH solution with and without blood or tissue. There was no significant leaching of the eight elements in the 2N-KOH solution for 1 week [10], except for recently developed Ti-loaded microspheres. In our recent experiments, we did not find leaching from the Ti-loaded microspheres in the 2N-KOH solution for 1 week (Fig. 2). We also monitored leaching of the elements in a 2N-H_2SO_4 solution for 1 hour for all the sets of microspheres, because some tissues such as the porcine liver and kidney can not be completely digested by 2N-KOH solution alone; we can achieve complete digestion of these organs by adding 2N-H_2SO_4 after the 2N-KOH. Seven of the nine sets of nonradioactive microspheres were stable in 2N-H_2SO_4. However, as shown in Fig. 2, Y- and In-loaded microspheres revealed a significant leaching of the elements from the microspheres in the 2N-H_2SO_4 solution.

In the near future, it might be possible to expand the number of different element labelled microspheres up to 18 sets. The PW 1480 spectrometer can measure elements with atomic numbers of 5 to 92, as described above. However, the sensitivity for detecting X-ray fluorescence for the heavier elements with an atomic number of more than 60 decreases, mainly due to an increase in the background level. In addition, most of the elements with an atomic number of more than 60 are scarce. On the other hand, the lighter elements with an atomic number of less than 21 are, in general, found in quantity in tissue. Therefore, we think that elements with atomic numbers of 22 to 60 are preferable as tracer elements for the microspheres. In order to develop new non-radioactive microspheres, insoluble particles of the tracer elements with a size of 0.1–0.5 μm

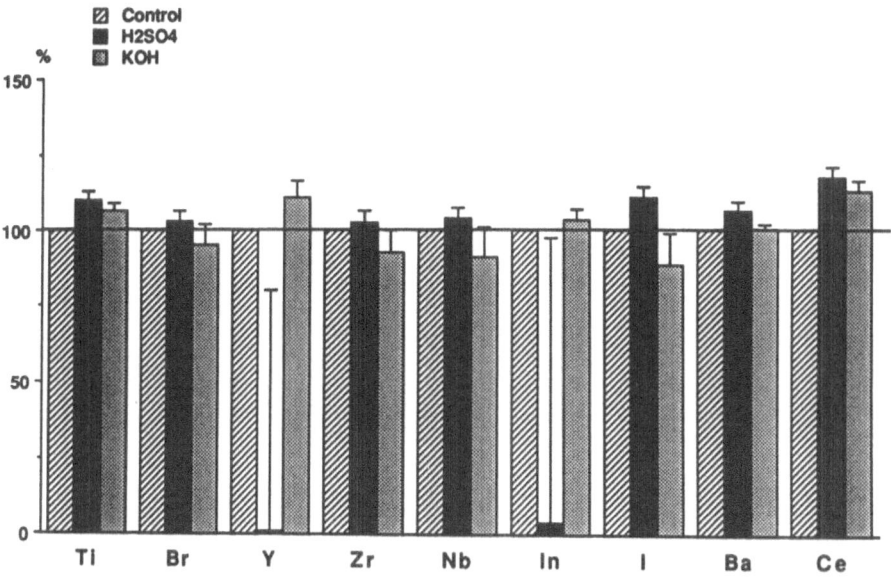

should be purchased commercially, and must be stable in 2N-KOH, tissue, and blood. These considerations indicate that another eight kinds of element loaded microspheres (V, Mn, Co, Ni, Cu, Sr, Mo, Ag) could become useful in measuring regional blood flow in the future.

Application of X-ray Fluorescence Spectrometry for Regional Blood Flow Measurement with the Non-Radioactive Microspheres in Various Physiologic Experiments

The accuracy of regional flow measured with the heavy element-loaded non-radioactive microspheres was assessed by comparison with the flow measured by radioactive microspheres in two anesthetized dogs. Br- and Zr-loaded micro-spheres (1×10^7 each) were injected into the left atrium together with radioactive microspheres labeled with ^{51}Cr and ^{95}Nb (1×10^6 for each). Radioactive counting was performed on minced myocardial and reference blood samples by using a 3 inch Na-I detector [1]. These samples were then dissolved in 2N-KOH, trapped on filter papers, and examined by X-ray fluorescence spectrometry. Comparing Br- and ^{95}Nb-flows (Table 2), the correlation coefficient was very close to 1.0 (0.9960), the standard deviation of the data from the regression line was small (0.148 ml/min per g), and the slope of the regression lines was close to 1.0 (1.04). Plotting the difference of the dual flows against mean of the dual flows

Table 2. Comparison of the flows measured with radioactive and nonadioactive microspheres

	$^{95}Nb-Br$	$^{95}Nb-Zr$	$^{51}Cr-Br$	$^{51}Cr-Zr$	$Zr-Br$	$^{51}Cr-^{95}Nb$
Regression analysis						
r	.9960	.9955	.9955	.9950	.9955	.9995
$s_{y|x}$	0.148	0.159	0.152	0.162	0.040	0.037
slope	1.04	1.08	1.07	1.11	0.97	1.03
$(b-1)/s_b$	3.946	6.663	6.124	8.663	12.218	9.456
p- value	<0.001	<0.001	<0.001	<0.001	<0.001	<0.001
Bland and Altman's analysis						
d (ml/min/g)	−0.036	−0.079	−0.071	−0.11	−0.04	−0.04
SD of d	0.17	0.21	0.19	0.24	0.07	0.05
d − 2SD	−0.37	−0.50	−0.46	−0.59	−0.18	−0.14
d + 2SD	0.30	0.34	0.31	0.36	0.10	0.07

Abbreviations: $(b-1)/s_b$, t- value to test whether the slope of the regression line is significantly greater than 1; $s_{y|x}$, standard deviation from regression; d, difference of flows measured by two technique; SD, standard deviation. The range between (d − 2SD) and (d + 2SD) means 95% confidence limits of the difference of flows measured by two technique. The other abbreviations are the same as in Table 1. (From [10] with permission)

(Bland and Altman's analysis) revealed that flows measured with Br were greater by 0.036 ml/min per g in mean than those measured with ^{95}Nb-flow. The range of the differences (mean ± 2SD) indicated that 95% of the differences (^{95}Nb − Br) were neither greater than 0.30 nor less than −0.37 ml/min per g. These results confirmed the validity of the flows measured with nonradioactive microspheres.

Recent investigations from Tokai University School of Medicine expanded the possible use of the present method in various animal experiments. Sakamoto et al. [11] investigated the accuracy of flow measurements of heart, liver, and kidney in long-term dog experiments. They confirmed that loss of microspheres from systemic organs and from the lung was small in experiments lasting 12 months, and that the relative distribution of dual microspheres injected 1–12 months before sacrifice correlated well with each other as well as the other two sets of nonradioactive microspheres injected just before sacrifice. For long term experiments, the application of nonradioactive microspheres reduced the difficulty in handling animals throughout the experiment because animals could now be kept in an ordinary animal care unit without any restrictions concerning radiation safety. Kuwahira et al. [12] confirmed the accuracy of the present method for evaluating the distribution of systemic and pulmonary blood flow in rats. They were able to measure flows in various systemic organs, with the smallest fractional distribution of 0.3% in the diaphragm, by left ventricular injection of 4×10^5 nonradioactive microspheres, and the flow distribution among 9 pieces from each lung by injection of 1×10^5 nonradioactive microspheres into the inferior vena cava. They also showed that they could measure regional flows repeatedly with the other sets of nonradioactive microspheres without significant changes in central and pulmonary hemodynamics.

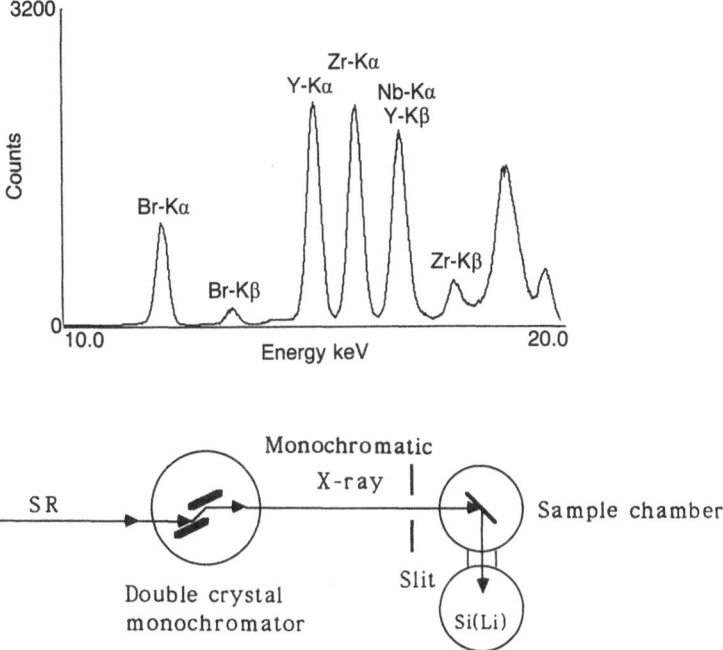

Fig. 3. A schematic representation of the synchrotron radiation-excited energy dispersive spectrometer is shown and representative K-spectra derived from samples containing ^{35}Br-, ^{39}Y-, ^{40}Zr-, and ^{41}Nb-loaded microspheres are shown. Abbreviations: *SR*, synchrotron radiation; *Si (Li)*, Si (Li) solid-state detector. The other abbreviations are the same as those in Figure 1

By applying a monochromatic synchrotron radiation as a primary excitation beam (Fig. 3), we could measure regional blood flow in contiguous small regions without cutting the tissue. The main reasons for the high sensitivity of synchrotron radiation-excited X-ray fluorescence spectrometry are the tunability of monochromatic X-rays, increasing the efficiency of X-ray fluorescence, and the high brightness and polarized nature of synchrotron radiation, increasing the probability of fluorescence excitation and decreasing background. In addition, the nearly parallel beam nature of synchrotron radiation allowed us to obtain high photon flux density in small regions (down to 1 μm^2). Thus, synchrotron radiation excited X-ray fluorescence spectrometry is useful to quantitate flows in small contiguous regions. The main problem is the stochastic nature of microsphere distribution, which makes it difficult to obtain the statistically required number of microspheres in small regions, rather than a limitation in the detection of small numbers of microspheres by the synchrotron radiation excited system. The acceptable minimum size for blood flow measurement with microspheres seems to be 10–20 mg or less.

In a previous study [10], we showed that synchrotron radiation-excited energy dispersive spectrometry increased the signal-to-background ratio of X-ray fluorescence spectrometry markedly (50 dB or roughly 300 times for Zr) compared to that with the wavelength dispersive system by measuring in duplicate the same microsphere samples. The measurement of X-ray fluorescence by synchrotron radiation-excited energy dispersive spectrometry was done at the Photon Factory, the National Laboratory for High Energy Physics, Tsukuba, Japan, because synchrotron radiation is currently unavailable in most laboratories. However, a much smaller storage ring for accelerated positrons is being developed, and it will expand the medical and industrial use of synchrotron radiation in the future.

In Fig. 4, we show the accuracy of blood flow measurements in small contiguous regions with synchrotron radiation-excited X-ray fluorescence spectrometry. In this experiment on an anesthetized dog, we injected dual nonradioactive microspheres (Ba- and I-, 40 million each) into the left atrium in 4 divided doses, spending 2–5 min between each injection. The calculated amount of microspheres per 1 g myocardium was approximately 20 000 for each set. The dog was then killed and the heart removed. We obtained a short axis slice from the mid portion of the left ventricle with a thickness of 10 mm. The slice was compressed and dried, reducing the thickness to about 2.5 mm. We performed 2 dimensional mapping of X-ray fluorescence on the slice (109 small regions) by using monochromatic synchrotron radiation with an energy level of 50 Kev and with a beam size of 1 × 1 mm. Relative fluorescence activity (%) was converted to absolute flow (μl/min per mg) with reference flow rate (ml/min) and X-ray fluorescence activity of reference blood. The dual flows correlated well (r =

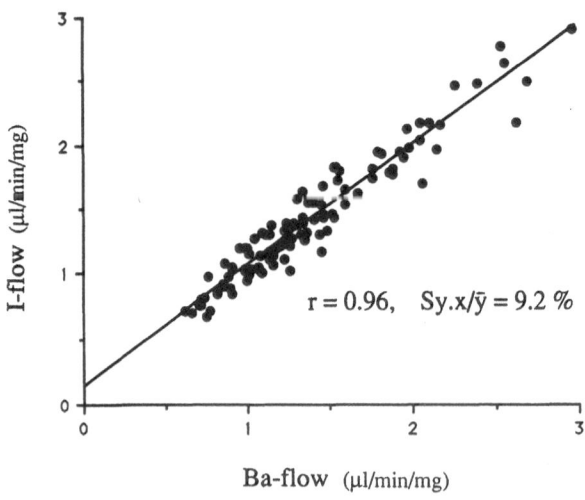

Fig. 4. Linear regession anaysis on dual flows measured with 2 sets of tracers (Ba- and I-) simultaneously injected

0.96) and the variability of the data from the regression line was small ($s_{y|x}/\bar{y}$ = 9.2%). Thus, we could measure accurately the blood flow distribution in small contiguous regions by synchrotron radiation-excited X-ray fluorescence spectrometry.

Comparison with Colored Microsphere Technique

As alternatives to radioactive microspheres for measuring regional blood flow, three methods using colored microspheres have been introduced: (1) direct counting of the colored microspheres under a microscope [2–3], (2) spectrophotometric determination of the amount of dye recovered from colored microspheres [4], and (3) fluorescence measurement of released dyes [5]. In this section, we discuss the advantages and the disadvantages of the three colored microsphere methods and the X-ray fluorescence spectrometry described above.

The advantages of X-ray fluorescence spectrometry are that 9 sets of microspheres can now be measured, with an increase to 18 sets in the near future. In addition, the sequence of measurement is fully automatic with a short counting time (90 s per element per sample) and the cost for nonradioactive microspheres is inexpensive (¥10 000 or US$75 for 1 g of microspheres of each set) without any serious cost for disposal. However, the cost of the wavelength spectrometer is high (approximately US$220 000 in the USA). We are now studying whether we can circumvent the process of tissue digestion by direct measurement on dried and compressed tissue samples using the wavelength dispersive X-ray fluorescence spectrometer.

The advantages of the direct counting method are that the costs of the colored microspheres and the measurement system are low, and that nine sets of colored microspheres are available to measure blood flow [2, 3]. The disadvantages are the tedious counting procedures, and the need for taking aliquots because only small pieces of tissue can be examined at any one time. The alternative colored microsphere method, determining the dye recovered from microspheres by spectrophotometry, reduced the tedious procedure of direct counting and improved the variability in duplicate measured flows [4]. However, the available numbers of colored microspheres are still fewer (5 colors) than for the direct counting method and the current X-ray fluorescence system. A recently introduced fluorescence system currently has seven different dyes, with at least five more potentially available, very high sensitivity and permits counting batches of 96 samples automatically [5].

References

1. Heymann MA, Payne BD, Hoffman JIE, Rudolph AM (1977) Blood flow measurements with radionuclide-labeled particles. Prog Cardiovasc Res 20:55–79
2. Shell W, Kligerman M, Chang BL, Meerbaum S, Corday E (1985) Measurement of myocardial blood flow with non-radioactive microspheres. Circulation 72:III-901

3. Hale SL, Alker KJ, Kloner RA (1988) Evaluation of nonradioactive, colored micro-spheres for measurement of regional myocardial blood flow in dogs. Circulation 78:428–434

4. Kowallik P, Schulz R, Guth BD, Schade A, Paffhausen W, Gross R, Heusch G (1991) Measurement of myocardial blood flow with multiple colored microspheres. Circulation 83:974–982

5. Glenny R, Bernard S, Brinkley M Validation of fluorescent-labeled microspheres for measurement of regional organ perfusion. J Appl Physiol 74:2585–2597

6. Iida A, Goshi Y (1990) Trace element analysis by X-ray fluorescence. KEK Reprint Tsukuba, National Laboratory for High Energy Physics. pp 89–193

7. Kaufman L, Deconinck F, Price DC, Guesry P, Wilson CD, Hruska B, Swann SJ, Camp DC, Voegele AL, Friesen, RD, Nelson JA (1976) An automated fluorescence excitation analysis system for medical applications. Invest Radiol 11:210–215

8. Kaufman LD, Shosa DW, Arbel A, Zender M (1982) Improved quantitation of low level tracers in X-ray fluorescent analysis. Nucl Inst Methods 193:105–110

9. Morita Y, Payne BD, Aldea GS, McWatters C, Husseini W, Mori H, Hoffman JIE, Kaufman L (1990) Local blood flow measured by fluorescence excitation of non-radioactive microspheres. Am J Physiol 258:H1573–H1584

10. Mori H, Haruyama S, Shinozaki Y, Okino H, Iida A, Takanashi R, Sakuma I, Husseini WK, Payne B, Hoffman JIE (1992) New nonradioactive microspheres and more sensitive X-ray fluorescence to measure regional blood flow. Am J Physiol 263: H1946–H1957

11. Sakamoto H, Tanaka Y, Mitomi T, Shinozaki Y, Haruyama S, Mori H (1992) Nonradioactive microspheres and X-ray fluorescence to measure regional blood flow in chronic animal experiments. FASEB Journal 6:A1473

12. Kuwahira I, Mori H, Moue Y, Shinozaki Y, Ohta Y, Yamabayashi H, Okino H, Gonzalez NC, Heisler N, Piiper J (1993) Cardiac output and regional blood flow measurement with nonradioactive microspheres by X-ray fluorescence spectrometry in rats. In: Vaupel P (ed) Advances in experimental medicine and biology, oxygen transport to tissue XV. Plenum Press, New York (in press)

Optimum Use of PET and SPECT for the Detection of Coronary Artery Disease

Nizar A. Mullani[1]

Summary. Detection of coronary artery disease is done by combining a non-invasive myocardial perfusion imaging at rest and stress with coronary angiography for direct visualization of the diseased artery. A review of the imaging modalities capable of myocardial perfusion imaging shows that positron emission tomography and single photon emission computed tomography provide clinically useful information on myocardial perfusion. Positron emission tomography is more accurate than single photon emission computed tomography and results in fewer false positive tests. However, the higher cost of positron emission tomography compared to single photon emission computed tomography requires a compromise between accuracy and the cost of detecting coronary artery disease. An analysis of the costs involved with each procedure shows that the most cost-effective method for detecting coronary artery disease in a population with a low probability of disease is with positron emission tomography and coronary angiography. In a population with moderate probability of disease, single photon emission computed tomography combined with coronary angiography yields slightly lower costs but a higher incidence of false positives. For a population with a high probability of disease, coronary angiography will yield the most cost-effective procedure for detecting coronary artery disease.

Key words: Positron emission tomography (PET)—Single photon emission computed tomography (SPECT)—Coronary angiography (Cath)—Coronary artery disease (CAD)—Cost-effective

Introduction

There are two major modes of operation for the detection of coronary artery disease (CAD) in humans: a direct invasive method of visualizing the artery lumen by coronary angiography (Cath) and the non-invasive detection in CAD by measuring myocardial perfusion at rest and stress [1]. The first method, Cath, has been the gold standard for the detection of CAD, since it not only shows the lumen of the coronary artery but also any obstructions or lesions within it which may be restricting blood flow to the myocardium being perfused by that artery. The second method, the non-invasive detection of CAD by myocardial perfusion imaging, has been the accepted noninvasive method of detecting the presence of a

---

[1] Division of Cardiology and Positron Diagnostic and Research Center, University of Texas Health Science Center, 6431 Fannin Street Houston, TX 77030, USA

stenosis in the coronary artery and is accomplished by measuring the change in perfusion to a segment of the myocardium at high flows induced by some form of stress (physical or chemical).

In the past, planar thallium-201 imaging has been used to image myocardial perfusion at rest and at exercise for the noninvasive detection of CAD. This technique images the heart from two or three different views; two-dimensional images of the uptake of thallium in the myocardium are compared at rest and stress for the detection of perfusion deficits under stress. Recently, planar imaging has been modified to take several different views around the body and to reconstruct the cross-sectional distribution of thallium in the myocardium. This technique, called single photon emission computed tomography (SPECT), provides an unobstructed view of the heart and minimizes the errors due to the overlapping and underlying tissue in planar imaging [2].

Positron emission tomography (PET), using nitrogen-13 ammonia or rubidium-82 chloride, has evolved as another noninvasive method of detecting CAD which is supposed to have higher accuracy for the detection of CAD in humans [3]. PET, which is based on different physical laws of radiation detection [4] than is SPECT, has some features which enable absolute quantitation of blood flow in the myocardium. Unfortunately, the better features of PET also result in increased cost for the equipment and operational expenses compared to such costs for SPECT. However, if PET is used appropriately, its increased cost can be offset by its better accuracy, which feature results in fewer unnecessary Cath procedures.

The main thrust of this chapter is to evaluate noninvasive methods of CAD detection, determine the impact of better quantitation towards improving CAD detection, evaluate cost-effective issues related to the use of noninvasive methods for CAD detection, and finally, to propose a possible scenario for optimizing the use of SPECT and PET for the detection of CAD.

Evaluation of Noninvasive Imaging Methods for CAD Detection

Noninvasive methods of detecting CAD rely on imaging myocardial perfusion during rest and stress. Thus, a major requirement for evaluating any instrumentation for CAD detection is its ability to measure myocardial perfusion. It is not necessary to measure absolute values of myocardial perfusion for the detection of CAD, since this detection is based on the relative distribution of blood flow to the myocardium during rest and stress. However, the accuracy with which myocardial perfusion is measured will influence the accuracy of CAD detection.

Besides PET and SPECT, there are some new imaging modalities which show promise for future applications towards the noninvasive detection of CAD. Echocardiography (Echo) with injection of microbubbles is being used to image relative myocardial perfusion. The use of Echo with microbubbles injected into the aortic root shows promise in early animal studies. However, this technique is in its early stages and will require further development and validation in patients [5]. Magnetic resonance imaging (MRI) using either the dynamic mode of operation or the fast echo planar imaging method, is being investigated for myocardial perfusion imaging. MRI does not currently measure myocardial perfusion and will require several years of development and validation before it can be useful clinically [6]. New developments in imaging the effect of paramagnetic contrast agents with high magnetic field strength systems show promise as indirect measures of myocardial perfusion at low flows, but fail to parallel the high flow necessary for assessing CAD. The digital fluoroscopy (DF) technique may be useful for perfusion imaging but has proven disappointing to date [7].

PET and SPECT, which use radioactive tracers to measure myocardial perfusion, are the most proven and accurate methods of detecting CAD noninvasively. SPECT can measure relative, but not absolute, perfusion due to its radiation detection physics, which results in inadequate attenuation correction [4]. PET, due to its unique radiation detection physics, is the only noninvasive imaging modality which can quantitatively measure myocardial perfusion at the present time [1]. Table 1 lists some essential information, from equipment cost to FDA approval, on these different imaging modalities and can be summarized thus: The only imaging techniques that are clinically validated and approved by the FDA at this time for assessing myocardial perfusion are SPECT and PET.

Does Improved Perfusion Measurement Relate to Improved CAD Detection?

The present method for noninvasive measurement of CAD with PET and SPECT is based on the theory that, at high coronary artery flows, there will be reduced

Table 1. Non-invasive imaging modalities considered for measurement of myocardial perfusion

	PET	SPECT	Echo	MRI	DF
Quantitative[a]	Yes	No	Maybe	Maybe	Possible
Perfusion capability[a]	Yes	Yes	Possible	Possible	Possible
Validation of quantitative flow	Yes	No	No	No	No
FDA-approved contrast agents	Yes	Yes	No	No	No
Cost of equipment	$1–2 million	$400–800 000	$150–300 000	$1–2 million	$1–2 million
Rest/stress test validation	Yes	Yes	No	No	No
Scan cost[b]	$1500	$1000	?	?	?
Sensitivity[c]	96%	75%	N/A	N/A	N/A
Specificity[c]	96%	67.5%	N/A	N/A	N/A
Scan time (rest/stress)	40–60 min	4 h	N/A	N/A	N/A

PET, Positron emission tomography; *SPECT*, single photon emission tomography; *Echo*, echocardiography; *MRI*, magnetic resonance imaging; *DF*, digital fluoroscopy
[a] Data compiled from [2, 3, 5, 6, 7, 10, 11, 18]
[b] Data obtained from [16]
[c] Data obtained from [13 and 14]

flow in a poststenotic region relative to a normal artery, due to the pressure drop across the stenosis. Therefore, the poststenotic myocardium will have decreased perfusion and will exhibit a perfusion defect with respect to the normally-perfused myocardium. The theory relating perfusion deficit to severity of CAD has previously been described by Mullani [8] and shows that the size of the myocardial defect and its severity are linked to the diameter of the coronary artery and the stenosis severity.

The ability of PET or SPECT to accurately detect a perfusion defect is elegantly described by Lim et al. [9] and is dependent on the image noise, reconstructed resolution, size of the perfusion defect, severity of the defect, and the uniformity of detection of myocardial perfusion in a normally-perfused heart. Image noise will depend on the radioactive tracer used, its extraction in the myocardium, scan times, image reconstruction techniques, and the reconstructed resolution. Resolution, size of the defect, and the severity of the defect are closely coupled. Small defects with mild perfusion defects will require higher resolution and low image noise for detection. Large defects that are severe can be detected with a poor resolution system, as long as the image noise is smaller than the measured defect severity.

Another important factor which plays a major role in lesion detectability is the uniformity of detection. In order to detect a defect with any confidence, the variation measured from region to region in a normally-perfused heart must be less than the measured perfusion deficit. If the threshold for detecting a perfusion deficit is set lower than the detection uniformity of the system, false perfusion deficits (false positives) will be created in the normal heart where the detection efficiency is low due to attenuation of the signal. Nienaber et al. [10] have shown that the variation in myocardial perfusion measured by PET is less than 10% following attenuation correction. Attenuation correction is routine with PET, but not with SPECT [11]. Variation of perfusion measurements for SPECT will depend on the size of the patient, position of the heart, proximity of the diaphragm, and the size of the breasts [12]. Variation of perfusion measurements obtained with SPECT during rest and stress will also depend on patient positioning, since the amount of tissue between the heart and the detector can change from the first scan to the second. The physical properties of PET, such as resolution and uniformity, are clearly better suited for quantitation of myocardial perfusion than those of SPECT. According to the theoretical work done by Lim et al. [9]. PET should provide better quantification than SPECT for the detection of CAD as demonstrated experimentally and clinically.

Comparison of PET and SPECT Accuracy for CAD Detection

The physics of PET and SPECT dictate that the higher resolution and more uniform detection capability of PET over SPECT should result in PET being able to detect smaller and less severe perfusion deficits than SPECT. The physics of lesion detection also predicts that this difference will disappear as the lesion becomes larger and more severe. In other words, PET is more suitable for detecting mild to moderate myocardial perfusion deficits caused by early CAD, while both PET and SPECT are equivalent for detecting large, severe perfusion deficits caused by advanced CAD.

Go et al. [13] compared PET and SPECT for the detection of CAD in the same patient population and documented the expected results predicted by the physics of the two detection systems. The overall accuracy of detecting CAD in 202 patients with a prevalence of disease of 75% (angiographically-determined lesions of greater than 50% diameter stenosis) was 90% for PET compared to 77% for SPECT. Twenty-seven of these patients were false negatives by SPECT but not by PET. Of these, 21 of the false negatives detected by SPECT were located in the inferior and posterior parts of the heart, where self-attenuation of the radiation is most severe. Three false positives by SPECT that were normal by PET were believed to be due to breast and diaphragmatic wall attenuation artifacts. The authors conclude that: "The improved contrast resolution of PET resulted in markedly superior images and a more confident identification of defects" [13]. In other words, lack of attenuation correction and deteriorating resolution towards the inferior part of the heart were the most likely causes of the errors in CAD detection by SPECT. Correction for attenuation correction and more uniform detection are the basis of the better performance of PET for the detection of CAD. Perhaps the most significant aspect of their conclusion is that they felt more *confident* in identifying a perfusion defect with PET, in other words, there was a better signal to noise ratio and less uncertainty in identifying the perfusion defect. Their observations are consistent with the theory of Lim et al. [9].

Several other independent studies show that the overall sensitivity and specificity of CAD detection by PET is better than these features in SPECT [14]. The average sensitivity and specificity of thallium stress testing corrected for referral bias, as reviewed by Diamond [15], is 75% and 67.5% respectively. Van Train et al. [2] in a multicenter trial, found the sensitivity and specificity of SPECT for the detection of CAD in non-myocardial infarction (MI) patients to be 91% and 45%, respectively. The multicenter averages for sensitivity and specificity by PET [14] are 96% and 96%, respectively. Allowing for conservative changes in these numbers over time, we can anticipate a sensitivity of 90%–95% for PET and 85%–90% for SPECT, and a specificity of 90%–95% for PET and 70%–75% for SPECT.

Cost-Effective Analysis of PET and SPECT for CAD Detection

The cost of detecting CAD is often confused with the cost of equipment and the cost of a particular procedure. The real cost of CAD detection should include the cost of all the tests necessary to confirm or rule out the presence of CAD. A detailed cost analysis that has been presented by Gould et al. [16] considers the long-term costs associated with misdiagnosing CAD. However, it is perhaps simpler and more expedient to consider the immediate costs involved with the detection of CAD and not the intangible long-term costs such as loss of productivity due to sudden death. The following analysis computes the cost of the major components of CAD detection in a patient who is suspected of having CAD, after routine preliminary tests such as electrocardiogram (EKG) and physical examination.

The most common procedure for the detection of CAD in a patient in whom CAD is suspected, after the performance of preliminary tests, is a noninvasive stress perfusion (nuclear test), followed by coronary catheterization (Cath) of those patients with positive nuclear tests. The cost of nuclear scans and Cath studies represents a major portion of the total cost of CAD detection and can be computed for a given population with a given prevalence of disease. Two imaging schemes can be compared for CAD detection: (a) the patient population is imaged by SPECT, followed by Cath for all positives and, (b) the patient population is imaged by PET, followed by Cath for all positives. The combined cost of nuclear and Cath for the two schemes will be a function of the prevalence of disease in the patient population, the sensitivity and specificity of the nuclear test, and the global cost of each test [17]. Table 2 lists the parameters and cost calculations for the two scenarios based on the Iskandrian study population [18] in which the use of SPECT for CAD detection was evaluated. We have excluded the MI patients from that study, leaving 336 non-MI patients imaged by SPECT for the detection of CAD. This study population had a 44% prevalence of disease. We used Eq. 1 to compute the cost of nuclear and arteriographic procedures for CAD detection for a given population (N), with a prevalence of disease (p), sensitivity (Sn), and specificity of the nuclear test (Sp).

$$\$CAD = [(N)(\$nuclear)] + [(N)(p)(Sn) + (N)(1 - p)(1 - Sp)][\$Cath] \quad (1)$$

where $CAD is the cost of CAD detection, $nuclear is the cost of nuclear study, and $Cath is the cost of catheterization.

The analysis in Table 2 shows that the combined cost for nuclear and Cath procedures for the detection of CAD in the Iskandrian study population is $4589 per patient for SPECT and $4161 for PET. Even though the cost of the nuclear scan for PET is greater than that for SPECT, the combined cost of detecting CAD by nuclear and Cath procedures is less for PET than for SPECT. The reason is that a large number of false positives are generated by SPECT, due to its low specificity; an expensive Cath is then required to rule out CAD. The higher accuracy of PET minimizes the number of normal Caths, so that the overall cost is reduced, even though the cost of the PET procedure, at $1500 per patient, is higher than the cost of the SPECT procedure, at $1000 per patient.

Some concern has been raised that the sensitivity and specificity of a test will change once the test is utilized more widely and, therefore, the cost of detecting CAD by SPECT and PET will change with time. To evaluate the effect of changes in sensitivity and specificity on our cost calculations, we reduced the sensitivity and specificity of PET from 95% to 90%, but improved the sensitivity

Table 2. SPECT and PET comparative cost of coronary artery disease (*CAD*) diagnosis based on the Iskandrian et al. SPECT study [18] population from which the myocardial infarction (*MI*) patients have been excluded

Iskandrian et al. study population (N = 336, Prevalence of Disease = 0.44)		
	Tl-SPECT	Rb-PET
Sensitivity	88%	95%
Specificity	62%	95%
True positives	129	140
False positives	72	9
Total Caths	201	149
Cost of Caths	$1 206 000	$894 000
Cost of scans	$336 000	$504 000
(Cath cost for false positives)	($432 000)	($54 000)
Cost of scan + Cath	$1 542 000	$1 398 000
Cost per Patient	$4 589	$4 161

Sensitivity and specificity for SPECT were obtained from the Iskandrian et al. study; these features of PET were obtained from [13]. PET is hypothetically substituted for SPECT for this calculation

Global costs: Cath, Coronary angiography, $6000; *Tl-SPECT*, $1000; Rb-PET, $ 1500

and specificity of SPECT to 90% and 70%, respectively. The new values for sensitivity and specificity were applied to the Iskandrian study population; the results are shown in Table 3. The cost of CAD detection by PET was $4214 per patient compared to $4393 per patient with SPECT as the nuclear test. Even with the use of favorable numbers for sensitivity and specificity for SPECT and less favorable numbers for PET, the cost of CAD detection in this patient population would have been lower for PET than for SPECT, and with fewer normal Caths.

PET is also more cost-competitive than SPECT when we apply similar calculations to the recent large multicenter trial study population described by Van Train et al. [2]. In this population of 243 non-MI patients scanned for CAD by SPECT, the prevalence of disease was quite high, at 74%, but the poor specificity of 45% resulted in a large number of normal Caths. The cost of CAD detection using SPECT for the nuclear scan is $5914 per patient compared to $5646 for PET (Table 4). The number of false positives in this study is so high that the cost of arteriograms to rule out CAD is almost as much as the cost of nuclear scans for all the patients. This is true even for the conservative cost estimates of Table 3, which shows that the cost of arteriograms of the false positives generated by SPECT is higher than the cost of SPECT scans for all the patients.

We are able to predict from the characteristics of SPECT and PET that, for a population with a low prevalence of disease, the number of normal Caths

Table 3. SPECT and PET comparative cost of CAD diagnosis based on optimistic sensitivity (Sn) and specificity (Sp) for SPECT but lower Sn and Sp for PET

High SPECT Sn and Sp and conservative PET Sn and Sp (N = 336; Prevalence of disease = 0.44)		
	Tl-SPECT	Rb-PET
Sensitivity	90%	90%
Specificity	70%	90%
True positives	133	133
False positives	57	19
Total Caths	190	152
Cost of Caths	$1 140 000	$912 000
Cost of scans	$336 000	$504 000
(Cath cost for false positives)	($342 000)	($114 000)
Cost of scan + cath	$1 476 000	$1 416 000
Cost per patient	$4 393	$4 214

The same patient population as in Table 2 is used but with modified values for sensitivity and specificity
Global costs: Cath, $6000; Tl-SPECT, $1000; Rb-PET, $1500

Table 4. SPECT and PET comparative cost of CAD diagnosis based on the multi-center SPECT study by Van Train et al. [2]

Van Train et al. multicenter non-MI SPECT study population (N = 243; prevalence of disease = 0.74)		
	Tl-SPECT	Rb-PET
Sensitivity	91%	90%
Specificity	45%	90%
True positives	470	450
False positives	35	6
Total Caths	199	168
Cost of Caths	$1 194 000	$1 008 000
Cost of scans	$243 000	$364 000
(Cath cost for false positives)	($210 000)	($36 000)
Cost of scan + cath	$1 437 000	$1 372 000
Cost per patient	$5 914	$5 646

Sensitivity and specificity for SPECT were obtained from the Van Train et al. study and conservative values for sensitivity and specificity are used for PET
Global costs: Cath, $6000; Tl-SPECT, $1000; Rb-PET, $1500

generated by PET should be lower than that generated by SPECT. Therefore, for this population, the number of normal Caths is reduced enough to make PET more cost-competitive than SPECT. Also, for a population with severe CAD and a high pretest probability of disease, it may be more cost-competitive to go directly to Cath for the diagnosis of CAD.

Table 5. Comparison of cost of nuclear scan and Cath cost for TI-SPECT and Rb-PET for hypothetical patient populations with different prevalence of disease[a]

Prevalence of disease	0.1	0.2	0.3	0.4	0.5	0.6	0.7	0.8
SPECT + Cath cost/ patient	$3160	$3520	$3880	$4240	$4600	$4960	$5320	$5680
PET + Cath cost/patient	$2580	$3060	$3540	$4020	$4500	$4980	$5460	$5940
SPECT-PET cost diff/ patient	$580	$480	$340	$220	$100	$(20)	$(140)	$(260)

[a] These costs are for the nuclear scans of every patient and Caths of all positive nuclear test findings. Global fees for the diagnostic tests are: Cath, $6000, TI-SPECT, $1000; Rb-PET, $1500. Conservative values of sensitivity and specificity are used for PET (Sn, 90%; Sp, 90%) while optimistic vales are used for SPECT (Sn, 90%; Sp, 70%)

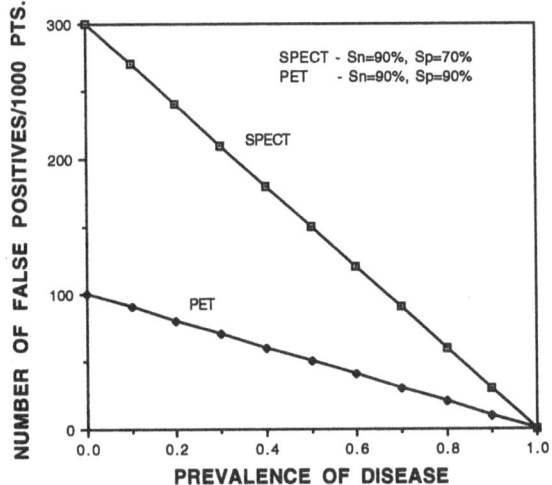

Fig. 1. A plot of the number of false positive test results generated by single photon emission computed tomography (*SPECT*) and positron emission tomography (*PET*) as a function of the prevalence of disease in a population. These false positives usually result in unnecessary coronary angiography (*Caths*). The values for specificity are conservatively set at 70% for SPECT and 90% for PET

To better understand the interplay of cost of CAD detection and prevalence of disease, we· have computed the cost per patient of SPECT and PET for hypothetical populations with different prevalence of disease (Table 5). Once again, conservative values of sensitivity (90%) and specificity (90%) were used for PET and optimistic values of sensitivity (90%) and specificity (70%) were used for SPECT. The number of false positives resulting from SPECT and PET are shown in Fig. 1 for a range of populations with different prevalence of disease. For patient populations which have a prevalence of disease of 60% or less, PET is more cost-effective than SPECT, due to the significantly lower number of unnecessary Caths of the false positive tests. For patient populations with higher than 60% prevalence of disease, SPECT may become more cost-competitive than PET. The breakeven point between PET and SPECT is a function of the imaging cost of PET, SPECT, and Cath and the values of sensitivity and specificity used for PET and SPECT. Details of how the use of PET and SPECT can be optimized from the cost-effective point of view are discussed in the following sections.

Is There an Optimum Utilization of PET and SPECT for CAD Detection?

An evaluation of the physical properties, cost of equipment, accuracy of lesion detection, and the cost-effectiveness of PET and SPECT shows that PET is more accurate for detecting CAD non invasively than SPECT. The higher accuracy of PET can reduce the number of unnecessary Caths and thereby lower the overall cost of CAD detection in humans for a given prevalence of disease. If the cost of equipment and operation of a PET or a SPECT camera is not an issue, it is clear that the better physical properties of PET would make it the better modality for detecting CAD. However, PET equipment is expensive and is not as readily available as SPECT. Therefore, we must attempt to optimize the use of PET and SPECT to reduce the cost of CAD detection while maintaining high detection accuracy.

The cost-effective analysis equation (Eq. 1) has parameters, besides the cost of a procedure, which influence the overall cost of CAD detection. These parameters are sensitivity, specificity, and the prevalence of disease, and can influence the breakeven point for the overall cost of detecting CAD. Assuming that a positive finding by PET or SPECT is automatically followed by a Cath, it is possible to examine the cost-effective analysis as a function of the prevalence of disease. An example of such a cost analysis is shown in Fig. 2; we can discuss the cost and accuracy implications of three groups of patients with different prevalence of disease.

Group 1: Low Prevalence of Disease

The number of people without disease in this group is much greater than the number with disease. Therefore, from the cost-effective equation, it is important for the nuclear test to have very high specificity to reduce the number of false positives and unnecessary Caths. As an example, if the average prevalence of disease is 20% and the specificity for PET and SPECT is 90% and 70%, respectively, SPECT will detect 24% of the population as false positives compared to 8% for PET. The poor specificity of SPECT results in a very high number of false positives and an overall increase in the cost of CAD detection. Using the SPECT system will this group can result in an increased cost of approximately 20% over that for PET. Compared to the cost of Cath of all the patients in this group, a PET scan followed by Cath of all the positive findings will reduce the cost of CAD detection by approximately one-half.

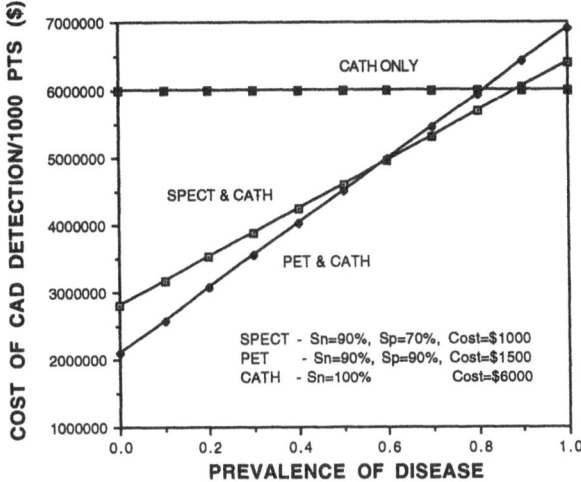

Fig. 2. Cost analysis of three modes of detecting coronary artery disease (*CAD*): Cath only, PET followed by Cath of all the positive tests, and SPECT followed by Cath of all the positive tests. The values for sensitivity and specificity were conservatively set at 90% and 70% for SPECT and 90% and 90%, respectively for PET. Cost of each test is $1000 for SPECT, $1500 for PET, and $6000 for Cath

Group 2: Moderate Prevalence of Disease

For the group of people with a prevalence of disease between 40% and 80%, the cost of catheterizing all of them is high because of the high number of normals who would be Cathed unnecessarily. A nuclear test, followed by a Cath of the positive findings from the nuclear test, results in significant reduction in the cost of CAD detection by both PET and SPECT. SPECT costs are slightly higher than those of PET, due to the higher number of false positive tests created by the lower specificity of SPECT. An example of the number of false positive tests and the cost associated with doing a Cath procedure on these patients in shown in Table 2, which shows a population with a disease prevalence of 44%. A similar cost analysis with a group of patients with a disease prevalence of 74% is computed in Table 4. In both cases the number of false positives created is five to six times higher for SPECT than for PET. The overall cost of CAD detection, including the cost of Caths, is less for PET, by less than 10%, in this group. Therefore, if cost were the major concern for CAD detection, SPECT could be used quite effectively in this group of people.

Group 3: High Prevalence of Disease

The group which has a high prevalence of disease, i.e., between 80% and 100%, will have a very large number of people with disease and very few without disease. The number of Caths performed on this group of patients will be dominated by the number of people with disease and the sensitivity of the nuclear test. The overall cost of CAD detection using a nuclear scan followed by a Cath in this group is higher than if the group were to be catheterized without a preceding nuclear scan. Therefore, the most cost-effective way of diagnosing disease in this group is to bypass the nuclear test and go directly to a Cath.

Patient Selection Criteria for Optimum Utilization of PET and SPECT for CAD Detection

The preceding analysis shows that the use of PET and SPECT for the detection of CAD can be optimized for cost and accuracy based on the performance characteristics of the nuclear imaging devices, the prevalence of disease, and the cost of the tests. Sensitivity, specificity, and prevalence of disease are statistical parameters which define the quality of a test or a population. A patient in group 1 would have a statistical average of 20% probability of having CAD. An analysis based on known prevalence of disease in a group of patients needs to be converted to a pretest probability of disease in a patient for it to be useful in selecting who should be tested with what method. Therefore, the key to the success of PET and SPECT utilization optimization depends on being able to define a pretest statistical probability of disease for each patient, depending on the symptoms, known probabilities of disease for each symptom, and patient history.

It is possible to construct a probability-based decision tree for the selection of the most appropriate test for each patient. This decision tree could be a knowledge-based system in which the probability of the best test for each patient can be computed based on patient history and symptoms. As an example of a simple decision tree, we know that, for the American population, the probability of dying from heart disease increases dramatically after the age of 50 years. Therefore, persons suspected of having CAD will have a larger pretest probability of CAD if they are over the age of 50 years than if they are younger than 50 years in age. The presence of angina can significantly increase the probability of CAD in a patient, such that a person over 50 years of age with confirmed angina should have a high enough pretest probability of disease to go directly for a Cath. A person with atypical chest pains who is over the age of 50 would have a moderate pretest probability of disease and could be scanned by SPECT. In order to avoid excessive false positives due to lack of attenuation correction in this group of patients, it would be better to image females by PET rather than by SPECT. Asymptomatic patients suspected of having heart disease would probably have a low probability of disease and therefore should be scanned by PET. An example of a simplified decision tree based on the symptoms and patient history discussed above is shown in Fig. 3. This decision tree could be expanded to include other risk factors and probabilities of CAD. It could also be automated to compute the highest probability of disease for each patient and select the optimum test for that patient.

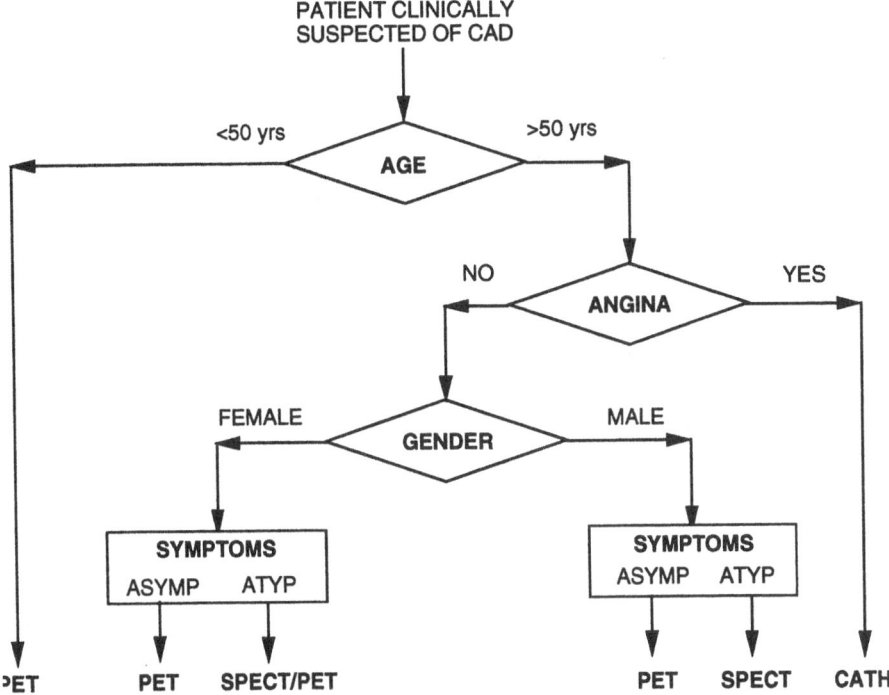

Fig. 3. A simplified decision tree for selecting the most cost-effective test (PET, SPECT or Cath for CAD detection) for a patient who is clinically suspected of coronary artery disease. Other parameters such as smoking, family history, high cholesterol, etc. can be added to this decision tree to improve its effectiveness

Conclusions

PET and SPECT are two very good noninvasive tests for the detection of CAD in humans. The physical properties of PET detection improve the accuracy of PET for CAD detection over SPECT. For the detection of CAD, with the exception of patients with a high pretest probability of disease, it is more cost-effective to screen a patient suspected of CAD with a nuclear scan such as SPECT or PET before subjecting the patient to a Cath. The high accuracy of PET reduces the number of false positive tests and unnecessary Caths so that the overall cost of detecting CAD by PET and Cath is lower than that that for SPECT and Cath. For the group of patients with a low probability of disease, PET screening is the most cost-effective, while for the group with moderate probability of disease, either PET or SPECT could be used from the cost point of view. The number of false positives detected by SPECT is always higher than by PET and is of concern in the clinical management of a patient. Patients with a high probability of disease should probably be imaged by Cath rather than pre-tested with a nuclear scan. A decision tree that might be useful for determining the optimum test for a patient needs to be developed to help reduce the cost of and the errors caused by these tests.

References

1. Gould KL (1990) Coronary artery stenosis. Elsevier Science, New York
2. Van Train KF, Maddahi J, Berman DS, et al (1990) Quantitative analysis of tomographic stress thallium-201 myocardial scintigrams: A multicenter trial. J Nucl Med 31:1168–1179
3. Hicks K, Ganti G, Mullani NA, Gould KL (1989) Automated quantitation of three-dimensional cardiac positron emission tomography for routine clinical use. J Nucl Med 30:1787–1797
4. Volkow ND, Mullani NA, Bendriem B (1988) Positron emission tomography instrumentation: An overview. Am J Physiol Imaging 3:142–153
5. Vandenberg BF (1991) Myocardial perfusion and contrast echocardiography: Review and new perspectives. Echocardiography 8:65–75
6. Le Bihan D (1990) Magnetic resonance imaging of perfusion. Magn Resonance Med 14:283–292
7. Detrano R, Haggman DL, Simpfendorfer C, Hobbs RE, Ramsey KJ, Strobl DJ, Salcedp EE, Sheldon WC, Friis R (1988) Digital fluoroscopy and intravenous cardiac angiography for the detection of coronary artery disease in selected subjects. A feasibility study. Cleve Clin J Med 55:129–135
8. Mullani NA (1984) Myocardial perfusion with rubidium-82: Theory relating severity of coronary artery stenosis to perfusion deficit. J Nucl Med 25:1190–1196
9. Lim CB, Kyung SH, Hawman EG, Jaszcak RJ (1982) Image noise, resolution, and lesion detectability in single photon emission CT. IEEE Trans Nucl Sci NS-29: 500–505
10. Nienaber CA, Ratib O, Gambhir SS, et al (1991) A quantitative index of regional blood flow in canine myocardium derived non-invasively with N-13 ammonia and dynamic positron emission tomography. J Am Coll Cardiol 17:260–269
11. Phelps ME, Hoffman EJ, Mullani NA, Ter-Pogossian MM (1975) Application of annihilation coincidence detection to transaxial reconstruction tomography. J Nucl Med 16:210–214
12. DePuey EG, Garcia EV (1989) Optimal specificity of thallium-201 SPECT through recognition of imaging artifacts. J Nucl Med 30:441–449
13. Go RT, Marwick TH, MacIntyre WJ, et al (1990) A prospective comparison of Rubidium-82 PET and Thallium-201 SPECT myocardial perfusion imaging utilizing a single dipyridamole stress in the diagnosis of coronary artery disease. J Nucl Med 31:1899–1905
14. Gould KL (1990) Agreement on the accuracy of thallium stress testing. J Am Coll Cardiol 16:1022–1023
15. Diamond GA (1990) How accurate is SPECT thallium scintigraphy? J Am Coll Cardiol 16:1017–1021
16. Gould KL, Goldstein RA, Mullani NA (1989) Economic analysis of clinical positron emission tomography of the heart with rubidium-82. J Nucl Med 30:707–717

17. Gould KL, Mullani NA, Williams B (1990) PET, PTCA, and economic priorities. Clin Cardiol 13:153–164
18. Iskandrian AS, Heo J, Kong B, Lyons E (1989) Effect of exercise level on the ability of thallium-201 tomographic imaging in detecting coronary artery disease: Analysis of 461 patients. J Am Coll Cardiol 14:1477–1486

Abstract

Uridine is a Marker of Viability in Myocardial Ischemia

*Hiroyuki Yaoita[1], Alan J. Fischman[2], Hiroshi Ohtani[2], H. William Strauss[2],
Michito Kanke, Tsukasa Asakura, Tomiyoshi Saito, Nobuaki Tsuchihashi,
and Yukio Maruyama[1]*

To test the hypothesis that uridine may be a marker of viability in myocardial
ischemia, myocardial uptake of uridine was compared to that of deoxyglucose, a
conventional marker of viability, in experimental myocardial ischemia. Fifteen
male Wistar rats had occlusion of the left coronary artery for 5, 10, or 60 min.
Four h after reperfusion, 60 uCi of ^{3}H uridine and 2 uCi of ^{14}C deoxyglucose were
administered intravenously and the animals were sacrificed 45 min later. Two
min before sacrifice, 100 uCi of thallium-201 was injected intravenously as a
marker of myocardial perfusion. After sacrifice, the heart was removed and the
left ventricle was sliced in a breadloaf fashion into 20 segments. Thallium-201
radioactivity in each segment was determined with a gamma counter; 1 month
later ^{3}H and ^{14}C radioactivity was determined with a beta counter. Then
the target to normal ratio (T/N) of the myocardial segment was calculated.

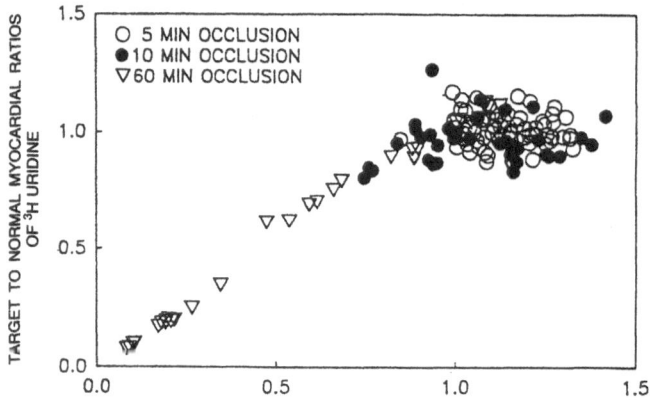

Fig. 1. Relationship of target to normal myocardial ratios between ^{3}H uridine and ^{14}C
Deoxyglucose

---

[1] First Department of Internal Medicine, Fukushima Medical College, Fukushima
[2] Massachusetts General Hospital, Boston, USA

Triphenyl tetrazolium chloride (TTC) staining was examined in neighboring myocardial specimens.

The absence of TTC staining and the reduction of deoxyglucose uptake was well correlated. In the infarcted segments (T/N of deoxyglucose < 1), deoxyglucose and uridine had a linear relationship in myocardial uptake. In the ischemic segments (T/N of deoxyglucose > 1), the mean T/N value for uridine was 1.07. These results suggest that retention of uridine may indicate viability in the ischemic myocardium.

B. Coronary Flow Dynamics

Transmural Distribution and Heterogeneity of Coronary Flow and Flow Reserve

Julien I.E. Hoffman, Arthur E. Flynn[ab]*, Richard E. Austin*[c]*, Dwain L. Coggins*[b]*, Gabriel S. Aldea*[b]*, Waleed K Husseini, Masami Goto*[a]*, and Joseph W. Doucette*[a][1]

Summary. Left ventricular subendocardial vulnerability to ischemia occurs in almost all forms of heart disease. The underlying mechanism of this vulnerability is due to the variation of intramyocardial pressures, which are highest under the endocardial surface of the left ventricle and lowest below the epicardium. At the onset of systole, pressures in the subendocardial vessels exceed those in the extramural coronary arteries. Consequently, flow from subendocardial arteries and arterioles is retrograde as well as anterograde. This retrograde flow accounts for retrograde systolic flow in the extramural arteries and also for forward systolic flow in the subepicardial vessels; little or no blood enters the myocardium from the extramural arteries in systole. At the end of systole, the subendocardial arteries and arterioles are narrowed, and therefore have a high resistance to refill in the next diastole. If diastolic time or perfusion pressure are decreased, then subendocardial flow will be compromised. Even in the subendocardium, however, there is variability of coronary flow reserve, so that decreased perfusion pressure affects first those small pieces of muscle that have the lowest reserve. Then additional pieces are recruited until, eventually, almost the whole subendocardium is ischemic. This variability, of unknown origin, probably accounts for the patchy necrosis seen in chronic ischemia.

Key words: Coronary flow reserve—Intramyocardial pressures—Systolic-diastolic interaction

Introduction

The vulnerability of the subendocardium of the left ventricle to reduced flow, flow reserve, and ischemia has been amply documented [1, 2]. It occurs with decreases in diastolic perfusion pressure, severe hypoxemia or polycythemia, aortic stenosis, any marked ventricular hypertrophy associated with ventricular dilatation, and very fast heart rates, as well as with narrowing of the large or small coronary arteries due to localized or diffuse disease. This review discusses

[1] Department of Pediatrics and Cardiovascular Research Institute, University of California, San Francisco, CA 94143, USA

Supported, in part, by Program Project Grant HL 25847 from the National Heart, Lung and Blood Institute. [a] Supported by the American Heart Association, California Affiliate [b] Supported by a National Institutes of Health Training Grant HL-07192 [c] Supported by a Sarnoff Fellowship

what is known about mechanisms causing this vulnerability, and introduces the additional complexity of heterogeneity of flow in each layer of the left ventricle.

Early Explanations for Subendocardial Ischemia

Early theories explained subendocardial ischemia as a consequence of systolic intramyocardial pressures being as high as intracavitary pressures at the endocardium and decreasing almost linearly to near atmospheric pressures under the epicardium. Investigators assumed that these pressure differences implied that in systole there would be systolic flow in the subepicardium but not in the subendocardium. Therefore the subendocardium would get flow only in diastole, and any decrease in diastolic duration or perfusion pressure would decrease flow mainly in the subendocardium. Many studies in animal models of heart disease seemed to fit this hypothesis [3], but it became untenable when investigators found that no or almost no flow entered the myocardium in systole [4].

The second theory was based on the suggestion by Bellamy [5] that there might be an intramyocardial waterfall, with a back pressure to flow that was higher than right atrial pressure. If back pressures were much higher in the subendocardium than in the subepicardium, then subendocardial flow would be selectively compromised at low perfusing pressures. Under normal circumstances, however, this back pressure does not seem to be different enough in different layers to explain selective subendocardial ischemia [6]. Nevertheless, there are occasions when back pressures may be important. Uhlig and colleagues [7] found a vascular waterfall in the coronary veins with resting pressures of about 12 mmHg in them, and also observed, in one experiment, that when left ventricular end-diastolic pressure rose during acute ischemia, there was an almost equivalent rise in coronary venous pressure. Takishima and his colleagues in Sendai (Watanabe et al. [8] and Satoh et al. [9]) studied this phenomenon more carefully. They found, in a maximally-vasodilated arrested canine isolated heart preparation, that raising left ventricular pressure to 30 mmHg increased zero flow pressure by about 20 mmHg. The mechanisms and regional effects of this change were studied in a similar preparation by Aldea et al. [10], who found that when left ventricular pressure was raised from zero to 30 mmHg the pressure in the great coronary vein increased from about 5 to 20 mmHg. They attributed this increased venous pressure to a vascular waterfall produced by compression of the veins between the distending ventricle and the overlying epicardial connective tissue; the fact that venous pressure was only two-thirds of cavity pressure was attributed to dissipation of cavity pressure in the ventricular wall. They also found that at a fixed low coronary perfusion pressure, raising left ventricular cavity pressure decreased flow in all layers of the left ventricle, but the decrease was most marked in the subendocardial layer; the possible mechanism is a high subendocardial radial stress or strain. The clinical application of these findings may be that with acute ventricular dilatation there is an increased coronary venous pressure (much higher than right atrial pressure) that decreases the driving pressure across the heart and jeopardizes subendocardial blood flow.

Current Explanation of Subendocardial Ischemia

Under most circumstances, however, subendocardial ischemia is due to another mechanism. To understand this, consider first the concept of intramyocardial capacitance that has been stressed by Spaan and his colleagues [11–13]. About 10%–15% of the normal left ventricular wall is blood, and this volume can change markedly throughout the cardiac cycle. In 1980 we observed that in the arrested, maximally-vasodilated dog heart, flow was greatest in the subendocardium [14], thus confirming what others had previously described [15]. In the same dog, however, when the heart was allowed to beat, not only did subendocardial flow decrease, as expected, but subepicardial flow increased. This phenomenon was studied in more detail by Flynn et al. [16], who studied the maximally-vasodilated dog left ventricle during cardiac arrest and then at heart rates of 60, 100, and 150 beats per min while the coronary arteries were perfused at constant pressure from a Gregg cannula. As heart rate increased, subendocardial flow decreased progressively, as previously described [17]. However, subepicardial flow increased by about 30% when the heart began to beat, and did not change much thereafter at different heart rates (Fig. 1). During these studies, a Doppler velocity meter on a small terminal branch of the left coronary artery showed that no flow entered the myocardium in systole; in fact, there was usually a small amount of retrograde flow in systole.

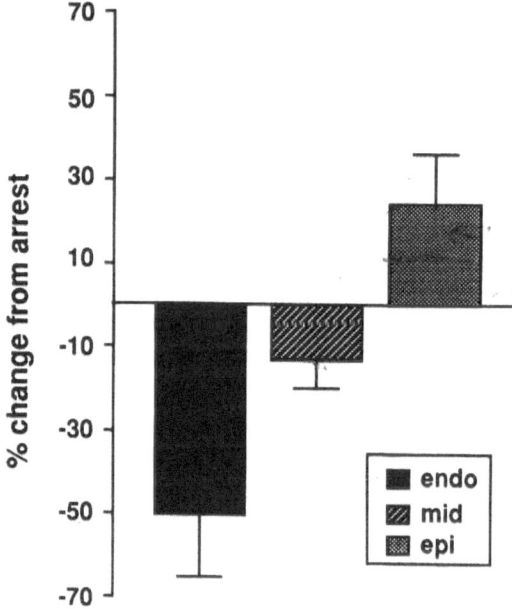

Fig. 1. Percentile changes in myocardial blood flow in subendocardial (*black column*), midmyocardial (*cross-hatched column*), and subepicardial (*gray column*) layers from cardiac arrest to a heart rate of 100 beats/min

These findings were interpreted as follows. When intramyocardial subendo-cardial pressure increases markedly at the beginning of systole, the intravascular pressure in subendocardial arteries becomes the sum of the old intravascular pressure (about 75 mmHg) and the additional intramyocardial pressure (about 100 mmHg) because the vessel wall is deformable; the transmural pressure remains constant as long as the intravascular volume does not change. The new intravascular pressure at the beginning of systole exceeds the pressure in the extramural coronary artery, thus permitting retrograde flow from the subendo-cardial arteries and arterioles during systole. The retrograde flow explains the retrograde flow noted by flowmeter in the terminal extramural coronary artery, and also accounts for the increased subepicardial flow, because some of the retrograde flow then moves anterograde in the subepicardial vessels. This concept fits well with the direct observation that there is continuous forward flow in subepicardial vessels, even though no blood enters the myocardium from the extramural vessels in systole [18, 19].

This hypothesis implies that, by the end of systole, there will be marked narrowing of all the subendocardial vessels, whereas those in the subepicardium, being exposed to low intramyocardial tissue pressures, will not be narrowed, Constancy of the diameters of the subepicardial vessels throughout the cardiac cycle has been observed directly [19]. In a study of small arterioles and capillaries in the rabbit left ventricle arrested in systole or diastole, Goto et al. [20] found that subendocardial vessels had half the diameter in systole that they had in diastole, whereas there was little change in the diameters of these subepicardial vessels throughout the cardiac cycle.

Based on these findings, a hypothesis for the vulnerability of the subendo-cardium to ischemia can be put forward. At the end of systole, subendocardial vessels are greatly narrowed, so that even though intramyocardial pressures are low throughout the myocardium in the next diastole, there is a higher resistance to reflow from the extramural vessels in the subendocardial than in the sub-epicardial vessels. If we could put tiny flowmeters on small arterioles in different layers of the left ventricle, then we would expect that flow from the extramural arteries would enter the subepicardial vessels first, then vessels in the midwall, and, finally, flow would enter the subendocardial vessels. Given enough perfusion pressure and time in diastole, the subendocardium will get enough blood flow, but if diastolic flow is reduced for any reason, then the subendocardial layer will be the first to become ischemic. If this hypothesis is true, then an increase in contractility would be expected to decrease subendocardial flow and a decrease in contractility would be expected to increase subendocardial flow. This is exactly what has been found experimentally [21].

There are many clinical implications of this hypothesis. One of them is that whenever there is a low cardiac output syndrome after cardiac surgery, support-ing the circulation by giving inotropes may be harmful to the subendocardium of the left ventricle. Not only may this happen because of the effect of inotropes on intramyocardial systolic pressures, but the impaired myocardium, often hypertrophied, will probably be dilated acutely so that coronary venous pressures will be high.

Heterogeneity of Flow and Flow Reserve

Heterogeneity of flow, however, applies to more than layers of the left ventricle. Recently, Austin et al. [22] examined flow at rest and during maximal vasodilatation in small portions (100–150 mg) of the left ventricle. At rest, there was moderate variability of flow per gram at rest, ranging from about 0.5 to 2.5 ml/min. During maximal vasodilatation there was also a wide range of flows per gram, from about 3 to 10 ml/min. What was particularly noteworthy was that there was no relation between resting and maximal flows in any piece; that is, a piece with a low resting flow could have a low, medium, or high maximal flow. If maximal flow per gram was divided by resting flow per gram to give a coronary flow reserve ratio, then that ratio could vary from 2 to about 20 (Fig. 2).

The distributions of resting and maximal flows, and the flow reserve ratio, were not random. They were highly correlated in different layers of the left ventricle. Within any layer, there were domains of high and low flows and flow reserve ratio; that is, a piece with a high flow or flow ratio was surrounded by pieces with high flows and flow ratios, whereas a piece with a low flow reserve ratio was surrounded by other pieces with similar low ratios. By autocorrelation analysis,

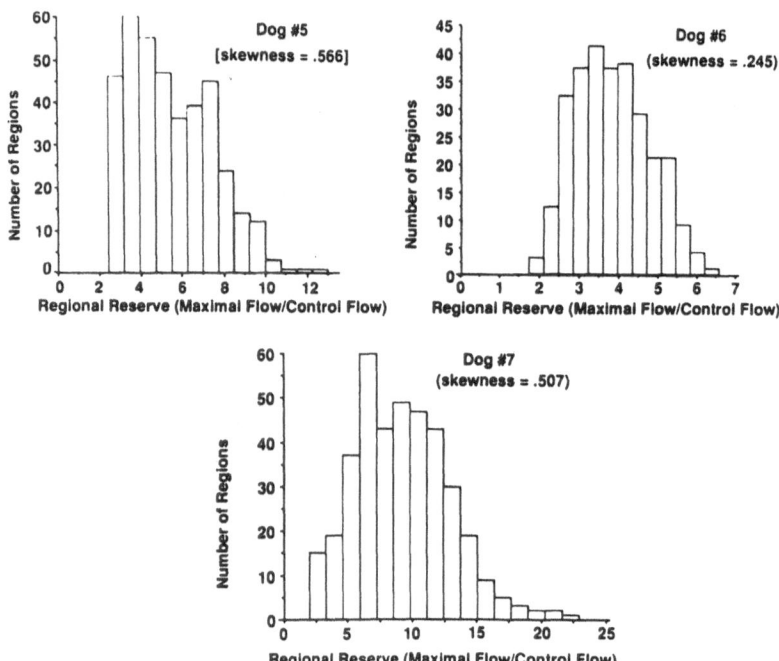

Fig. 2. Flow reserve ratio (ratio of maximal to resting flow) in three dogs under anesthesia. From [22] reproduced with permission. [Circulation Research] Copyright [1990] American Heart Association

the domains seemed to be about 4–8 mm in diameter. The reasons for the heterogeneity are unknown. Variation of flow at rest could be due to different wall forces and thus different oxygen consumption by each piece of tissue, but the variation of maximal flows cannot be so explained. It might be due to the closeness or remoteness of each piece from the main penetrating artery that supplies the region. The variation of flows is dependent upon the size of the piece of tissue that is examined, and Bassingthwaighte and his colleagues (Van Beek et al. [23]; Bassingthwaighte et al. [24]) have shown that the variability of flow as related to the size of the pieces follows a fractal distribution. They have not yet studied the heterogeneity of maximal flows or flow reserve, and the hypothesis does not explain the anatomy of the domains.

This demonstration of the heterogeneity of low reserve was used to investigate the problem of whether ischemia was a maximal vasodilator in the heart. Several groups of investigators [25–29] had observed independently that coronary flow reserve persisted despite severe myocardial ischemia, a finding that seemed to indicate that ischemia did not dilate vessels maximally. Because, until recently, most investigators had measured flows in pieces of the left ventricle that weighed from 1 to 10 g, Coggins et al. [30] reasoned that perhaps it was the heterogeneity of flow that caused the findings. They measured regional flows in three layers of the left ventricle at rest and after maximal vasodilatation at coronary perfusion pressures of 70, 50, 40, and 30 mmHg; each layer was cut into about 60 pieces, each weighing about 120 mg. At the lower perfusing pressures, with clear evidence of myocardial ischemia, some pieces showed no flow reserve (that is, their flows did not rise after adenosine was given), whereas other pieces retained some flow reserve. In general, the range of flow reserve at normal perfusing pressures was similar in all three layers, but, as perfusing pressures were lowered, more and more pieces lost their reserve, particularly in the subendocardial layer. Few pieces of the subepicardial layer lost their reserve, although the values for reserve decreased as perfusion pressure decreased. These investigators hypothesized that larger pieces of myocardium contained pieces with a high reserve ratio, as well as pieces with a low reserve ratio. Reducing perfusion pressure until some pieces lost reserve would cause ischemia in those pieces, which would not be able to increase their flow with adenosine, while adjacent pieces with retained reserve would increase their flows with adenosine. Because flows in all these small pieces were measured simultaneously when a big piece of tissue was examined, the retained reserve had been misinterpreted as failure of ischemia to cause maximal vasodilatation. It probably does cause maximal vasodilatation, but only in those pieces with a low flow reserve.

This concept of heterogeneity of flow reserve has important clinical impli- cations. Putting a needle into the myocardium to measure electrical potentials, ions, pH, oxygen tension, or to take a small biopsy may not give representative values; some form of averaging is needed. Furthermore, when perfusing pressures or flows are chosen, for example, when giving cardioplegia during cardiac sur- gery, an average flow and pressure that are adequate for pieces of myocardium with a high reserve may be inadequate for pieces with a low reserve, and a large factor of safety must be built in to avoid producing ischemic damage in pieces that have a low reserve.

References

1. Hoffman JIE (1987) Transmural myocardial perfusion. Prog Cardiovasc Dis 29: 429–464
2. Hoffman JIE (1990) Coronary physiology. In: Garfein OB (ed) Current concepts in cardiovascular physiology. New York, Academic pp 290–349
3. Hoffman JIE, Buckberg GD (1976) Transmural variations in myocardial perfusion. Prog Cardiol 5:37–89
4. Chilian WM, Marcus ML (1982) Phasic coronary flow velocity in intramural and epicardial coronary arteries. Circ Res 50:775–781
5. Bellamy RF (1978) Diastolic coronary pressure-flow relations in the dog. Circ Res 43:92–101
6. Aversano T, Klocke FJ, Mates RE, Canty JM Jr (1984) Preload-induced alterations in capacitance-free diastolic pressure-flow relationships. Am J Physiol 246 (Heart Circ Physiol 15):H410–H417
7. Uhlig PN, Baer RW, Vlahakes GJ, Hanley FL, Messina LM, Hoffman JIE (1984) Arterial and venous coronary pressure-flow relations in anesthetized dogs. Evidence for a vascular waterfall in epicardial coronary veins. Circ Res 55:238–248
8. Watanabe J, Maruyama Y, Satoh S, Keitoku M, Takishima T (1987) Effects of the pericardium on the diastolic left coronary pressure-flow relationship in the isolated dog heart. Circulation 75:670–675
9. Satoh S, Watanabe J, Keitoku M, Itoh N, Maruyama Y, Takishima T (1988) Influences of pressure surrounding the heart and intracardiac pressure on the diastolic coronary pressure-flow relation in excised canine hearts. Circ Res 63:788–797
10. Aldea G, Hoffman JIE, Husseini W, Mori H (1987) Determinants of regional myocardial blood flow. In: Brun P, Chadwick RS, Levy BI (eds) Cardiovascular dynamics and models; Proceedings of NIH-INSERM workshops. Les Editions INSERM, Paris, pp 44–47
11. Spaan JAE, Breuls NPW, Laird JD (1981) Diastolic-systolic coronary flow differences are caused by intramyocardial pump action in the anesthetized dog. Circ Res 49:584–593
12. Spaan JAE (1985) Coronary diastolic pressure-flow relations and zero flow pressure explained on the basis of intramyocardial compliance. Circ Res 56:293–309
13. Hoffman JIE, Spaan JAE (1990) Pressure-flow relations in coronary circulation. Physiol Rev 70:331–390
14. Verrier ED, Baer RW, Hickey RF, Vlahakes GJ, Hoffman JIE (1980) Transmural pressure-flow relations during diastole in the canine left ventricle. Circulation 62 [Suppl III]:62
15. Wüsten B, Buss DD, Deist H, Schaper W (1977) Dilatory capacity of the coronary circulation and its correlation to the arterial vasculature in the canine left ventricle. Basic Res Cardiol 72:636–650
16. Flynn AE, Coggins DL, Goto M, Aldea GS, Austin RE, Doucette JW, Husseini W, Hoffman JIE (1992) Does systolic subepicardial perfusion come from retrograde subendocardial flow? Am J Physiol 262 (Heart Circ Physiol 31):H1759–H1769

17. Domenech RJ, Goich J (1976) Effect of heart rate on regional coronary blood flow. Cardiovasc Res 10:224–231
18. Tillmanns H, Ikeda S, Hansen H, Sarma JSM, Fauvel J, Bing RJ (1974) Microcirculation in the ventricle of the dog and turtle. Circ Res 34:561–569
19. Ashikawa K, Kanatsuka H, Suzuki T, Takishima T (1986) Phasic blood flow velocity pattern in epimyocardial microvessels in the beating canine left ventricle. Circ Res 59:704–711
20. Goto M, Flynn AE, Doucette JW, Jansen CM, Stork MM, Coggins DL, Muehrcke DD, Husseini WK, Hoffman JI (1991) Cardiac contraction affects deep myocardial vessels predominantly. Am J Physiol 261 (Heart Circ Physiol 30):H1417–H1429
21. Marzilli M, Goldstein S, Sabbah HN, Lee T, Stein PD (1979) Modulating effect of regional myocardial performance on local myocardial perfusion in the dog. Circ Res 45:634–640
22. Austin RE Jr, Aldea GS, Coggins DL, Flynn AE, Hoffman JIE (1990) Profound spatial heterogeneity of coronary reserve. Discordance between patterns of resting and maximal myocardial blood flow. Circ Res 67:319–331
23. Van Beek JHGM, Roger SA, Bassingthwaighte JB (1989) Regional myocardial blood flow heterogeneity explained with fractal networks. Am J Physiol 257 (Heart Circ Physiol 26):H1670–H1680
24. Bassingthwaighte JB, King RB, Roger SA (1989) Fractal nature of regional blood flow heterogeneity. Circ Res 65:578–590
25. Aversano T, Becker LC (1985) Persistence of coronary vasodilator reserve despite functionally significant flow reduction. Am J Physiol 248 (Heart Circ Physiol 17): H403–H411
26. Canty JM Jr, Klocke FJ (1985) Reduced regional myocardial perfusion in the presence of pharmacologic vasodilator reserve. Circulation 71:370–377
27. Gold FL, Bache RJ (1982) Transmural right ventricular blood flow during acute pulmonary artery hypertension in the awake dog: Evidence for subendocardial ischemia during right ventricular failure despite residual vasodilator reserve. Circ Res 51:196–204
28. Grattan MT, Hanley FL, Stevens MB, Hoffman JIE (1986) Transmural coronary flow reserve patterns in dogs. Am J Physiol 250 (Heart Circ Physiol 19):H276–H283
29. Pantely GA, Bristow JD, Swenson LJ, Ladley HD, Johnson WB, Anselone CG (1985) Incomplete coronary vasodilation during myocardial ischemia in swine. Am J Physiol 249 (Heart Circ Physiol 18):H638–H647
30. Coggins DL, Flynn AE, Austin RE Jr, Aldea GS, Muehrcke D, Goto M, Hoffman JIE (1990) Nonuniform loss of regional flow reserve during myocardial ischemia in dogs. Circ Res 67:253–264

Effect of Vasomotor Tone on Coronary Flow Impediment in Systole

Nico Westerhof, Pieter Sipkema, and Paul Bouma[1]

Summary. We have studied the effect of cardiac contraction on coronary arterial flow, in the isolated blood-perfused donor-supported rat heat, with vasomotor tone present and after maximal vasodilation with adenosine ($n = 6$). A balloon in the left ventricle made it possible to change ventricular volume and thus pressure at a constant contractile state. Extra systolic potentiation was used to change cardiac contractility. Coronary pressure and flow measurements were performed at constant vasomotor tone as judged from stable diastolic coronary flow. The effects of developed left ventricular pressure, i.e., systolic minus diastolic pressure ($devP_{lv}$, in mmHg) and developed ventricular elastance i.e., systolic minus diastolic elastance ($devE_{lv}$, in mmHg/ml) on coronary blood flow (CBF) amplitude defined as diastolic minus systolic flow (ΔCBF, in ml/g.min) were determined through multiple regression of the form: $\Delta CBF = S_P.devP_{lv} + S_E.devE_{er} + I$, with S_P in ml/(g.min.mmHg), S_E in ml^2/(g.min.mmHg) and I in ml/(g.min). In the control vasoactive state we found (mean $\pm$ SD): $\Delta CBF = (0.015 \pm 0.005)$ $devP_{lv} + (0.0026 \pm 0.014)$ $devE_{lv} + (0.19 \pm 0.60)$. Maximal vasodilation increased coronary flow in diastole by a factor of 4.06 ± 0.70. The sensitivity to $devP_{lv}$ and $devE_{lv}$ increased by a factor of 5.40 ± 3.35 and 4.23 ± 0.61, respectively. These factors were not different. We therefore conclude contrary to the findings in the cat and rabbit and dog heart that left ventricular pressure has an effect on coronary flow impediment during maximal vasodilation and that vasodilation increases diastolic flow and the coronary flow amplitude by a similar factor.

Key words: Isolated rat heart—Blood perfused—Contractility—Left ventricular pressure—Coronary flow

Introduction

Coronary arterial flow is impeded in systole by cardiac contraction. We have found in the isolated maximally-dilated blood-perfused cat heart and Tyrode-perfused rabbit heart that coronary arterial flow impediment in systole is mainly determined by cardiac contractility and only affected little by left ventricular pressure [1–3]. The effect of left ventricular pressure in these hearts was about one order of magnitude less than the effect of contractility [2]. From these findings we proposed that it is the stiffness of the myocardium surrounding the vasculature that affects the blood vessels [1–3]. If one derives from the myo-

---

[1] Laboratory for Physiology, Free University of Amsterdam, van der Boechorst straat 7, 1081 BT Amsterdam, The Netherlands

:ardial stiffness i.e., the stress-strain or force-length relation, the elastance
$(\Delta P/\Delta V)$ of blood vessels it is found that elastance is higher in systole than in
diastole [4]. At a given constant perfusion pressure an increased elastance results
n a decreased vessel lumen and decreased flow. We performed these earlier
experiments in the isolated maximally-dilated cat [1, 2] and rabbit heart [3] to
avoid possible changes in flow resulting from alterations in vasomotor tone.
Naturally these extreme conditions are not very physiological. It was subsequently reported that the small contribution of ventricular pressure to coronary
flow impediment was also found in the much larger dog heart [4].

In the present study we studied the effect of cardiac contraction on coronary
flow in the blood-perfused rat heart to investigate whether or not the earlier
findings also apply to small hearts. We also compared the effects of cardiac
contraction on arterial flow during the presence of vasomotor tone and after
maximal vasodilation.

Methods

Experiments

Experiments were carried out on the isolated blood-perfused rat heart supported
by a donor rat. Perfusion pressure was kept constant at 105 mmHg under all
conditions. A latex balloon was placed in the left ventricle so that only isovolumic
contractions were allowed. The volume of the balloon could be changed to obtain
different values of left ventricular pressure. Balloon volume was chosen such that
over the ventricular volume range encountered, transmural balloon pressure was
negligible. The hearts were paced at 5 Hz. Short periods (about 2 s) of high
frequency pacing followed by a sudden return to a heart rate of 5 Hz were used to
increase contractility due to post extrasystolic potentiation. Coronary flow was
measured with a cannulated ultrasonic transit time measuring system (2N probe
and T101 Transonic Systems, Ithaca, N.Y.). Pressure in the balloon was
measured with a catheter tip manometer system (Millar Instruments, Houston,
Tex.). Maximal vasodilation was accomplished by infusion of adenosine (1 mg/hr).
Further details of the method have been reported before [5].

Data Analysis

Data were analyzed as follows. The flow amplitude (ΔCBF, i.e., diastolic minus
systolic flow) was determined and related to developed left ventricular pressure
defined as systolic minus diastolic pressure (devP$_{lv}$) and also related to developed
elastance (devE$_{lv}$) being the slope of the end-systolic pressure-volume relation.
devE$_{lv}$ is considered as a measure of contractility [6]. By assuming that the
end-systolic pressure-volume relation is going through the origin, elastance for
isovolumic beats can be calculated as the ratio of systolic pressure and volume.
Developed elastance is the difference of elastance between systole and diastole.
The relation between flow amplitude, and developed left ventricular pressure,

and developed elastance was determined with multiple regression to obtain the following relation: $\Delta CBF = S_P\ devP_{lv} + S_E\ devE_{lv} + I$. The coefficients S_P, S_E, and I were then averaged over the six hearts and presented as mean $\pm$ SD.

For details of the calculations the reader is referred to [5].

Results

With constant perfusion pressure, ventricular pressure was changed by changes in balloon volume, and contractility was changed with potentiation. A representative example of left ventricular pressure and coronary arterial flow during the presence of vasomotor tone and after maximal vasodilation with adenosine is presented in Fig. 1. It may also be seen that both diastolic flow and the amplitude of coronary flow (diastolic minus systolic flow) increase after vasodilation. It may be seen that after resumption of the normal pace frequency (5 Hz) left ventricular pressure is increased due to enhanced contractile state and that after a limited number of beats the pressure returns to a steady state. Moreover, diastolic flow is not changing during the measurement period indicating that vasomotor tone remains constant. Finally it should be noted that systolic flow increases with vasodilation.

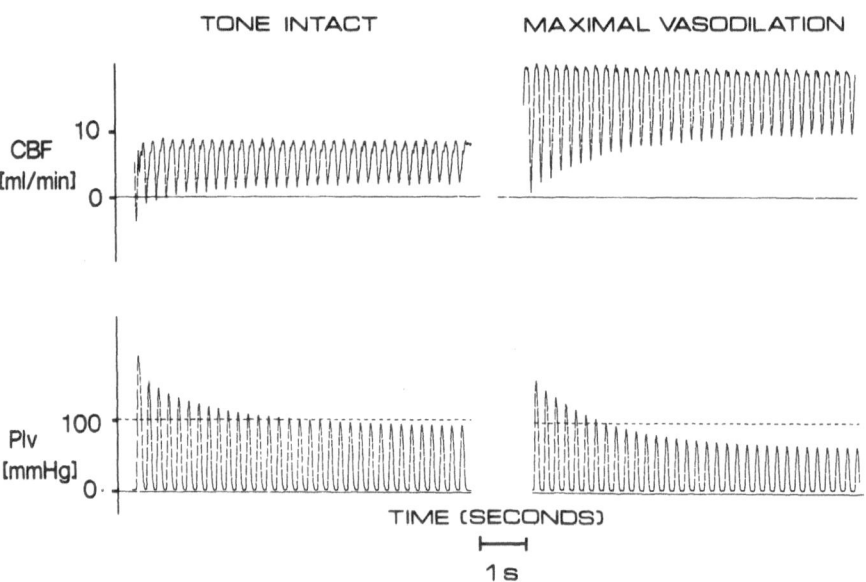

Fig. 1. An example of coronary arterial flow (*CBF*) and left ventricular pressure (P_{lv}) during changes in contractility obtained by potentiation with vasomotor tone intact and after maximal vasodilation. Both diastolic flow and the amplitude of the oscillations in coronary flow increase with vasodilation. (Adapted from [5] with permission)

By repeating this intervention for different volumes, basic data for the multiple regression are obtained. It was found that both contractility and left ventricular pressure have an effect on the coronary flow. The average relation between the flow amplitude and developed left ventricular pressure and developed elastance with vasomotor tone present is: $\Delta CBF = (0.015 \pm 0.005) \cdot devP_{lv} + (0.0026 \pm 0.0014) \cdot devE_{lv} + 0.19 \pm 0.60$.

After maximal vasodilation we found the sensitivities to ventricular pressure and contractility to be increased by a factor of 5.40 ± 3.35 and 4.23 ± 0.61, respectively. The diastolic coronary flow was increased by a factor of 4.06 ± 0.70. These three factors were not significantly different.

Discussion

We found that in the isolated blood-perfused rat heart the effect of cardiac contraction on the coronary arterial flow results from both the varying stiffness of the myocardium and from the effect of left ventricular pressure. We also found that with maximal vasodilation both the flow in diastole and the sensitivities of the flow amplitude for developed left ventricular pressure and developed elastance are similarly augmented (by about a factor of 4) as compared to the situation with intact vasomotor tone. The ratio of the sensitivities to developed pressure and developed elastance both increased by a factor of about 4 with vasodilation. The findings imply also that systolic flow is affected by vasomotor tone.

We previously measured in the isolated blood-perfused cat heart [1, 2] and the isolated Tyrode-perfused rabbit heart [3] under maximal vasodilation that the effect of left ventricular pressure on coronary flow impediment was negligible with respect to the effect of elastance. In the blood-perfused cat heart we found under maximal dilation that the sensitivity for pressure was about 0.01 ml/min per g per mmHg and for elastance 0.1 ml^2/min per g per mmHg [2]. In other words, the sensitivity to pressure appears to be similar in the vasodilated cat heart (0.01 ml/min per g per mmHg) and the rat heart (0.015 ml/min per g per mmHg) under control conditions. With vasodilation, the sensitivity to pressure is increased by a factor of four so that in similar conditions of the vasculature, i.e., maximal dilation, the sensitivity to pressure is much stronger in the rat heart than in the cat or rabbit heart.

In the dog heart, we observed that ventricular pressure had very little effect on endocardial flow [4]. However, Kouwenhoven and Spaan [7] found that the effect of left ventricular pressure was related to the initial decline of coronary flow in systole, and that later in systole flow impediment was mainly the result of elastance. Why the effect of ventricular pressure on coronary flow in the rat heart during vasodilation is stronger than in larger hearts is not clear. A possible explanation could be that pressure acts to a certain depth in the endocardium and that in the rat heart this layer contributes relatively more to overall flow than in larger hearts.

The sensitivities of coronary arterial flow, in the blood-perfused rat heart and the blood-perfused cat heart, to elastance during maximal vasodilation are 0.01 (see results 0.0026×4.23) and 0.1 [2], respectively. This difference may be explained as follows. Elastance, and therefore also developed elastance, is a heart-size-dependent measure of contractility. The rat ventricle is about ten times smaller than the rabbit or cat heart. This implies that with similar pressures as found in these species [8] elastance, which is a pressure-flow ratio, is about ten times larger in the rat heart than the cat or rabbit heart. Indeed the sensitivity of the flow amplitude for elastance is about 10 times larger in the cat heart [2] than in the rat hearts reported here. In other words the product of sensitivity and elastance is similar in both preparations. Another interpretation is that the myocardial stiffness variation, which depends on muscle properties, is similar in the differently-sized hearts and the elastance is a heart-size-dependent parameter. For the even larger dog heart [4] the effect of the varying elastance is therefore predicted to be even smaller than found in the rabbit and cat heart.

When vasodilation is compared with the situation where vasomotor tone is present we find a similar relative contribution of the effect of developed left ventricular pressure and developed elastance. In other words, vasomotor tone does not grossly change the contribution of contractility. Therefore we expect that in larger hearts with vasomotor tone present the effect of left ventricular pressure on coronary arterial flow is, as in vasodilation, negligible.

From the finding that, in systole, coronary flow is still depending on vasomotor tone it is to be concluded that even in systole the vessel lumen is not solely determined by the properties of the (stiff) myocardium but also by the vessel wall itself. Since vasomotor tone affects the sensitivity to pressure and elastance in a quantitatively similar manner it may be suggested, by extrapolation, that in the larger hearts left ventricular pressure will also not play an important role when vasomotor tone is present. Contractility is therefore expected to have an overriding effect on coronary flow in larger hearts, not only during maximal vasodilation, but also when vasomotor tone is present.

Since vasodilation affects systolic coronary flow the findings suggest also that the vessel wall properties (wall elasticity) contribute to vessel elastance in systole.

References

1. Krams R, Sipkema P, Westerhof N (1989) Varying elastance concept may explain coronary systolic flow impediment. Am J Physiol 257:H1471–H1479
2. Krams R, Sipkema P, Zegers J, Westerhof N (1989) Contractility is the main determinant of coronary systolic flow impediment. Am J Physiol 257:H1936–H1944
3. Krams R, van Haelst ACTA, Sipkema P, Westerhof N (1989) Can coronary systolic-diastolic flow differences by predicted by left ventricular pressure or time-varying intramyocardial elastance? Basic Res Cardiol 84:149–159

4. Van Winkle DM, Swafford AN Jr, Downey JM (1991) Subendocardial coronary compression in beating dog hearts is independent of pressure in the ventricular lumen. Am J Physiol 261:H500–H505
5. Bouma P (1991) Coronary flow and cardiac contraction. PhD Dissertation, Free University of Amsterdam
6. Suga H, Sagawa K, Shoukas AA (1973) Load dependence of the instantaneous pressure-volume ratio of the canine left ventricle and effects of epinephrine and heart rate on the ratio. Circ Res 32:314–322
7. Kouwenhoven HJ, Spaan JAE (1992) Retrograde coronary flow is limited by time-varying elastance. Am J Physiol 263:H484–H490
8. Westerhof N, Elzinga G (1991) Normalized input impedance and arterial decay time over heart period are independent of animal size. Am J Physiol 261:R126–R133

Heart Contraction and Coronary Blood Flow

Jos. A.E. Spaan[1]

Summary. Heart contraction is the cause of pulsatile coronary arterial and venous flow. It also reduces time-averaged flow through the myocardium, also referred to as myocardial perfusion. The impediment of myocardial perfusion is predominantly subendocardial, which for a long time resulted in the paradigm that this impediment was caused by a tissue pressure coupled to left ventricular pressure. However, it has recently been shown that heart muscle contraction directly affects coronary blood flow. A unique concept of contraction-perfusion interaction is lacking. Both contractility and left ventricular pressure do affect coronary perfusion, but are mutually dependent. An important determinant for the contraction-perfusion interaction appears to be the filling of the intramyocardial vasculature and the structure of the connections between vessel walls and myocytes. Different aspects of the contraction-perfusion interaction are considered, including the effect of contraction on lymph pressure.

Key words: Coronary mechanics—Intramyocardial compliance—Cardiac lymph—Elastance—Tissue pressure—Contractility

Introduction

Because of the embodiment of the coronary microcirculation within the myocardium, heart contraction does affect coronary blood flow. Notwithstanding some centuries [1–6] of research, a satisfying concept of this impeding effect has not yet been provided. The basic observations are simple. Contraction impedes arterial systolic flow, accelerates systolic venous flow, and reduces time average flow. The impediment of time-averaged coronary blood flow is not equal over the wall but is most pronounced at the subendocardium and almost absent in the subepicardium [7].

The mechanism by which cardiac contraction interacts with coronary blood flow has not yet been resolved. In the early decades of this century emphasis was placed on the analysis of coronary arterial inflow to understand coronary flow mechanics. In many studies the idea that the decrease in systolic coronary arterial flow would imply a decrease in myocardial perfusion was discussed [4,

---

[1] Department of Medical Physics and Informatics, Cardiovascular Research Institute, University of Amsterdam, The Netherlands

6]. This was either explained by an increase in systolic resistance or by systolic collapse of vessels. In 1981, the concept of the intramyocardial pump was introduced, relating arterial inflow variations and venous outflow variations to variations in myocardial blood volume [8, 9]. The term "intramyocardial pump" was selected because, in systole, the heart muscle would expel blood out of the myocardium. Subsequently, relaxation forces would contribute to restoration of the intramyocardial blood volume in diastole. Essential to that model was a large time constant for changing intramyocardial blood volume [10]. It was suggested that this time constant would be in the order of 1.6 s. Hence, intramyocardial blood volume would never be in equilibrium during a normal heart beat, neither in systole nor in diastole [11]. However, according to the paradigm of that time, the energizer for the myocardial pump was thought to be tissue pressure.

In the concept of tissue pressure it was implicitly assumed that the coronary microcirculation was embedded within extracellular fluid and was subjected to so-called tissue pressure. Within models of wall mechanics, not taking into account the microscopic structure of tissue, including collagen [12], there had to be an equilibration between tissue pressure and left ventricular pressure at the endocardium (e.g., [13, 14]). For the subepicardium this equilibration pressure was atmospheric pressure. Furthermore, tissue pressure would decrease linearly from subepicardium to subendocardium. However, it was recently reported that, in an empty beating heart, coronary arterial flow was still pulsatile [15] and time-averaged subendocardial flow was still impeded [16]. Consequently one was forced to conclude that the tissue pressure concept was too simple.

More recently, an alternative has been presented in which the effect of contraction on intramyocardial blood flow is determined by the change in elastance when the heart contracts [17]. This model in itself has beauty, but at first sight it is hard to see how the difference in effect at subepicardium and subendocardium can be explained. Hence, in the present state of science related to cardiac contraction and coronary flow there is no ready explanation for this interaction. The purpose of this chapter is to review the different mechanisms involved and to contribute to the direction of research that will bring about a solution in the end.

The Elastance Concept

The elastance concept will be elucidated first, since it is the feeling of the author that at the moment it is the best start for discussing interaction between heart contraction and myocardial perfusion. Application of the elastance concept to the coronary circulation mechanics has been put forward by Westerhof and his group (Krams et al. [15]). It is a simple analogy to the left ventricular function analysis put forward by Suga et al. [18]; the principle is illustrated in Fig. 1.

The Suga concept is too simple to explain the mechanics of the left ventricle in full. However, demonstrates the role of two major determinants of left ventricular

62   J.A.E. Spaan

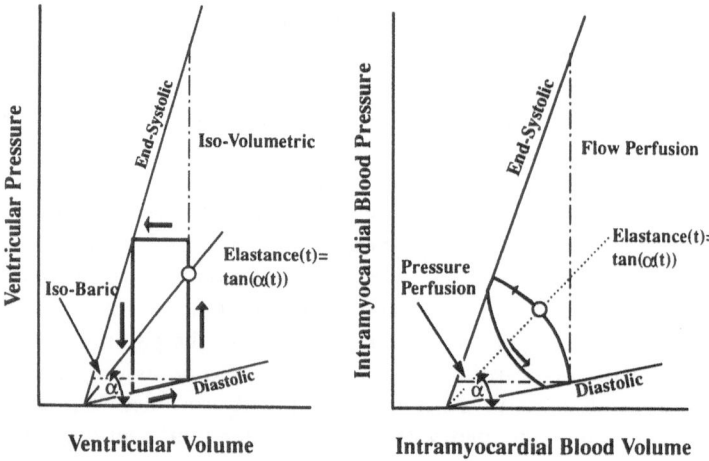

Fig. 1. The elastance concept. *Left panel.* Diastolic and systolic pressure volume relationships for left ventricular function. The slope of the curve indicated by α is referred to as elastance. For the left ventricle, a pressure-volume relationship can be approximated by a simple loop [18]. *Right panel.* Elastance model applied to the coronary circulation. For further discussion see text

function, namely, contractility and filling of the left ventricle. We will consider these effects at constant afterload. A higher level of filling of the left ventricle will result in a higher stroke volume. An increased contractility at constant diastolic left ventricular volume will result in a larger ejection fraction. The suggestion of Westerhof and his group was to consider the intramyocardial vascular space to be subjected to the same forces as the left ventricular cavity. Hence, intramyocardial blood pressure would rise, not because of an increased left ventricular pressure, but it would be directly related to contraction of myocytes and, obviously, to the filling of the intramyocardial blood vessels. The analogy between intramyocardial pump function and left ventricular pump function stops here. The absence of valves and the distribution of the blood, as well as resistance, over different vessels complicates the application of the elastance model to the coronary circulation.

For the intramyocardial blood space, the pressure volume relation is more complicated. The absence of valves makes isovolumetric phases unlikely. Moreover, the distribution of blood and resistance over different vessels makes it unclear at what level the different volumes will stabilize. The right hand panel in Fig. 2 demonstrates that if intramyocardial pressure remained constant, intramyocardial volume would vary during the heart cycle. As a result, coronary arterial and venous flows would pulsate. If intramyocardial volume were constant, arterial and venous pressures would be pulsatile instead. Neither condition can be realized in practice, but they may be reflected by perfusion at constant flow and constant pressure, respectively. Both perfusion conditions are illustrated by Fig. 2.

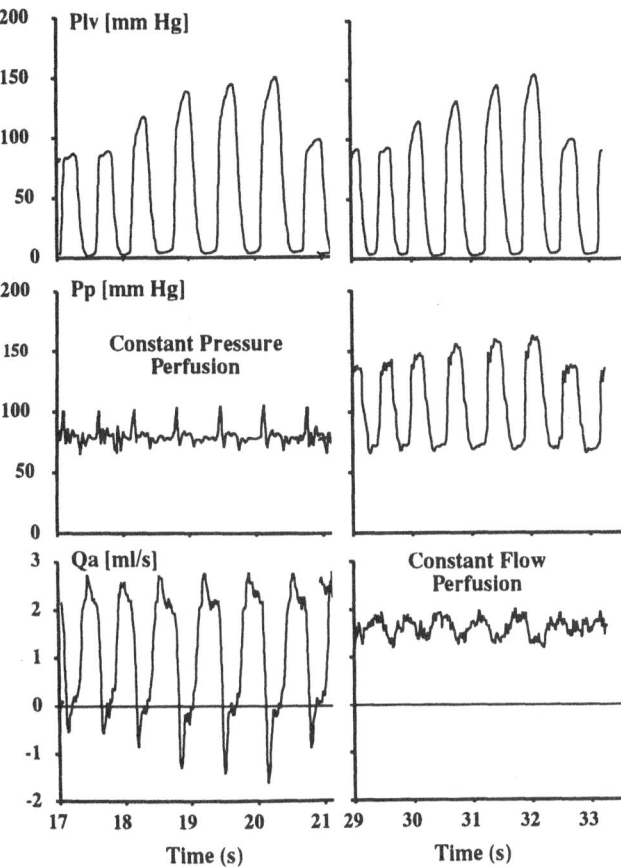

Fig. 2. Effect of aortic clamping on coronary arterial flow at constant pressure perfusion (*left panels*) and at constant flow perfusion (*right panels*). Note the persistent increase in coronary pressure during whole systole in the *right hand panel* and only an early systolic decrease in flow in the *left panel*. (Data from [23])

The elastance concept interferes with the intramyocardial pump model put forward earlier only in relation to the generator of the pump. It does not affect the concept of volume displacement, although it may affect the concepts of time constants involved, as will be discussed below.

Several observations are in accordance with the elastance model. Krams et al. [15] showed convincingly that pulsations in coronary arterial blood flow at constant arterial pressure and constant contractility were relatively independent of left ventricular pressure. Furthermore, they demonstrated that pulsations in coronary arterial flow increased with the level of arterial pressure [19]. This is easily explained by the elastance model by the assumption that intramyocardial blood volume is higher at higher coronary arterial pressure. The third observation is that coronary arterial flow pulsations increased with increasing contractility [20].

Effect of Left Ventricular Pressure

The elastance model, however, cannot explain everything. It is well known that the effect of cardiac contraction on time-averaged flow is predominantly sub-endocardial. If there were a direct elastance effect, why would such an effect not hold for the subepicardium? We and others demonstrated that the coronary flow and pressure signals are not insensitive to left ventricular pressure [21–23]. This is illustrated in Fig. 2. In anesthetized goats with cannulated and perfused main stem, the aorta was clamped for a few beats to rapidly increase left ventricular pressure. When perfused at constant pressure, coronary arterial flow was pulsatile and the effect of increasing left ventricular pressure had an effect only early in systole, when elastance was still low but increasing. However, in mid-systole, the flow was rather insensitive to left ventricular pressure. It was reasoned that, in this phase, elastance would be dominant and would shield the intramyocardial vessels from the compressive effects of left ventricular pressure. However, when the perfusion system was switched into constant flow perfusion mode, it appeared that the pulsations in coronary pressure were sensitive throughout systole, even when elastance was high.

In general, the pressure source of a complicated network is best reflected when perfused with a system having a high internal load. In principle, constant flow perfusion means an infinite load for flow variations. Hence, the pulsations in coronary arterial pressure would reflect the compression force on the vessels. This same force could generate the flow variations at constant pressure perfusion. However, the flow variations were not a mirror image of these pressure variations. Apparently, the possibility for volume displacement at constant pressure perfusion has the consequence that the systolic flow wave is different from the systolic pressure wave with constant flow perfusion.

The difference in transmission between constant flow and constant pressure perfusion might point to the direction that, with the latter, the systolic time constant for proximal volume variations is small. This time constant is increased with the resistance on the perfusion line. Such an effect may also occur with stenoses in a coronary artery. However, this should be further investigated.

The results at constant pressure perfusion indicate that left ventricular pressure may contribute to volume reductions of intramyocardial vessels. Hence, left ventricular pressure still may be a factor to reckon with in describing the effect of heart contraction on myocardial perfusion.

Cardiac Contraction and Lymphatic Pressure

The complicated nature of interaction between cardiac contraction and fluids within the heart muscle is further demonstrated by waves in epicardial lymphatic pressure as recently measured by our group [24, 25]. A lymph vessel alongside the left anterior descending (LAD) coronary artery was cannulated with a small tube with an outside diameter of 1 mm. Via a special connector, the pressure wave was measured by a catheter tip manometer. The compliance of the system was such that the estimated frequency response of the measurement was better than 50 Hz. A typical registration is demonstrated in Fig. 3.

In the first place, one may appreciate that the lymph pressure is low; in systole it is generally not higher then 20% of left ventricular pressure. The sensitivity of systolic lymphatic pressure to left ventricular pressure is much lower than the coronary arterial pressure wave at constant flow perfusion. Obviously, one might argue that epicardial lymph pressure does not really reflect the intramural lymphatic pressure or interstitial pressure because the epicardial lymph is occluded and shifts of lymphatic fluid between subendocardium and subepicardium may occur. Occlusion of flow, however, normally results in an increase of pressure. As far as the argument of lymphatic displacement from endo to epi is considered, one should note the fact that such an argument holds for measurements in large epicardial arteries and veins as well.

The lymphatic pressure wave consists of different phases. First is the wave corresponding to the atrial wave in left ventricular pressure. Considering the low pressure level of this pressure wave, the sensitivity of lymphatic pressure to diastolic left ventricular pressure is high. In the beginning of systole, lymphatic pressure rises, but then suddenly reduces again. This behavior is similar to the observed systolic arterial flow wave, where, initially, the transmission of left ventricular pressure to flow was shielded off in mid-systole by elastance. The middle phase of the systolic lymphatic pressure wave would then be determined

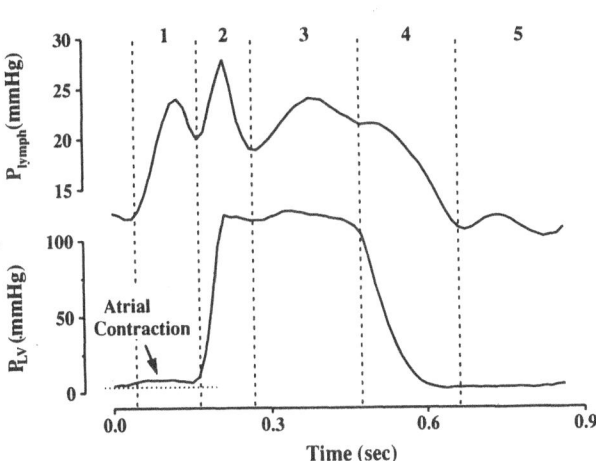

Fig. 3. Typical lymphatic pressure measured in a cannulated epicardial lymph vessel of an anesthetized dog. Note the relatively large sensitivity of lymph pressure to left ventricular pressure in diastole. In systole, different phases may be distinguished. Note the initial rapid rise at early systole. The effect of left ventricular pressure is shielded in mid-systole due to a stiffer myocardium

by the elastance effect on intramyocardial lymphatic pressure. After reduction of elastance, lymphatic pressure increases again, probably because the left ventricular pressure effect dominates the elastance effect in the end-systolic phase.

Intramyocardial Pressures and Cardiac Contraction

In this chapter, it has been demonstrated that there is not a single mechanism explaining the effect of cardiac contraction on coronary blood flow. Apart from this, the two concepts at hand both have conceptual problems. Tissue pressure is not a simple concept, since the pressure should be allocated to a fluid space. At least a few of these fluid spaces should be distinguished: blood in different types of vessels, fluid within the myocytes, interstitial fluid, and lymph. All these fluids are within compartments divided by mechanical structures that can withstand a pressure difference. Consequently, there is no primary reason why all these pressures would be the same. Hence, the question is: what is tissue pressure? Is it the pressure in one of these compartments or is it the average fluid pressure over the compartments, or could it be a weighed average?

Conceptually, tissue pressure is often equated to interstitial pressure. However, histological studies reveal that the interstitium is very small and that the concept of blood vessels suspended within liquid is just much too simple. In recent years, there has been a growing insight that interaction between contraction and coronary blood flow cannot be understood without taking into account the anchoring of collagen to blood vessels and myocytes. We have to conclude for the moment that the two concepts, tissue pressure and elastance, as compressing force for the intramyocardial pump, combined or not, are only partly satisfying for explaining the contraction effect on the coronary circulation. The elastance concept provides a framework in which phenomena such as the pulsatility of coronary flow independent of left ventricular pressure and also the effect of contraction on lymphatic pressure can be explained. It explains nicely that pressure generation is related to the volume within a certain compartment. However, it is not clear whether elastance, as determined from left ventricular function, is the factor that should be taken as a measure of direct action of local muscle stiffness on intramyocardial blood vessels.

One may expect that, in the near future, the distribution of blood volume, and how this is affected by contraction, will receive ample attention. This is because of the relation between volume and resistance, but, more especially, because vascular volume is a major determinant of the effect of contractility on the intramyocardial vessels. In this respect, it is noteworthy that most of the coronary blood volume is within the myocardial microcirculation [26]. There is an urgent need for the development of a concept that explains why contraction reduces vascular volume predominantly at the subendocardium [27].

References

1. Scaramucci J (1695) De motu cordis, theorema sextum. In: Theoremata familiaria de physico-medicis lucubrationsibus Iucta leges mecanicas (in Latin) pp 70–81
2. Porter WT (1898) The influence of the heartbeat on the flow of blood through the walls of the heart. Am J Physiol 1:145–163
3. Gregg DE, Greem HD (1940) Registration and interpretation of normal phasic inflow into a left coronary artery by an improved differential manometric method. Am J Physiol 130:114–125
4. Sabiston DC Jr, Gregg DE (1957) Effect of cardiac contraction on coronary blood flow. Circulation 15:14–20
5. Wiggers CJ (1954) The interplay of coronary vascular resistance and myocardial compression in regulating coronary flow. Circ Res 2:271–279
6. Downey JM, Kirk ES (1975) Inhibition of coronary blood flow by a vascular waterfall mechanism. Circ Res 36:753–760
7. Hoffman JIE, Spaan JAE (1990) Pressure-flow relations in the coronary circulation. Physiol Rev 70:331–390
8. Spaan JAE, Breuls NPW, Laird JD (1981) Diastolic-systolic coronary flow differences are caused by intramyocardial pump action in the anesthetized dog. Circ Res 49:582–593
9. Spaan JAE, Breuls NPW, Laird JD (1981) Forward coronary flow normally seen in systole is the result of both forward and concealed back flow. Basic Res Cardiol 76:582–586
10. Spaan JAE (1985) Coronary diastolic pressure-flow relation and zero flow pressure explained on the basis of intramyocardial compliance. Circ Res 56:293–309
11. Vergroesen I, Noble MIM, Spaan JAE (1987) Intramyocardial blood volume change in first moments of cardiac arrest in anesthetized goats. Am J Physiol 253 (Heart Circ Physiol 22):H307–H316
12. Borg TK, Caulfield JB (1981) The collagen matrix of the heart. Fed Proc 40: 2037–2041
13. Bruinsma P, Arts T, Dankelman J, Spaan JAE (1988) Model of the coronary circulation based on pressure dependence of coronary resistance and compliance. Basic Res Cardiol 83:510–524
14. Chadwick RS, Tedgui A, Michel JB, Ohayon J, Levy BI (1990) Phasic regional myocardial inflow and outflow: Comparison of theory and experiments. Am J Physiol 258 (Heart Circ Physiol 27):H1687–H1698
15. Krams R, Sipkema P, Westerhof N (1989) Varying elastance concept may explain coronary systolic flow impediment. Am J Physiol 257 (Heart Circ Physiol 26): H1471–H1479
16. VanWinkle DM, Swafford AN, Downey JM (1991) Subendocardial coronary compression in beating dog hearts is independent of pressure in the ventricular lumen. Am J Physiol 261 (Heart Circ Physiol 30):H500–H505
17. Westerhof N (1990) Physiological hypotheses: Intramyocardial pressure. A new concept, suggestions for measurement. Basic Res Cardiol 85:105–119
18. Suga H, Sagawa K, Shoukas AA (1973) Load independence of the instantaneous pressure-volume ratio of the canine left ventricle and effects of epinephrine and heart rate on the ratio. Circ Res 32:314–322
19. Krams R, Sipkema P, Westerhof N (1990) Coronary oscillatory flow amplitude is more affected by perfusion pressure than ventricular pressure. Am J Physiol 258 (Heart Circ Physiol 27):H1889–H1898

20. Krams R, Sipkema P, Zegers J, Westerhof N (1989) Contractility is the main determinant of coronary systolic flow impediment. Am J Physiol 257 (Heart Circ Physiol 26):H1936–H1944
21. Chilian WM, Marcus ML (1982) Phasic coronary blood flow velocity in intramural and epicardial coronary arteries. Circ Res 50:775–781
22. Chilian WM, Marcus ML (1985) Effects of coronary and extravascular pressure on intramyocardial and epicardial blood velocity. Am J Physiol 248 (Heart Circ Physiol 17):H170–H178
23. Kouwenhoven E, Vergroesen I, Han Y, Spaan JAE (1992) Retrograde coronary flow is limited by time-varying elastance. Am J Physiol 263 (Heart Circ Physiol 32): H484–H490
24. Han Y, Vergroesen I, Spaan JAE (1993) Stopped-flow epicardial lymph pressure is affected by left ventricular pressure in anesthetized goats. Am J Physiol 264 (Heart Circ Physiol 33):H1624–H1628
25. Han Y, Vergroesen I, Goto M, Dankelman J, VanderPloeg CPB, Spaan JAE (1993) Left ventricular pressure transmission to myocardial lymph vessels is different during systole and diastole. Pflügers Arch 423:448–454
26. VanderPloeg CPB, Dankelman J, Spaan JAE (1993) Functional distribution of coronary vascular volume in the beating goat heart. Am J Physiol 264 (Heart Circ Physiol 33):H770–H776
27. Goto M, Flynn AE, Doucette JW, Jansen CMA, Stork MM, Coggins DL, Muehrcke DD, Husseini WK, Hoffman JIE (1991) Cardiac contraction affects deep myocardial vessels predominantly. Am J Physiol 261 (Heart Circ Physiol 30):H1417–H1429

Blood Velocity Profiles Along Poststenotic Coronary Artery and Stenotic Intramyocardial Flow

Fumihiko Kajiya, Osamu Hiramatsu, Akihiro Kimura, Masami Goto, Yasuo Ogasawara, and Katsuhiko Tsujioka[1]

Summary. To evaluate the effect of coronary stenosis on coronary flow, we evaluated: (1) the blood velocity profiles across the vessel at portions distal to stenosis, (2) the velocity waveform in the septal artery during left main coronary artery stenosis with or without vasodilators, and (3) transmural flow distribution with low perfusion pressure as in stenosis before and after intracoronary nitroglycerin. The experiments were performed in 29 dogs using our 80-channel 20 MHz ultrasound velocimeter. The poststenotic velocity configuration was characterized by a narrow region of high velocity with diastolic reverse flows near the wall which may dissipate energy. Septal artery blood flow velocity which reflected myocardial inflow showed a diastolic-predominant waveform always accompanied by a systolic retrograde blood velocity component. Coronary artery stenosis enhanced the systolic retrograde flow with a decrease in diastolic flow, reducing myocardial inflow. The systolic retrograde flow was augmented further by coronary vasodilation (intracoronary adenosine or nitroglycerin) and did not improve myocardial inflow (or decreased it). Intracoronary nitroglycerin increased epimyocardial flow, but did not increase endomyocardial flow. In conclusion, an augmented retrograde flow which is increased by vasodilators plays an important role in disturbing myocardial inflow during coronary artery stenosis. This could be called a "coronary slosh phenomenon". Thus, increased systolic retrograde flow and the decreased diastolic flow are closely related in reducing myocardial flow, especially subendocardial flow.

Introduction

Detailed assessment of flow dynamics in coronary artery stenosis is essential for understanding the pathophysiology of obstructive coronary disease. In the presence of moderate to severe coronary artery stenosis, a significant pressure loss develops across the stenosis which causes reductions in coronary flow reserve and effective coronary flow to the myocardium. Gould [1] investigated the relation between diastolic pressure gradient across the coronary artery stenosis and blood velocity. By referring to Young and Tsai's reports [2, 3], he adopted a simplified equation of pressure loss consisting of two terms, the first term relating to viscous friction pressure loss in the stenotic segment and the second term to flow separation pressure loss. For the study of stenotic hemodynamics, it is necessary to measure velocity fields in the poststenotic region in detail. However,

---

[1] Department of Medical Engineering and Systems Cardiology, Kawasaki Medical School, 577 Matsushima, Kurashiki-shi, Okayama, 701-01 Japan

69

in vivo measurements of phasic blood velocity, especially velocity profiles across the vessel, were hampered by methodological limitations.

In the presence of a coronary stenosis, the systolic-to-diastolic flow ratio (S/D), which is a useful parameter for estimating the endomyocardial-to-epimyocardial flow ratio [4], increased with increasing severity of stenosis [5–7]. However, these observations were the results of blood flow measurements in relatively large epicardial coronary arteries which have large capacitance [8], and therefore obscure the true phasic flow pattern of the arterial inflow into myocardium [9, 10]. Accordingly, several investigators have studied the phasic pattern of the intramyocardial artery (mainly the septal artery) and/or a small epicardial artery just before penetration into the myocardium [9–13]. The phasic pattern in these arteries is characterized by systolic retrograde flow compared with that in epicardial large coronary arteries.

In this study, we first measured blood velocity profiles at various positions along stenotic arteries for in vivo assessment of viscous friction pressure loss and flow separation pressure loss. Second, to evaluate myocardial inflow disturbance for various degrees of coronary artery stenosis, we measured the velocity waveform in the septal artery following left main coronary artery stenosis. Third, we evaluated the change in the stenotic coronary velocity waveform and intramural flow distribution following coronary vasodilation with adenosine or nitroglycerin.

Methods

Measurement of Coronary Blood Velocity Waveforms on or Near the Central Axial Region and Velocity Profiles Across Vessels

We measured blood velocity waveforms and velocity profiles with our 80-channel 20 MHz pulsed Doppler velocimeter, which has been described previously in detail [14, 15]. In short, the transducer consists of a $\pi \times (0.5)^2$ mm^2 piezoelectrical crystal with a 20 MHz carrier frequency. Since the depth resolution is 0.2 mm, the sample volume for each sampling point is approximately $\pi \times (0.5)^2 \times 0.2$ mm^3. This system has 80 sampling gates. Doppler signals from the multicircuit are analyzed by a zero-cross method and the signals from an optional channel were analyzed by a fast Fourier transform (FFT) method, both in real time. Fourier analysis was performed on 128 data points. The frequency resolution, temporal resolution, and maximum detectable frequency of our system were 290 Hz, 2.56 ms, and 25 kHz, respectively. The low frequency cut-off was at 375 Hz.

Animal Preparation

Protocol 1

Analysis of blood velocity profiles at various positions along stenotic arteries [16].

Five adult mongrel dogs (15–25 kg) were anesthetized with sodium pentobarbital (25 mg/kg i.v.). After intubation, the animals were ventilated by a respirator pump with room air, which was supplemented with 100% oxygen at a rate sufficient to maintain arterial oxygen tension at physiological level. The chest was opened by both a median thoracotomy and a left thoracotomy of the fourth or fifth intercostal space. The heart was exposed and suspended in a pericardial cradle. The left anterior descending coronary artery (LAD) was exposed for a 3 or 4 cm length. All side branches along the segment of velocity measurement were ligated, and a snare occluder was placed around the proximal portion of the segment to produce a stenosis. The blood velocity waveforms and velocity profiles in the poststenotic region were measured for an approximately 90% diameter stenosis as assessed radiographically (AHA criteria). To produce an asymmetric stenosis, we applied an external weight to the stenotic segment from the outer wall of the LAD in addition to the snare occluder.

Figure 1 shows the measuring position and the probe holder in this protocol. Specially designed half cylinder probe holders of various size were used to fix the transducer at appropriate positions along the vessel. This probe holder was made of silicon and a transducer was mounted in it at an angle of 60° to the bottom surface. The phasic patterns of the velocity profiles across the vessel were observed in the plane perpendicular to the myocardial surface at distances which were nominally 0.5–1D, 1–2D, and 3D (D, outer diameter) distal to the exit of stenosis by the 80-channel zero-cross method. In the case of the 0.5–1D measuring site, the ultrasound beam was located on or near the central axial region of the vessel between 0.5 and 1.0D. The blood velocity waveforms with velocity spectra on or near the central axial region were displayed for axisymmetric

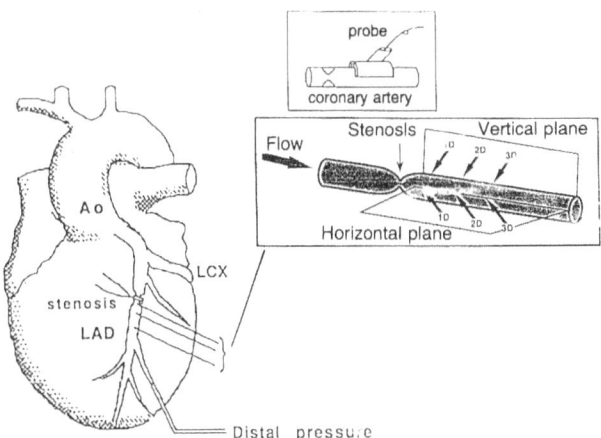

Fig. 1. The sites of poststenotic blood velocity measurement in the left anterior descending coronary artery (*LAD*) (*left*) and the positioning of velocity probes on the vessels (*right*). The phasic patterns of the velocity profiles across the LAD were observed at 0.5–1D, 1–2D, and 3D (*D*, outer diameter) distal to the exit of stenosis. (This figure was originally presented by Kajiya et al. in [16]. With permission of the American Society of Mechanical Engineers)

stenosis and in the case of asymmetric stenosis the region corresponding to peak velocities within the vessel. In two additional dogs, the velocity profiles in the horizontal plane were also measured as well as those in the vertical plane (Fig. 1). This procedure required greater exposure of the arteries.

Electrocardiograms were recorded from standard leads. Aortic pressure and left ventricular pressure were measured by an 8F pigtail manometer catheter(model PC-470, Millar, Houston, Texas) which was inserted into the ascending aorta and left ventricle through the right carotid artery. A 21-gauge Teflon catheter was introduced into a side branch of the LAD, 4 to 5 cm distal from the stenosis. Measurement of poststenotic pressure was taken with a fluid-filled pressure transducer (Model DHC, Nihon Kohden Tokyo, Japan).

Protocol 2

Analysis of blood velocity waveforms following different degrees of coronary artery stenosis with or without vasodilation [17, 18].

Twelve mongrel dogs (15–25 Kg) were premedicated and anesthetized in the similar manner to protocol 1. After the heart was exposed and suspended in the pericardial cradle, the left main coronary artery (LMCA) was carefully isolated.

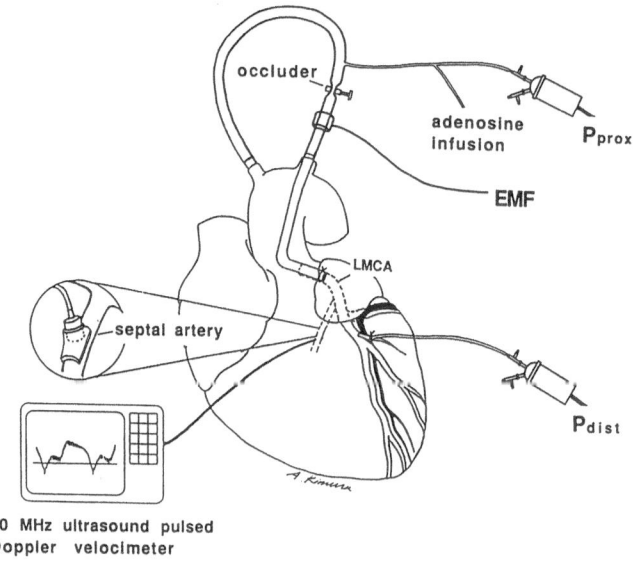

Fig. 2. The experimental preparation. Septal arterial blood flow velocity was measured by our 20 MHz multichannel ultrasound pulsed Doppler velocimeter. *EMF*, Electromagnetic flow transducer; *LMCA*, left main coronary artery; P_{prox}, proximal coronary arterial pressure measured through the auxiliary tube; P_{dist}, distal pressure measured in the diagonal branch. (This figure was originally presented by Kimura et al. in [17]. With permission of the American Physiological Society)

Heparin (5000 units i.v. and 3000 units hourly) was injected intravenously. The LMCA was perfused directly from the right carotid artery by a direct shunt circuit (Fig. 2). A cannula placed in the right carotid artery was connected by plastic tubing to a Gregg cannula. The Gregg cannula was introduced into the aorta through a left subclavian arteriotomy, advanced into the ostium of the LMCA, and secured with a ligature. An occluder was placed just proximal to the Gregg cannula near the LMCA. The coronary arterial pressure was measured both proximal (P_{prox}) and distal (P_{dist}) to the occluder; the proximal coronary arterial pressure was measured in the cannula, and the distal coronary arterial pressure was measured through a stiff cannula which was inserted into a diagonal branch of the left anterior descending coronary artery or at the tip of the Gregg cannula (Nihon Kohden DHC pressure transducer Tokyo, Japan). The velocity waveform analyzed by the FFT method, was measured for the systolic forward component (SF), reverse velocity component (SR), and the diastolic velocity component (DF) under the following conditions: (1) control condition with no stenosis (N), (2) moderate stenosis with a P_{dist} of up to 60 mmHg (S1), (3) severe stenosis with a P_{dist} of up to 35 mmHg (S2), and (4) LMCA complete occlusion.

Using the same canine preparation, adenosine (0.3–0.7 mg/min) was infused continuously into the LMCA through the external lumen of the cannula under three conditions, i.e., N, S1 and S2. The infusion rate of adenosine was increased until there was no further increase in LMCA flow. Measurements of the septal artery blood velocities were obtained for the three conditions. The degree of stenosis was kept constant between the S1 and the S1 + adenosine administration (A), and between the S2 and the S2 + A.

In an additional 12 dogs, the effect of nitroglycerin on the phasic pattern of the septal artery was evaluated. After the measurements of the coronary pressures and the phasic coronary arterial flow during control conditions, nitroglycerin (0.2 mg) was injected into the coronary arteries through the Gregg cannula, and the measurements were repeated when the LMCA flow reached its maximum value which was observed within 5 s after the nitroglycerin administration in all the animals. After allowing flow to return to its control level, a stenosis was made. Then the measurements were performed before and after administration of nitroglycerin.

Protocol 3

Transmural flow distribution during coronary artery stenosis [18].

To evaluate transmural flow distribution during severe stenosis (P_{dist} = 40 mmHg) and during stenosis (the same value of P_{dist}) + nitroglycerin, the intramyocardial blood flow distribution was measured in 6 among the additional 12 dogs with radionuclide-labeled microspheres (3M Company, St. Paal, Minn; New England Nuclear, Boston, Mass). Each dog received two intracoronary injections of 3 to 5×10^5 microspheres (15 ± 1 μm in diameter, mean ± standard deviation) selected from ^{153}Gd, ^{57}Co, ^{114}In, ^{51}Cr, ^{113}Sn, ^{85}Sr, ^{95}Nb, ^{54}Mn or ^{65}Zn.

Results and Discussion

Blood Velocity Profiles and Waveforms in the Coronary Artery at the Distal Portion to the Stenosis: Protocol 1

Figure 3 shows a normal coronary blood velocity waveform near the central axial region of the LAD without stenosis, obtained by FFT (top panel), and the velocity profiles in the vertical plane, obtained by the 80-channel zero-cross analysis (bottom), using the coordinates of radial position (vertical axis), and velocity and cardiac cycle (longitudinal axis). The velocity waveform in the time domain exhibited a diastolic-predominant pattern, which is characteristic of normal coronary artery flow. A brief reverse flow was frequently observed during the early systolic phase. The velocity profiles across the vessel were almost parabolic without any flow irregularity at the vicinity of the vessel wall, indicating the absence of flow disturbances in the coronary artery without stenosis.

Figure 4 shows the blood velocity waveform near the central axial region obtained by the FFT and velocity profiles in the vertical plane at the position immediately distal (0.5–1.0D) and 3D distal to the axisymmetric stenosis of about 90%. The peak diastolic velocity became high (over 1 m/s) and early systolic

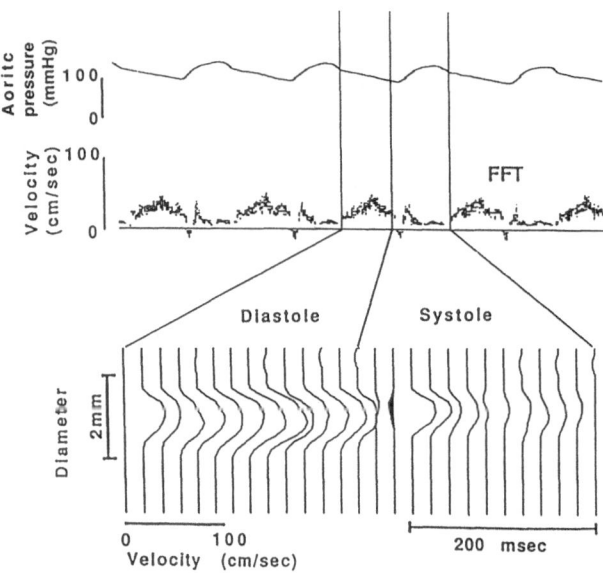

Fig. 3. Typical blood velocities in the left anterior descending coronary artery without stenosis. The *middle panel* is the fast Fourier transform (*FFT*) display of the blood velocity waveform near the central axial region of the LAD. The *bottom panel* shows the velocity profiles in the vertical plane by the multichannel zero-cross method. (This figure was originally presented by Kajiya et al. [16]. With permission of the American Society of Mechanical Engineers)

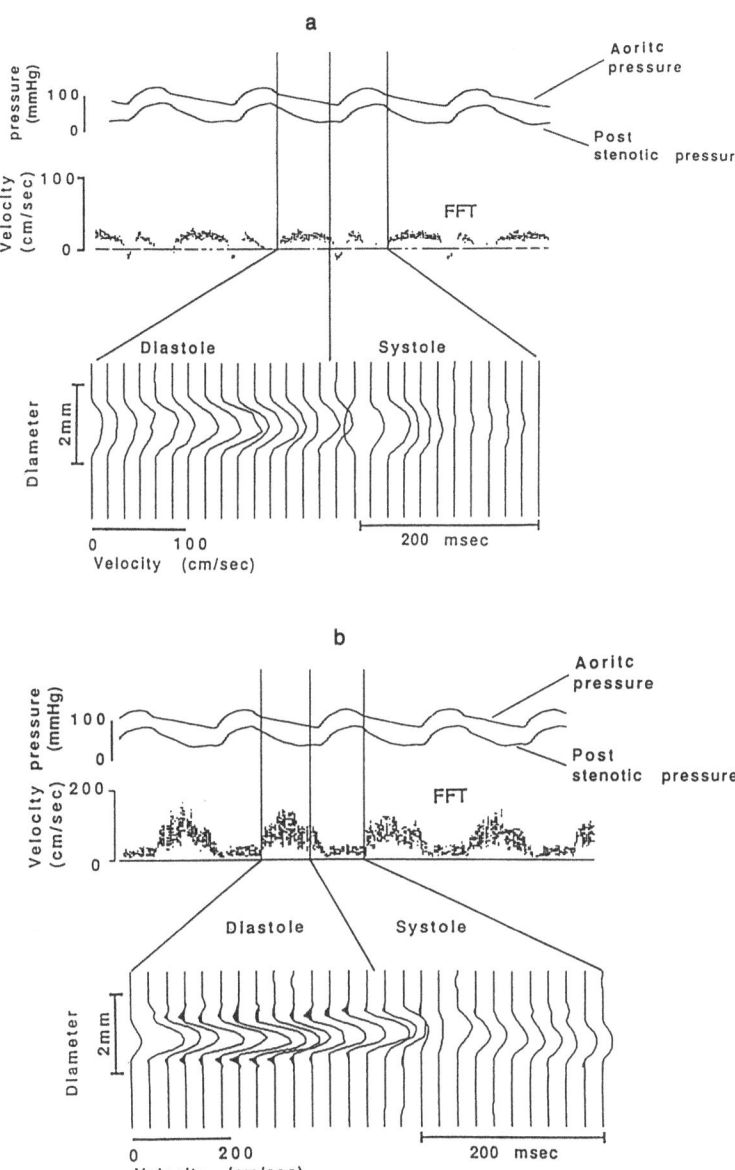

Fig. 4A,B. Blood velocity waveform and velocity profiles in the vertical plane at a position 0.5–1.0 diameters distal (**a**) and 3.0 diameters distal (**b**) to the stenotic portion. Dark shadow in **a** indicates the reverse flows during diastole. Note that at the position just distal to the stenosis, the blood velocities near the central axial region are high and there are reverse flows indicating flow separation near the vascular wall, but at the position 3 diameters distal to the stenosis the central velocities are relatively low and there is no detectable reverse flow. (This figure was originally presented by Kajiya et al. in [16]. With permission of the American Society of Mechanical Engineers)

reverse flow disappeared following coronary artery stenosis. The velocity spectra were relatively broad. In the display of blood velocity profiles, we consistently observed reverse flows near both side walls (dark shading) for axisymmetric stenoses, indicating flow separation in this region. In the blood velocity at 3D distal, the peak diastolic velocity had decreased significantly in this region compared with that just downstream of the stenosis. The velocity spectra became narrower. The velocity profiles across the vessel showed that reverse flow near both side walls was absent, indicating disappearance of flow separation in this region. The velocity profiles in the horizontal plane exhibited similar patterns to those in the vertical plane.

Figure 5 shows the blood velocities in the vertical plane at a position immediately (0.5–1.0D) distal to the asymmetric stenosis. The blood velocity waveform by the FFT (top) was obtained at the position corresponding to peak velocity across the vessel. The diastolic velocity was high with relatively broad spectra as shown in Fig. 4. The reverse flows (dark shading) were always observed near the stenotic side wall during diastole, again indicating flow separation near the stenotic side wall. Unlike the velocity profiles with flow reversals at the stenotic side wall in the vertical plane, those in the horizontal plane showed an almost symmetric pattern. In conclusion, the poststenotic velocity waveform was characterized by the existence of flow separation and recirculation, and a high stenotic viscous flow, which may contribute to pressure loss by dissipating energy.

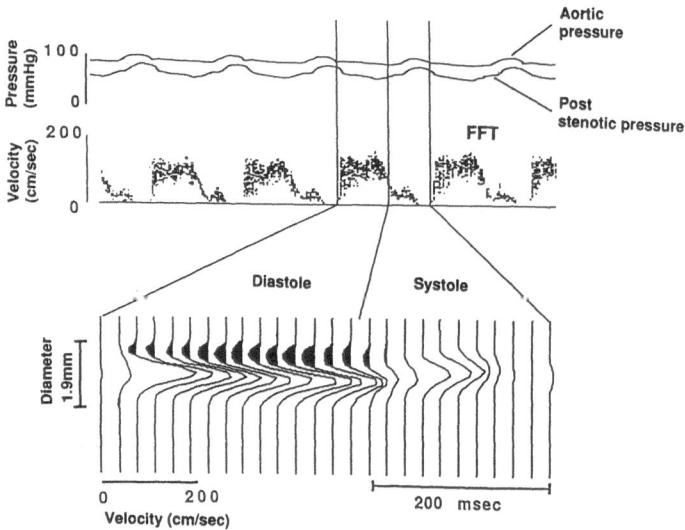

Fig. 5. Example of a blood velocity waveform in the vertical plane at 0.5–1.0 diameters distal to an asymmetric stenosis and velocity profiles across the vessel. Note that there are reverse flows near the stenotic-side wall. (This figure was originally presented by Kajiya et al. in [16]. With permission of the American Society of Mechanical Engineers)

Blood Velocity Waveforms in the Septal Artery during Coronary Artery Stenosis
with or without vasodilation: Protocol 2

Figure 6 shows typical recordings of blood velocity in the septal artery, measured
by FFT, with increasing severity of stenosis. Septal arterial velocity was measured
under control condition with no stenosis (N), moderate stenosis (S1), severe
stenosis (S2), and complete coronary artery occlusion. In the absence of a
stenosis, the blood velocity in the septal artery was predominantly diastolic
and early systolic retrograde flow was observed. The diastolic antegrade flow
decreased almost linearly with increasing severity of stenosis, whereas the
systolic retrograde flow increased. After coronary occlusion, the blood velocity
waveform in the septal artery exhibited a "to-and-fro" pattern. As a result, the
total inflow into the myocardium through a cardiac cycle decreased progressively
with the increase in the severity of stenosis. A decrease in the backpressure to the

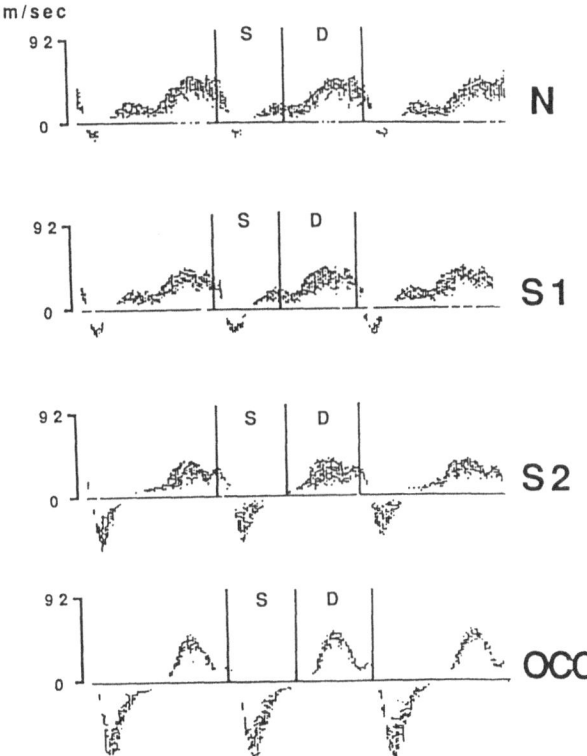

Fig. 6. Typical recordings of the septal artery blood flow velocities for no stenosis (N),
moderate stenosis (S1), severe stenosis (S2), and complete coronary artery occlusion
(OCC) (left). Note that systolic retrograde flow increased with augmentation of stenosis.
(This figure was originally presented by Kimura et al. in [17]. With permission of the
American Physiological Society)

retrograde flow and an increase in the epicardial capacitance distal to the stenosis may contribute to the increase in the systolic retrograde flow.

Figure 7 shows typical recordings of blood velocity in the septal artery before and after intracoronary adenosine (A) administration under three conditions, i.e., no stenosis (N + A), moderate stenosis (S1 + A), and severe stenosis (S2 + A). Adenosine increased systolic retrograde blood velocity component significantly, but there was no significant increase in the diastolic antegrade blood velocity component (Fig. 8). Thus, in the presence of coronary artery stenosis, vasodilation caused by adenosine administration enhanced systolic retrograde movement of the blood in the myocardial arteries and does not improve the flow reduction caused by stenosis. Therefore, the systolic retrograde flow may be an important factor in explaining the intramural steal phenomenon [19]. Following nitro-glycerin administration, diastolic antegrade velocity increased significantly, while systolic retrograde velocity area increased as well. In conclusion, coronary artery stenosis enhanced systolic retrograde flow with reduction of diastolic flow, resulting in the decrease in total myocardial inflow. The coronary vasodilation by adenosine augmented the systolic retrograde flow and did not improve the net myocardial inflow.

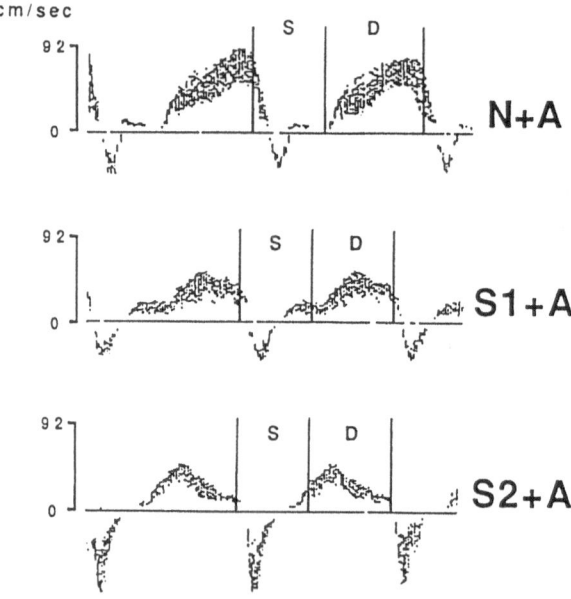

Fig. 7. Typical recordings of septal artery blood flow velocities for coronary stenosis after adenosine administration that were measured in the same dog as shown in Fig. 6. Note that adenosine increased systolic retrograde flow, but not diastolic antegrade flow for mild and severe stenosis. (This figure was originally presented by Kimura et al. in [17]. With permission of the American Physiological Society)

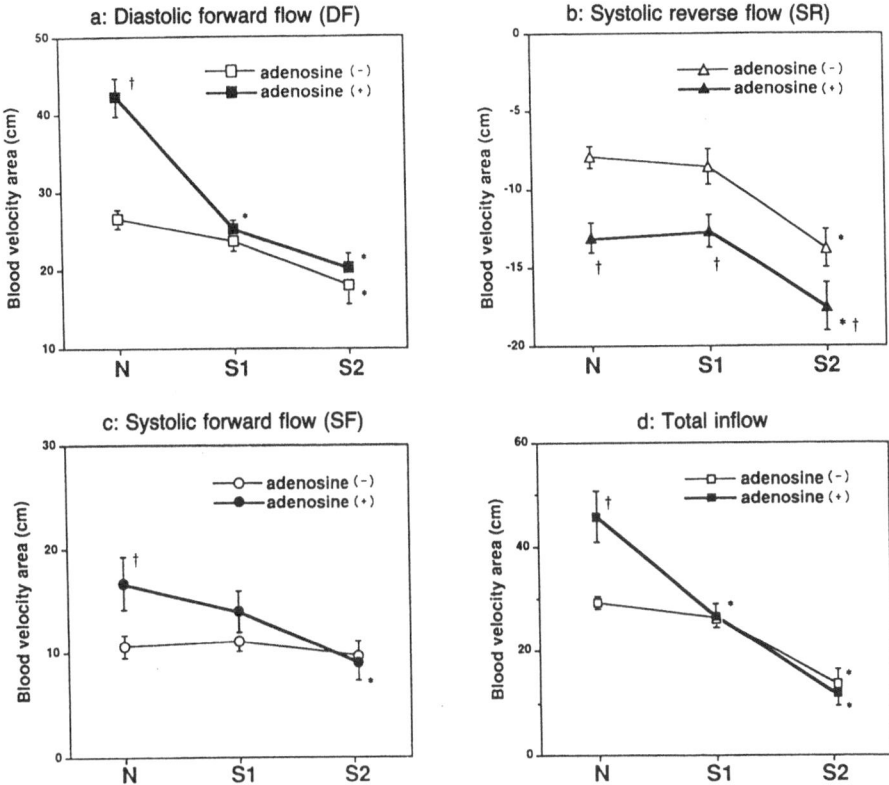

Fig. 8A–D. Blood velocity area of diastolic antegrade flow (**a**), systolic retrograde flow (**b**), systolic antegrade flow (**c**), and total inflow (**d**) with no (*N*), moderate (*S1*), and severe (*S2*) stenoses before and after adenosine administration. *$P < 0.05$ vs N; †$P < 0.05$ adenosine (−) vs adenosine (+). *Open symbols*, Adenosine (−); closed symbols, adenosine (+) (This figure was originally presented by Kimura et al. in [17]. With permission of the American Physiological Society)

Transmural Flow Distribution during Coronary Artery Stenosis with or without nitroglycerin: Protocol 3

After injection of the radionuclide-labeled microspheres under control condition, nitroglycerin (0.2 mg) was injected into the coronary arteries. The mean LMCA pressure was maintained at 40 mmHg. As soon as the LMCA flow reached its maximum value, measurement of the myocardial flow by the microspheres was repeated. Figure 9 shows the transmyocardial distribution of coronary flow before and after intracoronary nitroglycerin administration when the perfusion pressure was kept constant at a low level. Before nitroglycerin administration, the transmural myocardial flows were low due to low perfusion pressure of 40 mmHg, especially in the subendocardium. With the administration of nitroglycerin, the subepicardial flow increased significantly from 0.71 ± 0.34 ml/min per g to 1.27

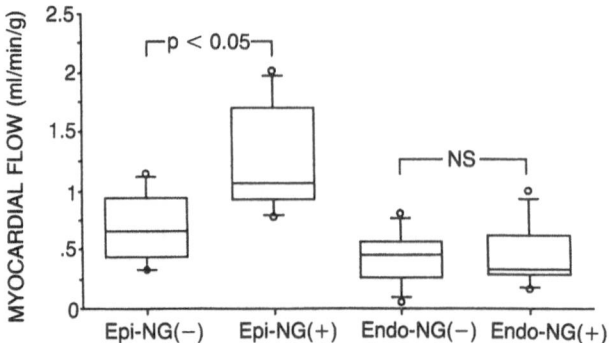

Fig. 9. Myocardial flows in subepicardium and subendocardium while maintaining perfusion pressure at a low level (40 mmHg) before and after intracoronary administration of nitroglycerin. *Epi-NG(−)*, Subepicardial flow under basal condition; *Epi-NG(+)*, subepicardial flow with nitroglycerin; *Endo-NG(−)*, subendocardial flow under basal condition; *Endo-NG(+)*, subendocardial flow with nitroglycerin. Because of low perfusion pressure, subepicardial and subendocardial flows are low even before nitroglycerin (about 0.7 and 0.5 ml/min per g, respectively). Note that with administration of nitroglycerin, the subepicardial flow increased significantly, but the subendocardial flow did not. (This figure was originally presented by Goto et al. in [18]. With permission of the American Heart Association)

$\pm$ 0.51 ml/min per g ($P < 0.05$). However, the subendocardial flow did not change significantly, being 0.48 $\pm$ 0.20 ml/min per g and 0.42 $\pm$ 0.30 ml/min per g before and after administration of nitroglycerin, respectively. With the administration of nitroglycerin, the subendocardial-to-subepicardial flow ratio decreased significantly, from 0.73 $\pm$ 0.19 to 0.32 $\pm$ 0.14 ($P < 0.01$). This may be mainly due to the augmented systolic reverse flow from the subendocardium and/or a relative increase in the resistance of subendocardial vessels to diastolic flow or a regional steal [19].

In conclusion, intracoronary nitroglycerin administration which enhances systolic retrograde flow, as well as the increase in diastolic flow, increased epimyocardial flow, but not endomyocardial flow.

Acknowledgments. We are grateful to Prof. JIE Hoffman and his colleagues for their collaboration on the nitroglycerin experiment [18]. We thank the Society of Mechanical Engineers, the American Physiological Society, and the American Heart Association, Inc. for permission to reproduce in part our paper published in Journal of Biomedical Engineering 114, 1992 [16], American Journal of Physiology 262, 1992 [17], and Circulation 85, 1992 [18].

This work was supported, in part, by a grant in aid for Scientific Research from the Ministry of Education, Science and Culture of Japan (No 01480253), by a research grant for Cardiovascular Disease (1A-1) from the Ministry of Health and Welfare, Japan, and by a grant from the Japan Society for the Promotion of Science, the Japan-US Cooperative Science Program.

References

1. Gould KL (1978) Pressure-flow characteristics of coronary stenosis in unsedated dogs at rest and during coronary vasodilation. Circ Res 43:242–253
2. Young DF, Tsai FY (1973) Flow characteristics in models of arterial stenosis: 1. Steady flow. J Biomech 6:395–410
3. Young DF, Tsai FY (1973) Flow characteristics in models of arterial stenosis: 1. Unsteady flow. J Biomech 6:547–559
4. Buckberg GD, Fixler DE, Archie JP, Hoffman JIE (1972) Experimental subendo-cardial ischemia in dogs with normal coronary arteries. Circ Res 30:67–81
5. Furuse A, Klopp EH, Brawley RK, Gott VL (1975) Hemodynamic determinations in the assessment of distal coronary artery disease. J Surg Res 19:25–33
6. Gould KL (1978) Pressure-flow characteristics of coronary stenosis in unsedated dogs at rest and during coronary vasodilation. Circ Res 43:242–253
7. Kajiya F, Tsujioka K, Ogasawara Y, Hiramatsu O, Wada Y, Goto M, Yanaka M (1989) Analysis of the characteristics of the flow velocity waveforms in left atrial small arteries and veins in the dog. Circ Res 65:1172–1181
8. Douglas JE, Greenfield JC, Jr (1970) Epicardial coronary artery compliance in the dog. Circ Res 27:921–929
9. Chilian WM, Marcus ML (1982) Phasic coronary blood flow velocity in intramural and epicardial coronary arteries. Circ Res 50:775–781
10. Chilian WM, Marcus ML (1985) Effects of coronary and extravascular pressure on intramyocardial and epicardial blood velocity. Am J Physiol 248 (Heart Circ Physiol 17):H170–H178
11. Carew TE, Covell JW (1976) Effect of intramyocardial pressure on the phasic flow in the intraventricular septal artery. Cardiovasc Res 10:56–64
12. Eckstein RW, Moir TW, Driscol TE (1963) Phasic and mean blood flow in the canine septal artery and an estimate of systolic resistance in deep myocardial vessels. Circ Res 12:203–211
13. Kajiya F, Tomonaga G, Tsujioka K, Ogasawara Y, Nishihara H (1985) Evaluation of local blood flow velocity in proximal and distal coronary arteries by laser Doppler method. J Biomech Eng 107:10–15
14. Ogasawara Y, Hiramatsu O, Kagiyama M, Tsujioka K, Tomonaga G, Kajiya F, Yanashima T, Kimura Y (1984) Evaluation of blood velocity profile by high frequency ultrasound pulsed Doppler velocimeter by a multigated zerocross method together with a Fourier transform method. IEEE Comput Cardiol 447–450
15. Kajiya F, Ogasawara Y, Tsujioka K, Nakai M, Goto M, Wada Y, Tadaoka S, Matsuoka S, Mito K, Fujiwara T (1986) Evaluation of human coronary blood flow with an 80-channel pulsed Doppler velocimeter and zero-cross and Fourier transform methods during cardiac surgery. Circulation 75 [Suppl III]:III53–III60
16. Kajiya F, Hiramatsu O, Kimura A, Yamamoto T, Yada T, Ogasawara Y, Tsujioka K (1992) Blood velocity patterns in poststenotic regions and velocity waveforms for myocardial inflow associated with coronary artery stenosis in dogs. J Biomech Eng 114:385–390

17. Kimura A, Hiramatsu O, Yamamoto T, Ogasawara Y, Yada T, Goto M, Tsujioka K, Kajiya F (1992) Effect of coronary stenosis on phasic pattern of septal artery in dogs. Am J Physiol 262:H1690–H1698
18. Goto M, Flynn AE, Doucette JW, Kimura A, Hiramatsu O, Yamamoto T, Ogasawara Y, Tsujioka K, Hoffman JIE, Kajiya F (1992) Effect of intracoronary nitroglycerin administration on phasic pattern and transmural distribution of flow during coronary artery stenosis. Circulation 85:2296–2304
19. Feigl EO, Buffington CW, Nathan HJ (1987) Adrenergic coronary vasoconstriction during myocardial underperfusion. Circulation 75 [Suppl I]:I1–I5

Morphometric Measurement of Subendocardial Vessel Dimensions in Systolic and Diastolic Arrested Rat Heart

Bernard I. Levy[1], *Jane Lise Samuel*[2], *Victor E. Kotelianski*[3], *Francoise Marotte*[2], *Pierre Poitevin*[1], and *Richard S. Chadwick*[4]

Summary. After maximal vasodilation and KCl or $BaCl_2$ heart arrest (groups D and S, respectively), vessel wall fibronectin was labelled in the *in situ* frozen myocardium of ten Wistar rats. Capillary and venular dimensions and densities were measured in transverse sections of the inner third of the left ventricular free wall for both groups. The measured capillary diameters and density were, respectively, $6.86 \pm 1.73\,\mu m$ and $1434 \pm 233/mm^2$ in group D, and $4.97 \pm 1.12\,\mu m$ and $1316 \pm 400/mm^2$ in group S. The venular cross-sectional areas were $181 \pm 27\,\mu m^2$ in group S and $415 \pm 65\,\mu m^2$ in group D, while the respective venular densities were $15.4 \pm 3.5/mm^2$ and $17.1 \pm 2.9/mm^2$. These morphometric measurements correspond to an intramyocardial blood volume of $6.0\,ml/100\,g$ of left ventricle (*LV*) in group D and $2.9\,ml/100\,g$ LV in group S. These results suggest: that (1) even in the group with $BaCl_2$ arrest, there are very few subendocardial capillaries smaller than the critical value for red cell passage, and (2) the systolic-diastolic changes in intramyocardial blood volume may form a significant part of the total vasodilated intramyocardial blood volume.

Key words: Intramyocardial blood volume—Myocardial microcirculation—Capillary density—Capillary diameter

Introduction

During the last 10 years, it has become apparent that intramyocardial blood volume and vessel compliance, thought to be located mainly in the capillaries and small veins, play a major role in the phasic myocardial blood flow pattern [1]. No direct measurement of the total intramyocardial vessel volume in vivo has been reported. Some studies, using the corrosion cast technique, have reported diameters of capillaries and capillary density in the left ventricular wall [2, 3]. Other authors have observed in vivo subepicardial capillaries by optical methods

Institut National de la Santé et de la Recherche Médicale, [1] Unit 141 and [2] Unit 127, Hôpital Lariboisiere, 75010 Paris, France
[3] CNRS URA 230, 46 rue d'Ulm, 75005 Paris, France
[4] Theoretical Biomechanics Group, Biomedical Engineering Branch, Division of Research Service, National Institutes of Health, Bethesda, MD 20892, USA
The authors wish to acknowledge the support of the NIH-INSERM collaborative agreement.

83

[4–7]; they reported significant differences between systolic and diastolic epicardial capillary dimensions.

There is a lack of information concerning the dimensions of deep intramyocardial vessels, especially their ability to be modified during the cardiac cycle by myocardial contraction and relaxation. The purpose of this work was to estimate the blood volume of capillaries and venules in the inner part of the rat left ventricle after systolic or diastolic arrest.

Methods

Experiments were performed on ten Wistar rats weighing $360 \pm 27\,g$. After anesthesia was induced by intraperitoneal injection of pentobarbital (50 mg/kg), the trachea was cannulated and then connected to a rodent respirator (Harvard model 680 South Natick, MA, USA). A mid-sternal thoracotomy was performed; the beating heart was exposed and introduced into a plastic container placed into the thorax. The heart was then arrested in diastole (group D) by IV injection of KCl (saturated solution), or in contracture (group S) induced by injection of barium chloride solution (30 mM). An IV perfusion of carbochromene was performed for 5 min (20 mg/kg) before heart arrest in order to induce maximal coronary vasodilation and to counterbalance the vasoconstrictive effects of barium and potassium. Immediately after cardiac arrest, liquid nitrogen and precooled isopentane were simultaneously poured into the container. The heart was then rapidly removed and the left ventricle trimmed free. Tissue was excised from the region midway between the apex and the base of the ventricle, and the ventricle was oriented to be sectioned roughly perpendicular to the long axis. The tissue samples were attached to a thin sheet of cork with embedding medium (RUA Instruments IVRY, France) and kept frozen in cooled isopentane. Sections (5-μm-thick) from the two experimental groups were cut on a cryostat at $-20\,°C$ and collected on glass slides. Slides were incubated for 30 min at $37\,°C$ with fibronectin antibodies, at a dilution of 1:50 in phosphate buffered saline (PBS) pH 7.2, containing 2% bovine serum albumin. After being washed in PBS, the sections were incubated with fluorescein-labelled anti-rabbit immunoglobulins (Miles Laboratories, Rueil Malmaison, France). The antibodies used were raised in a rabbit against human fibronectin and were then purified by immunoaffinity chromatography on the same fibronectin. The specificity of the antibodies toward fibronectin and their cross-reactivity with fibronectins of other species have been described elsewhere [8]. Sections used as controls were incubated with non-immunized rabbit serum.

After another washing, the sections were mounted with elvanol and examined with a Leitz Dialux microscope with epifluorescence optics equipped with a Ploemopack filter set (Rueil Malmaison, France). Photomicrographs were made on TriX pan Kodak film and developed in Rodinal (Agfa). Prints were enlarged to obtain 1 mm per μm on the paper print. Photographs were then mounted to represent a continuous view of the left ventricular free wall myocardium from the endocardium to the epicardium. Figure 1 shows typical examples of montages

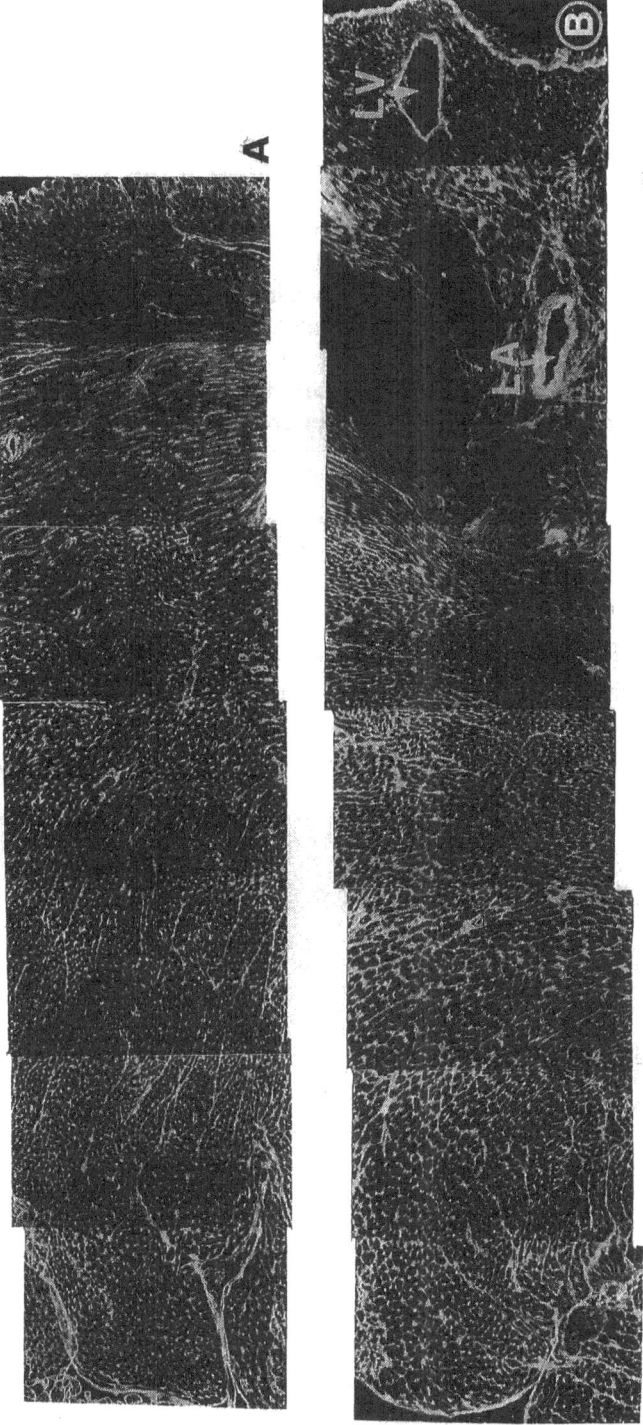

Fig. 1 A,B. Photomicrographic montage (×40) of transverse sections of the fibronectin-labelled left ventricular free wall **A** after KCl arrest and **B** after BaCl₂ arrest. View from the endocardium to the epicardium from left to right. *LA*, Large epicardial artery; *LV*, large epicardial vein

obtained during diastolic (A) or systolic (B) arrest. On the enlarged photomicrographs, arterioles, capillaries, and venules are easily identified from the labelled fibronectin of their walls (Fig. 2). Vessel cross-sectional dimensions taken from points midway in the vessel walls were measured using a HP 9873 digitizer connected to a HP 9836 microcomputer. The maximal resolution of the digitizer was 25 μm, corresponding to 0.025 μm in the actual dimension measurements.

In every heart studied, the dimensions of all capillaries located in the inner third of the myocardium were measured "blind" by two independent operators. In the same way, the dimensions of all subendocardial veins and venules were measured. From these measurements, we calculated the density and the mean cross-sectional area of capillaries and venules.

Results are expressed as means and standard deviations. A one way analysis of variance was performed. Differences between groups were then evaluated using the Newman Keuls test.

Results

Capillaries

The mean subendocardial capillary diameter was 6.86 ± 1.73 μm in K^+-arrested hearts and 4.97 ± 1.12 μm in Ba^{2+}- arrested hearts (F = 359, DF = 1199, $P <$ 0.001). Frequency distributions of individual capillary diameters (Fig. 3) were similar in groups D and S, with a shift to the lower diameter values for group S.

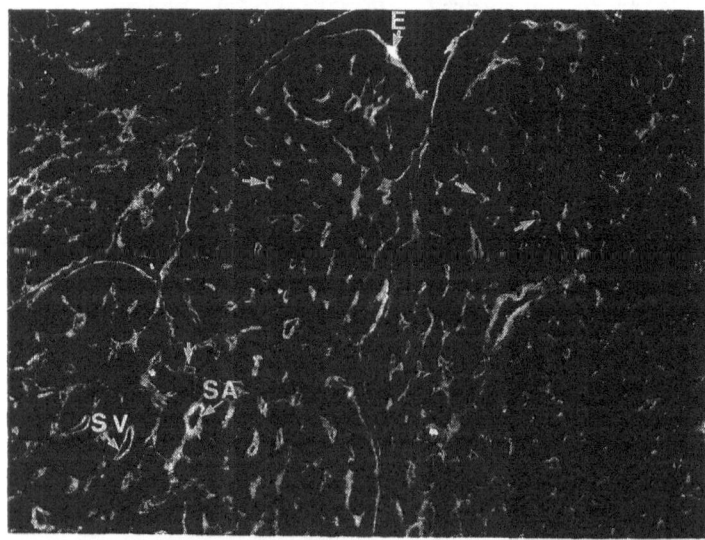

Fig. 2. Enlarged view (×500) of the subendocardium area. *Arrows* indicate some capillaries *E*, Endocardium; *SV*, small vein; *SA*, small artery

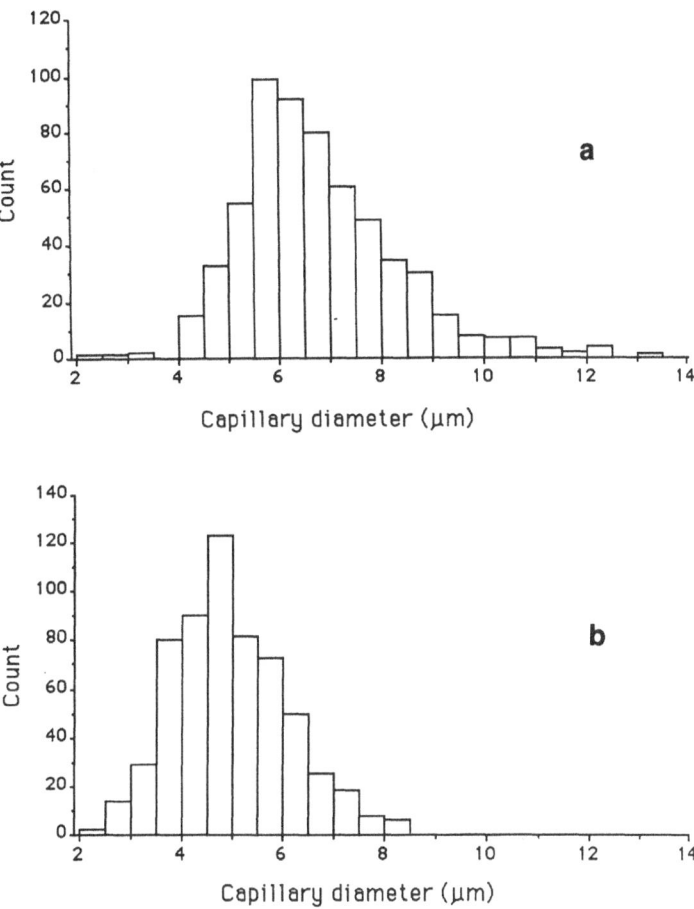

Fig. 3 A,B. Frequency distribution of capillary diameters in the left ventricular subendo-cardium after **a** diastolic arrest and **b** systolic arrest

Capillary density was measured by counting the total number of capillaries in adjacent subendocardial surface areas of $200 \times 200 \,\mu m$. In systolic and diastolic arrested hearts, the capillary density was $1316 \pm 400/mm^2$ and $1434 \pm 233/mm^2$, respectively (non significant difference, $P = 0.49$). The capillary volume/100 g, calculated as the product of the capillary density by the mean capillary cross-sectional area was thus $2.6 \pm 0.8 \,ml/100 \,g$ of subendocardial left ventricle in group S and $5.3 \pm 0.9 \,ml/100 \,g$ in group D ($P < 0.01$).

Venules

The venular density was much lower than the capillary density for both groups. The densities were, respectively, $15.4 \pm 3.5/mm^2$ and $17.1 \pm 2.9/mm^2$ in systolic and diastolic arrest (non-significant difference, $P = 0.38$). The venular cross-sectional areas were $181 \pm 27\,\mu m^2$ in group S and $415 \pm 65\,\mu m^2$ in group D. This corresponds to venular volumes of $0.28 \pm 0.06\,ml/100\,g$ for group S and $0.71 \pm 0.16\,ml/100\,g$ for group D ($P < 0.001$).

Discussion

The morphological method used in our study allowed us to count and to measure all capillaries, even those empty of red cells, in the subendocardial layers. Measurements in the midwall and epicardial layers, where fibers are not perpendicular to the section plane, did not prove to be reliable. Also, very few arterioles were seen in the subendocardial sections, so that their diameter and density could not be reported. We hypothesize that two effects contribute to their low number density: (a) the number of branching levels between large coronary arteries (those with a diameter of approximately 0.2 mm in rats and 2–3 mm in dogs and humans) and capillaries is smaller in the rat than in the dog or in humans, while capillary diameters are about the same. (b) Those arterioles that are present are not oriented perpendicular to the section plane. We do not feel that a significant flow redistribution took place during cooling. With the surface held at $-150\,^{\circ}C$, the freezing rate can be estimated at $10\,\mu m/ms$ [9]; thus, the time to cool the core of a rat heart from $37\,^{\circ}C$ down to $10\,^{\circ}C$ (when the blood becomes very viscous) can be estimated to be less than 0.5 s, i.e., probably less than the time constant for microcirculatory flow in the rat.

The previously reported capillary densities were obtained from in vitro measurements made from fixed sections of tissue, following coronary perfusion with India ink or casting medium. These measurements have to be corrected for the effects of the high injection pressure that may dilate the capillaries and for the effects of shrinkage, which may be up to 50% [2]. It is likely that the dye or casting medium may induce hypoxia and thus vasodilation, and, furthermore, the media may not fill all the capillaries. For these reasons, there are large variations in previously reported capillary densities (from $340/mm^2$ [10] to $3170/mm^2$, [2]). For rat hearts, Buss et al. [11] reported $2100/mm^2$ and Martini and Honig [12] $2200/mm^2$. Our results of $1316/mm^2$ in systole and $1434/mm^2$ in diastole are lower than those previously reported in rats, probably because the present method did not induce hypoxia, thus not producing recruitment [13]. Our systolic-diastolic values for capillary density are consistent with muscle fiber thickening during contraction, but, nevertheless, are not statistically significant.

Results obtained for mean capillary diameter in the present study (4.97–6.86 μm) are in the range of those reported by different methods. Using casting, reported mean values were 5.6 μm in dog [2], and 5.14 μm [3], 5.3 μm [14], 6–6.4 μm [6], and 7.2–8.9 μm [11] in rat left ventricle.

Our reported capillary diameter distribution ranged from 2 to 13 μm. The lower limit is an interesting point to discuss in terms of the ability of red cells to enter. Henquell et al. [5] have shown that 2.45 μm is the lower limit for a glass tube diameter through which a red cell can pass. They suggest that the smallest capillaries must be channels for plasma flow alone, especially during systole in the subendocardium. However, the lower limit for deformable capillaries is probably smaller than that of a rigid tube. Such "white" capillaries have been described in various tissues, but have never been observed in the epicardial surface of the beating heart. Our results show that even in the subendocardial layers, during systole when tissue pressure is high, there are very few capillaries (about 2%) smaller than 3 μm. This suggests that "white" capillaries are not likely to exist in the beating heart [15].

Our reported venular volumes, 0.28 ml/100 g for group S and 0.71 ml/100 g for group D, were only 10% of the corresponding capillary volumes. We have found no other subendocardial measurements of venular volumes for comparison. These low values are probably best explained by the same arguments given previously concerning arterioles. Furthermore, since it is difficult to differentiate between large capillaries and small venules, venous volume might be underestimated. The diastolic-systolic capillary cross-sectional area ratios calculated from previously published data are 2.36 for dog, 2.43 for turtle [4], and 1.56 for rat [5] in the epicardial surface of the beating heart. From our data, we obtained a similar ratio, of 2.04. This corresponds to a blood volume variation between the diastolic and the systolic arrested hearts of 2.7 ml/100 g tissue.

While this absolute volume variation is in agreement with an estimate made by Spaan [1], it represents a larger fraction of the intramyocardial blood volume (45%) than the value he inferred for vasodilated heart (25%). The main reason for this discrepancy is the difference between the venular volume estimated by Spaan in the dog and the much smaller venular volume measured by the present technique in the rat.

References

1. Spaan JAE (1985) Coronary diastolic pressure-flow relation and zero flow pressure explained on the basis of intramyocardial compliance. Circ Res 56:293–308
2. Bassingthwaighte JB, Yipintsoi T, Harvey RB (1974) Microvasculature of the dog left ventricular myocardium. Microvasc Res 7:229–249
3. Potter RF, Groom AC (1983) Capillary diameter and geometry in cardiac and skeletal muscle studied by means of corrosion casts. Microvasc Res 25:68–84
4. Tillmanns H, Ikeda S, Hansen H, Sarma JSM, Fauvel JM, Bing RJ (1974) Microcirculation in the ventricle of the dog and turtle. Circ Res 34:561–569
5. Henquell L, LaCelle PL, Honig CR (1976) Capillary diameter in rat heart in situ: Relation to erythrocyte deformability, O_2 transport, and transmural O_2 gradients. Microvasc Res 12:259–274
6. Steinhausen M, Tillmanns H, Thederan H (1978) Microcirculation of the epimyocardial layer of the heart. A method for in vivo observation of the microcirculation of superficial ventricular myocardium of the heart and capillary flow pattern under normal and hypoxic conditions. Pflügers Arch 378:9–14

7. Nellis SH, Liedtke AJ (1982) Pressures and dimensions in the terminal vascular bed of the myocardium determined by a new free-motion technique. In: Tillmanns H, Kubler W, Zebe H (eds) Microcirculation of the heart. Springer, New York Berlin Heidelberg Tokyo, pp 61–74

8. Kotelianski VE, Arsenyeva EL, Bogacheva GT, Chernousov MA, Glukhova MA, Ibrighimov AR, Metsis ML, Petrosyan MN, Rokuin OV (1982) Identification of the species-specific antigenic determinant(s) of human plasma fibronectin by monoclonal antibodies. FEBS Lett 42:199–202

9. Clarks AJR, Clarks PAA (1983) Capture of spatially homogenous chemical reactions in tissue by freezing. Biophys J 42:25–30

10. Reynold SRM, Kirsch M, Bing RJ (1958) Functional capillary beds in the beating, KCI-arrested and KCI-arrested-perfused myocardium of the dog. Circ Res 6:600–611

11. Buss DD, Hyde DM, Stoval MY (1981) Application of stereology to coronary micro-circulation. Basic Res Cardiol 76:411–415

12. Martini J, Honig CR (1969) Direct measuremant of intercapillary distance in beating rat heart in situ under various conditions of O_2 supply. Microvasc Res 1:244–256

13. Duran WN (1982) Myocardial capillary recruitment studied by indicator dilution curves. In: Tillmanns H, Kubler W, Zebe H (eds) Microcirculation of the heart. Springer, New York Berlin Heidelberg Tokyo, pp 109–117

14. Tomanek RJ, Searls JC, Lachenbruch PA (1982) Quantitative changes in the capillary bed during developing peak and stabilized cardiac hypertrophy in the spontaneously hypertensive rat. Circ Res 51:295–304

15. Vetterlein F, Hemeling H, Sammler G, Pettio A, Schmidt G (1989) Hypoxia-induced acute changes in capillary and fiber density and capillary red cell distribution in the rat heart. Circ Res 64:742–752

Evaluation of Dynamic Mechanical Properties of Coronary Arterial System Using Multi-Channel Random Noise Technique

Kenji Sunagawa[1], Yasuhiko Harasawa, and Koji Todaka[2]

Summary. We evaluated, in ten isolated, cross-circulated canine hearts, the dynamic mechanical properties of the coronary arterial system using a multichannel random noise technique. We perturbed coronary arterial pressure with a high speed servo-pump and altered ventricular pressure by random pacing. We then determined admittance spectra where both coronary arterial pressure and left ventricular pressure are the inputs and coronary arterial flow is the output. The coronary arterial admittance from coronary arterial pressure to the flow indicated that the dynamic mechanical properties of the coronary arterial system approximated a windkessel model like the other arterial systems. The coronary admittance from left ventricular pressure to coronary arterial flow indicated that, in the low frequency range, the characteristics of attenuation of coronary arterial flow by ventricular contraction were consistent with the vascular waterfall mechanism. We conclude that the instantaneous coronary arterial flow could be described as a function of coronary arterial pressure and left ventricular pressure in our tested condition.

Key words: Two-channel admittance model—Multichannel random noise technique—Vascular waterfall—Intramyocardial pump—Time varying elastance

Introduction

Hydraulic impedance has often been used to characterize mechanical properties of arterial systems. It allows us to estimate details of visco-elastic properties and pulse wave reflectiveness of the target vascular system. This technique, in principle, is applicable to any vascular system as long as it is linear and time invariant. In case of the coronary artery, however, its apparent time varying nature, that is, coronary arterial flow decreases in systole and increases in diastole, has given an impression that the coronary impedance analysis of the beating heart is invalid. But whether the coronary arterial system is truly time varying or not has not been well established.

The purpose of this investigation was to characterize the mechanical properties of the coronary arterial system using hydraulic admittance. The admittance is

---

[1] Department of Cardiovascular Dynamics, National Cardiovascular Center, Research Institute 5-7-1 Fujishirodai, Suita, Osaka, 565 Japan
[2] Research Institute of Angiocardiology and Cardiovascular Clinic, Kyushu University, 3-1-1 Maidashi, Higashi-ku, Fukuoko, 812 Japan

the reciprocal value of impedance. Thus it carries similar information to the impedance and has the same limitations as the impedance. Therefore, central to the validity of the hydraulic admittance analysis of the coronary arterial system, is the system linearity and time invariance. We hypothesized that the coronary arterial system is linear and time invariant. We assumed then that the instantaneous coronary arterial flow is a function of coronary arterial pressure. This is the same assumption of conventional vascular impedance measurement. We further assumed that the instantaneous coronary arterial flow is a function of the left ventricular pressure as well. The second assumption was required to take into account the effect of ventricular contraction on coronary arterial flow. Thus the instantaneous coronary arterial flow ($F(f)$) is expressed in the frequency domain as

$$F(f) = H1(f)Pca(f) + H2(f)Plv(f) \qquad (1)$$

where $H1(f)$ is the admittance from coronary arterial pressure, $Pca(f)$, to coronary arterial flow, and $H2(f)$ is the admittance from left ventricular pressure, $Plv(f)$, to coronary arterial flow. Both $H1(f)$ and $H2(f)$ can be accurately estimated only if coronary arterial pressure and left ventricular pressure are sufficiently independent. Under normal conditions, however, the coronary arterial pressure is tightly coupled with left ventricular pressure. This consistent coupling makes identification of $H1(f)$ and $H2(f)$ inaccurate. To circumvent this problem, we separately and randomly perturbed the coronary arterial pressure and left ventricular pressure, and identified the coronary admittance. The results indicated that despite the apparent time-varying nature of the coronary arterial system, the multi-channel random perturbation technique allowed us to accurately estimate coronary arterial admittance, suggesting that the coronary arterial system is linear and time invariant.

Experiments

A pair of dogs were anesthetized with an intravenous injection of sodium pentobarbital (25 mg/kg). We thoracotomized one dog and used it as a donor heart. The other dog was used to support the excised hearts. We removed the heart while retrogradely perfusing it by the support dog through the aorta. We then inserted a latex balloon in the left ventricle. The balloon was filled with tap water and hooked up to a volume servo-pump. The servo-pump enabled us to control ventricular volume as we desired. We separated the left circumflex coronary artery and placed an electromagnetic flow probe on it. The coronary arterial pressure was servo-controlled. We abolished the coronary autoregulation by continuously infusing adenosine into the coronary artery. We used ten isolated hearts for this series of experiments.

In order to measure hydraulic admittance from coronary arterial pressure to coronary arterial flow and that from left ventricular pressure to coronary arterial flow, we randomly perturbed coronary arterial pressure using the high fidelity servo-pump. We also perturbed left ventricular pressure by irregularly pacing the heart with a computer controlled pacemaker. We used different, independent random command signals for the servo-pump and the pacemaker to make the ventricular pressure and the coronary arterial pressure reasonably independent. We digitized analog signals of coronary arterial flow, coronary arterial pressure, left ventricular pressure, and left ventricular volume at 200 Hz with 12 bit resolution. The acquired data were saved in a hard disk for subsequent analyses.

Data Analysis

If we multiply the conjugate of frequency transformed coronary arterial pressure, Pca(f)*, with Eq. 1, we obtain the following equation.

$$F(f)Pca(f)^* = H1(f)|Pca(f)|^2 + H2(f)Plv(f)Pca(f)^* \quad (2)$$

Similarly multiplying the conjugate of left ventricular pressure, Plv(f)*, with Eq. 1 yields

$$F(f)Plv(f)^* = H1(f)Pca(f)Plv(f)^* + H2(f)|Plv(f)|^2 \quad (3)$$

Since all power terms (i.e., $|Pca(f)|P^2$ and $|Plv(f)|^2$) and cross power terms (i.e., $F(f)Pca(f)$, $Pca(f)Plv(f)$, and $F(f)Plv(f)$) appearing in Eqs. 2 and 3 are measurable, simultaneous solution of Eqs. 2 and 3 yields Hl(f) and H2(f) as follows.

$$H1(f) = \frac{\begin{vmatrix} F(f)Pca(f) & Plv(f)Pca(f) \\ F(f)Plv(f) & |Plv(f)|^2 \end{vmatrix}}{\begin{vmatrix} |Pca(f)|^2 & Plv(f)Pca(f) \\ Pca(f)Plv(f) & |Plv(f)|^2 \end{vmatrix}} \quad (4)$$

$$H2(f) = \frac{\begin{vmatrix} F(f)Pca(f) & Plv(f)Pca(f) \\ F(f)Plv(f) & Plv(f)^2 \end{vmatrix}}{\begin{vmatrix} |Pca(f)|^2 & Plv(f)Pca(f) \\ Pca(f)Plv(f) & |Plv(f)|^2 \end{vmatrix}} \quad (5)$$

Tight coupling of coronary arterial pressure and left ventricular pressure results in an ill-conditioned matrix inversion. This is equivalent to saying that the denominator is close to zero. That is why we must dissociate coronary arterial pressure from left ventricular pressure. We determined the admittance over the frequency range from 0.02 Hz through to 20 Hz. To test the accuracy of the estimated admittance model, we determined the multiple coherence function. The coherence function varies from 0 to 1. Unity coherence indicates that coronary flow variation is totally attributable to changes in coronary arterial pressure and left ventricular pressure. If the coherence function is 0.5, 50% of coronary flow variation is linearly attributable to coronary arterial pressure and left ventricular pressure. To test the accuracy of the two channel admittance model in the time domain, we predicted instantaneous coronary arterial flow for a given coronary arterial pressure and left ventricular pressure using the estimated admittance values. To be stringent, the data segment used to test the accuracy was not used to estimate the admittance values.

Results

Illustrated in Fig. 1 is the average admittance spectrum. The left panel is the admittance from coronary arterial pressure to coronary arterial flow (H1), and the right is from left ventricular pressure to coronary arterial flow (H2). As can be seen in the left panel, the coronary admittance was rather low in the low frequency range and increased with frequency. The corner frequency was about 6 Hz. Since H1 represented coronary arterial admittance when the effect of ventricular contraction was removed, the reciprocal of the admittance should resemble the arterial impedance. This was indeed the case.

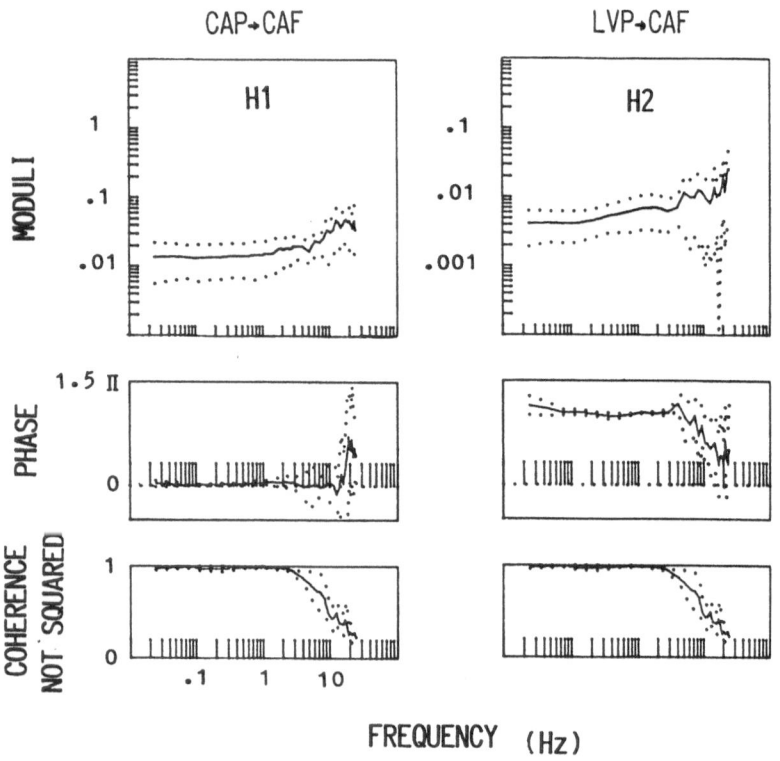

Fig. 1. Coronary arterial admittance spectrum. *Left panel*, Coronary arterial admittance spectrum from coronary arterial pressure (*CAP*) to coronary arterial flow (*CAF*) (*H1*). *Right panel*, Coronary arterial admittance spectrum from left ventricular pressure (*LVP*) to CAF, (*H2*). See text for details

In the right panel of Fig. 1, the admittance from left ventricular pressure to coronary arterial flow was shown. The admittance was low in the low frequency range and increased above 1 Hz. This is to say that the effect of ventricular contraction on the coronary arterial flow was less significant in the low frequency range and became more significant above 1 Hz. The phase analysis indicated that the ventricular contraction resulted in a decrease of the coronary arterial flow. The fact that the effect of ventricular pressure on coronary arterial flow was fairly constant below 0.1 Hz suggests that the vascular waterfall mechanism might take place in this frequency range.

The bottom panels of Fig. 1 represent the multiple coherence function. As can be seen, the coherence was close to unity in the frequency range between 0.02 Hz

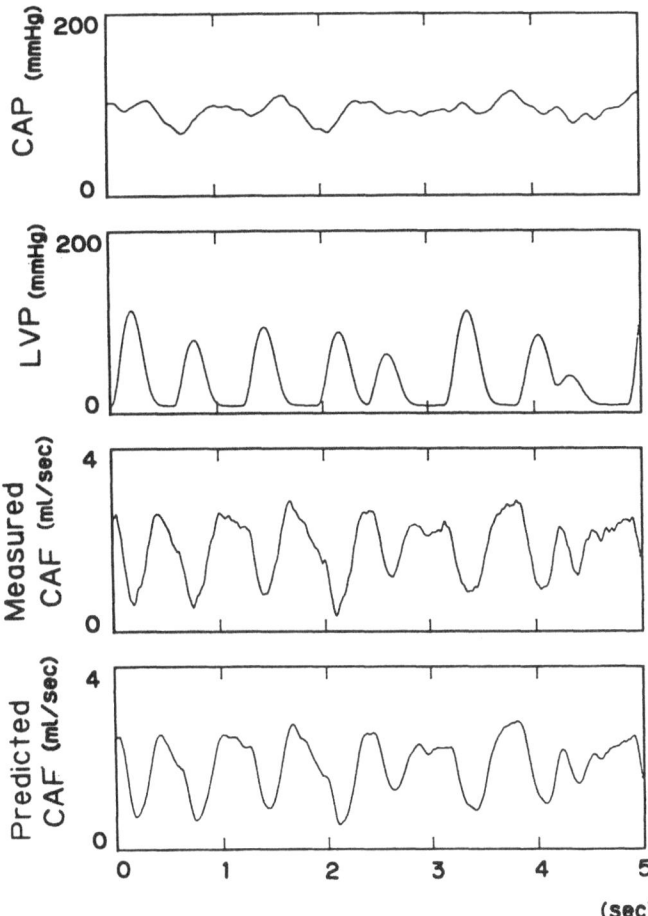

Fig. 2. Prediction of coronary arterial flow using two-channel coronary admittance, coronary arterial pressure (*CAP*) and left ventricular pressure (*LVP*). Predicted coronary arterial flow (*CAF*) was close to measured CAF

and 3.0 Hz. Thus in the frequency range below 3.0 Hz, the two channel linear admittance model we developed was capable of describing instantaneous coronary arterial flow as a function of left ventricular pressure and coronary arterial pressure. Above 3.0 Hz, however, the coherence quickly decreased. At 10 Hz, the two channel admittance model explained only 50% of coronary flow variations. The provided model may no longer be valid in the frequency range above 10 Hz.

Illustrated in Fig. 2 are coronary arterial pressure, left ventricular pressure, measured coronary arterial flow, and predicted coronary arterial flow on the basis of estimated two channel admittance model. As can be seen, the predicted instantaneous coronary arterial flow was quite similar to the observed one suggesting the validity of the admittance model. Linear regression analysis indicated that the correlation coefficient between the predicted and measured coronary arterial flow was higher than 0.90 in all animals.

Discussion

We have demonstrated that the changes in instantaneous coronary arterial flow could be explained by a two channel admittance model as a function of coronary arterial pressure and left ventricular pressure. As we indicated, the admittance model is valid only when the coronary arterial system is linear and time invariant. The recent investigation by Krams et al. [1] indicated that what affected instantaneous coronary arterial flow was not left ventricular pressure but time varying elastance. Since we used isovolumic contraction and changed left ventricular pressure by irregular pacing, we perturbed left ventricular pressure by changing ventricular contractility, and thus time varying elastance. Therefore our protocol was not able to tell whether left ventricular pressure itself or time varying elastance was the determinant of coronary arterial flow.

Many models of the coronary arterial system have been proposed. Illustrated in Fig. 3 are well known coronary arterial models [2]. If the coronary arterial flow was determined through the mechanism of the vascular waterfall [3] (the left panel of Fig. 3), the admittance from coronary arterial pressure to coronary arterial flow should be fairly flat over the wide frequency range as long as coronary arterial pressure was reasonably higher than the extra vascular compression pressure. Similarly the admittance from the left ventricular pressure to coronary arterial flow should also be flat over the wide frequency range. As shown in Fig. 1, our data were not consistent with this model.

If the coronary arterial system was equivalent to the intramyocardial compliance model [4], we expect that the admittance from the coronary arterial pressure to the flow is somewhat similar to what we experimentally obtained. On the contrary, the admittance from the left ventricular pressure to coronary arterial flow would be different. Because of the effect of a serial capacitance, the effect of ventricular contraction on the coronary arterial flow should decrease when frequency was decreased. Our data indicated that in the frequency range below 0.1 Hz, the admittance was rather flat. Therefore the intramyocardial pump model was inconsistent with our observations.

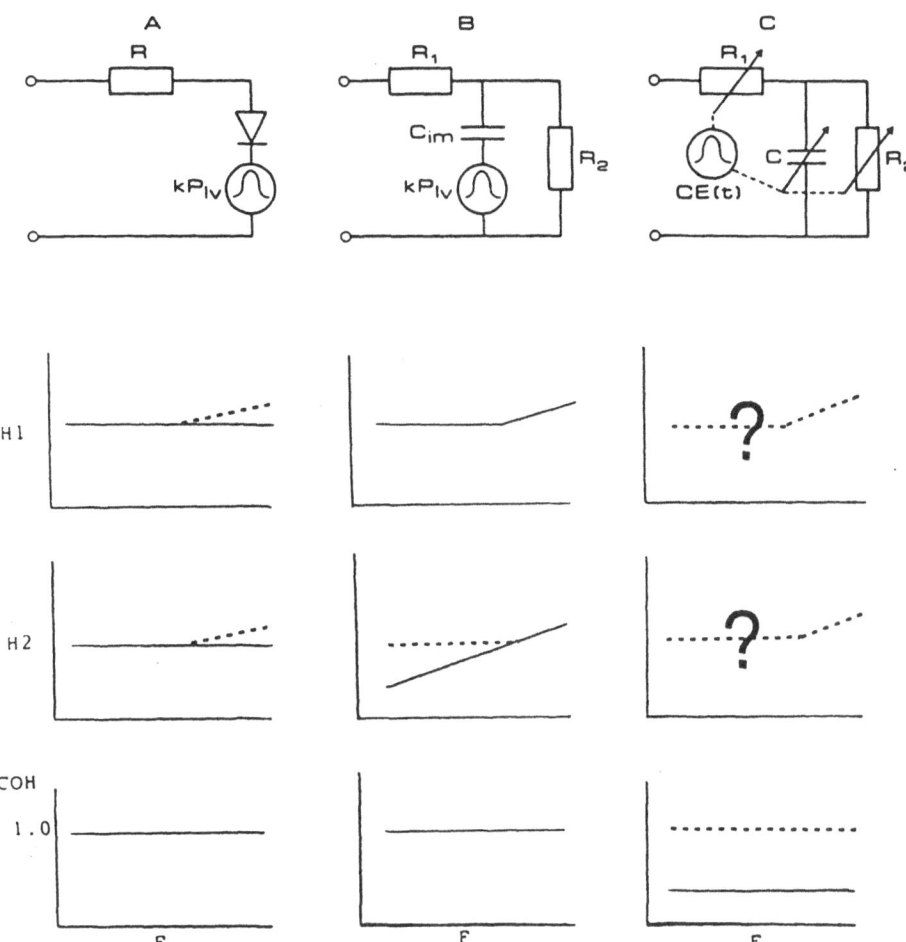

Fig. 3. Models of coronary arterial coupling with left ventricular mechanics. *Left panel,* Vascular waterfall model; *middle panel,* intramyocardial pump model; *right panel,* time varying elastance model R, R_1, R_2, resistance; KP_{lv}, scaled left ventricular pressure; C_{im}, intramyocardial compliance; $CE(t)$, left ventricular contraction; COH, coherence. [modified from [2] with permission]

In the case of time varying elastance model, the complexity of the model made the prediction of admittance spectrum prohibitive. As we discussed, however, the left ventricular pressure we used in this experiment was analogous to the time varying elastance. Therefore even though in the two channel admittance model the left ventricular pressure was considered to be a determinant of coronary arterial flow, some of the apparent left ventricular pressure effect may well be attributable to time varying elastance. Obviously investigation into quantitatively separate time-varying pressure effects and elastance effects remains to be done.

References

1. Krams S, Sipkema P, Westerhof N (1989) Varying elastance concept may explain coronary systolic flow impediment. Am J Physiol 257 (Heart Circ Physiol 26): H1471–H1479
2. Krams R, Sipkema P, Westerhof N (1990) Coronary oscillatory flow amplitude is more affected by perfusion pressure than ventricular pressure. Am J Physiol 258 (Heart Circ Physiol 27):H1889–H1898
3. Downey JM, Kirk ES (1975) Inhibition of coronary blood flow by a vascular waterfall mechanism. Circ Res 34:251–257
4. Spaan JAE (1985) Coronary diastolic pressure-flow relation and zero flow pressure explained on the basis of intramyocardial compliance. Circ Res 56:293–309

Myogenic and Flow-Induced Responses in Coronary Arterioles

Lih Kuo, William M. Chilian, and Michael J. Davis[1]

Summary. The coronary vascular bed exhibits an intrinsic ability to autoregulate blood flow over a wide range of arterial perfusion pressures. Despite numerous studies suggesting the involvement of myogenic and flow-induced responses in the regulation of blood flow in many organ systems, the role of these two mechanisms in the control of vasomotor tone in the coronary microcirculation is not clear. The aim of this paper is to describe myogenic and flow-induced responses of isolated porcine subepicardial arterioles to physiological changes in intraluminal pressure and flow. Experiments are also described which quantitate the interaction of these two responses in vitro. Subepicardial arterioles, 40–100 μm in diameter, were isolated and cannulated with two glass micropipettes connected to independent reservoir systems. Initially, myogenic responses were studied by moving both reservoirs in the same direction to alter myogenic tone in the absence of flow. Flow-induced responses were studied by simultaneously moving the reservoirs in equal but opposite directions thus generating a pressure gradient (ΔP) to initiate flow without changing the mean intraluminal pressure (IP). Myogenic constrictions and dilations were observed when IP was increased ($> 60 \, cmH_2O$) and decreased ($< 60 \, cmH_2O$), respectively. The threshold and maximum flow-induced dilation occurred at $\Delta P = 4 \, cmH_2O$ (flow = 4 nl/s) and $20 \, cmH_2O$ (flow = 13 nl/s), respectively. Flow-induced dilation competed with myogenic constriction when flow and pressure were elevated simultaneously. In addition, flow potentiated myogenic dilation when IP was decreased. The magnitude of the flow-induced dilation was greatest at intermediate levels of tone (IP = $60 \, cmH_2O$) but was attenuated at higher and lower levels of tone. In the presence of flow ($\Delta P = 4 \, cmH_2O$), pressure-diameter relationships were shifted upward, and the magnitude of myogenic responsiveness was attenuated. After mechanical removal of the endothelium, spontaneous tone and myogenic responses were preserved, but flow-induced dilation was abolished. Therefore, both pressure-dependent and flow-induced responses occur in isolated coronary arterioles, but only flow-induced responses require an intact endothelium. These two responses closely interact either competitively or additively, depending on the direction of local vascular pressure changes.

Key words: Endothelium—Autoregulation—Endothelium-derived relaxing factor—Pressure—Endothelium-dependent

---

[1] Department of Medical Physiology and Microcirculation Research Institute, College of Medicine, Texas A&M University Health Science Center, College Station, TX 77843, USA

Introduction

Autoregulation and metabolic coronary dilation are the two most important physiological mechanisms that modulate myocardial perfusion [1]. These regulatory adjustments are primarily confined to arterioles less than 150 μm in diameter where the majority (> 50%) of coronary vascular resistance is located [2, 3]. Traditionally, coronary blood flow regulation has been attributed to metabolic control mechanisms, although the unequivocal identification of the causal metabolite still remains unknown [4]. In addition to the metabolic mechanism, it has been shown in many organ systems other than heart that the myogenic response plays a potentially important role in autoregulation of blood flow [5]. The myogenic response is defined as the constriction or dilation of a vessel in response to an increase or decrease in intraluminal pressure, respectively. Some investigators have provided indirect evidence to support coronary myogenic mechanisms in the intact heart [6]. However, precise documentation of a myogenic contribution to coronary autoregulation has been difficult to unequivocally establish in vivo because the methods employed to alter intraluminal pressure also modify the flow perfusion to the tissue, which could potentially activate metabolic mechanisms of flow control. In addition to pressure-dependent myogenic responses, it has been shown recently that velocity of blood flow is able to modulate vascular caliber in many organ systems [7–11]. In the coronary circulation, large conduit arteries dilate in response to an increase in flow [7, 8]. However, it remains unknown if this mechanism also influences tone in coronary resistance vessels. Furthermore, a plethora of evidence indicates that endothelium is able to modulate vascular tone by releasing vasoactive substances in response to specific pharmacological stimuli, but it is unclear whether coronary microvascular tone is also influenced by the transduction of hemodynamic forces such as intravascular pressure (myogenic response) and shear stress (flow-induced response) through the endothelial cells. In this context, we summarize some of our recent studies in this area and attempt to address the following questions: (1) Do coronary arterioles demonstrate myogenic and flow-induced responses?, (2) Do these responses depend upon an intact endothelium? and (3) Do myogenic and flow-induced responses interact with each other to regulate coronary microvascular tone? In order to have precise and independent control of intraluminal pressure and flow without the involvement of metabolic control mechanisms, arterioles (40–100 μm) from porcine subepicardium were isolated, cannulated, and studied in vitro. Myogenic and flow-induced responses were investigated in the presence and absence of endothelium. Experiments were also designed to quantitatively study the interaction of these two responses in vitro.

Material and Methods

General Preparations

Pigs (6–10 weeks old of either sex) were sedated with ketamine (2.5 mg/kg, i.m.) and Rompun (2.25 mg/kg, i.m.), anesthetized with pentobarbital sodium

(20 mg/kg, i.v.), intubated, and ventilated with room air. After a left thoracotomy, heparin (1000 units/kg) was administered into the left atrium, and the heart was electrically fibrillated, excised, and immediately place in cold (4 °C) saline solution.

The techniques for localization and dissection of coronary arterioles were reported in detail previously [12]. In brief, a mixture of India ink and gelatin in physiological salt solution (PSS) was perfused into the left anterior descending artery and the circumflex artery to visualize the coronary arterioles. At 4 °C, arteriolar branches from left anterior descending or circumflex arteries (0.8–1.2 mm in length and 40–100 μm internal diameter) were selected and dissected from the surrounding cardiac tissue and transferred for further dissection to a dish (4 °C) containing filtered PSS-albumin solution at pH 7.4. After careful removal of any remaining cardiac tissue, an arteriole was then transferred for cannulation to a Lucite vessel chamber containing PSS-albumin solution equilibrated with room air at ambient temperature. One end of the arteriole was cannulated with a glass micropipette (40-μm tip internal diameter and filled with filtered PSS-albumin solution) and the arteriole was securely tied to the pipette with 11-O ophthalmic suture. The ink-gelatin column inside the vessel was flushed out at low perfusion pressure (< 20 cmH₂O). Any small branches were tied off, then the other end of the vessel was cannulated with a second micro-pipette and secured with suture. Electrical resistances (measured by LCR Bridge Circuit, model LCR-740, Leader Electronics Corp., Japan) of the two pipettes were matched (±0.5%).

Instrumentation

After the vessel was cannulated, the preparation was then transferred to the stage of an inverted microscope (model IM35, Carl Zeiss, Thornwood, N.Y.) coupled to a Dage TV camera (67M Newvicon, Michigan City, Ind.), video micrometer, and video recorder. Internal diameters were measured continuously throughout the experiment by videomicroscopic techniques.

To study the pressure- and flow-induced responses, we developed a dual-reservoir system that allowed these two responses to be examined independently [13]. In brief, the micropipettes were connected to two independent reservoir systems, and intraluminal pressures were measured through side-arms of the two reservoir lines by low-volume displacement strain gauge transducers (Statham P23 Db, Gould, Cleveland, Ohio). Leaks were detected by differences between reservoir pressure and luminal pressure. Any preparations with leaks were excluded from the data analysis. If the reservoirs were set at the same hydrostatic level, the isolated arteriole could be pressurized without flow by simultaneously moving both reservoirs in the same direction. Thus, the myogenic response could be studied without the interference of flow. Flow could be initiated by simultaneously moving the reservoirs in opposite directions, generating a pressure gradient, ΔP. Because the resistances of both cannulation pipettes were equivalent, simultaneous movement of the reservoirs in equal and opposite directions did not alter mid-point luminal pressure [13]. A constant intraluminal pressure during variation of ΔP (i.e., flow rate) was confirmed by a direct measurement of intraluminal pressure using the servo-null technique (IPM model 4A) [13].

Endothelium Denudation Procedure

The endothelial layer of the cannulated arteriole was removed by a mechanical abrasion technique which was described previously [14]. Briefly, a concentric glass abrasive micropipette with a long shank (5 mm) and an irregular tip (25 µm in diameter) was advanced into the lumen of the cannulated vessel. A negative intraluminal pressure (-20 cmH$_2$O) was produced to collapse the vessel and to allow the abrasive pipette to closely contact the endothelial cells. The endothelial cells were disrupted by passing the pipette back and forth several times through the vessel lumen. After this procedure, the cellular debris inside the lumen was flushed out by perfusion ($\Delta P = 40$ cmH$_2$O) with warm (37 °C) PSS-albumin solution for 5 min. The vessels were allowed to equilibrate at 60 cmH$_2$O intra-luminal pressure for 40–60 min to regain vessel tone. The efficacy of endothelial denudation was re-examined as previously described [14]. If endothelial functional responses persisted (i.e., the vessel relaxed to bradykinin), the denudation procedure was repeated until the endothelium-dependent relaxation was abolished. If the vessel constriction to Acetylcholine (ACh, an endothelium-independent constrictor in porcine coronary arterioles) was not comparable to the control (indicating damage of vascular smooth muscle caused by abrasion), the vessel was excluded from the data analysis.

Experimental Protocols

In the chamber, the cannulated vessel was bathed in PSS-albumin solution, and the temperature was maintained at 36°–37 °C by an external heat exchanger. The vessel was set to its in situ length and allowed to develop spontaneous tone at 60 cmH$_2$O luminal pressure without flow [12]. The reason for selecting this internal pressure was that intraluminal pressures were found to be 40–50 mmHg in vivo in arterioles of this size [3].

Protocol 1: Do Isolated Coronary Arterioles Exhibit Myogenic and Flow-Induced Responses?

After the vessel developed spontaneous tone, the relation between intraluminal pressure and vessel diameter (myogenic response) was examined. Initially, vessel diameter was measured at the control pressure (60 cmH$_2$O), then pressure was increased to 140 cmH$_2$O in steps of 20 cmH$_2$O, then reduced in the same steps to 20 cmH$_2$O, and finally returned to the control. At each step, the pressure was maintained until a stable lumen diameter was obtained (2–4 min). After the vessel redeveloped tone at control intraluminal pressure, flow was initiated by opposite movements of the two reservoirs. Diameter was measured at each level of flow corresponding to a pressure gradient (ΔP) of 4, 10, 20, 40, and 60 cmH$_2$O, and again at zero flow ($\Delta P = 0$). In arterioles of this size, this range of ΔP, 4 to 60 cmH$_2$O, corresponds to volumetric flows of 4.1 ± 0.3 to 36 ± 2 nl/sec as determined in a previous study [13].

Protocol 2: Do Myogenic and Flow-Induced Responses Depend Upon an Intact Endothelium?

To answer this question, the above mentioned protocols for examination of the pressure-diameter and flow-diameter relationships were repeated in the absence of endothelium. The efficacy of endothelial denudation was verified phar-macologically as previously described [14].

Protocol 3: Interaction Between Pressure- and Flow-Induced Responses

The effect of pressure on flow-induced responses was studied by randomly setting the luminal pressure to 20, 60, or 100 cmH$_2$O, which produced different levels of myogenic tone. After each new level of myogenic tone was established, flow was initiated by equal and opposite movements of the two reservoirs. The diameter was measured at each level of flow corresponding to a pressure gradient (ΔP) of 4, 10, 20, 40, and 60 cmH$_2$O, and again at zero flow (ΔP = 0). To study the influence of flow on pressure-induced responses, the pressure-diameter relation-ship between 20 to 140 cmH$_2$O was studied at a constant ΔP (4 cmH$_2$O).

At the end of each experiment, the vessels were maximally dilated with nitroprusside (10^{-4} M), and the sequence of pressure and flow changes described previously was then performed to obtain the passive pressure-diameter and flow-diameter relationship. All drugs used in this study were obtained from Sigma Chemical (St. Louis, Mo.).

Data Analysis

For analysis of myogenic responses, the internal diameters of the vessels were normalized to the maximally dilated diameters at 60 cmH$_2$O luminal pressure (in the presence of nitroprusside). In the flow protocol, each vessel diameter was normalized to the passive diameter at the control luminal pressure (60 cmH$_2$O) in the presence of nitroprusside. To analyze the effect of myogenic tone on flow-induced responses, the internal diameter of each vessel was normalized to the passive diameter at the corresponding intraluminal pressure (20, 60, or 100 cmH$_2$O) in the presence of nitroprusside. To analyze the effect of flow on myogenic responsiveness, the vessel diameters were normalized to their maximally dilated state at 60 cmH$_2$O luminal pressure in the presence of nitroprusside. Normalized diameters were averaged at each pressure or flow step. All data are reported as mean $\pm$ SEM. Statistical comparisons between groups (with and without endothelium) and within groups were made with factorial or repeated-measures analysis of variance tests with Fisher least significant difference multiple-range tests when appropriate. Significance was accepted at $P \leq 0.05$.

Results

At 60 cmH$_2$O intraluminal pressure and 37 °C bath temperature, subepicardial arterioles developed spontaneous tone within 40 min after cannulation. Figure 1a shows the development of spontaneous tone, i.e., constriction from an initial

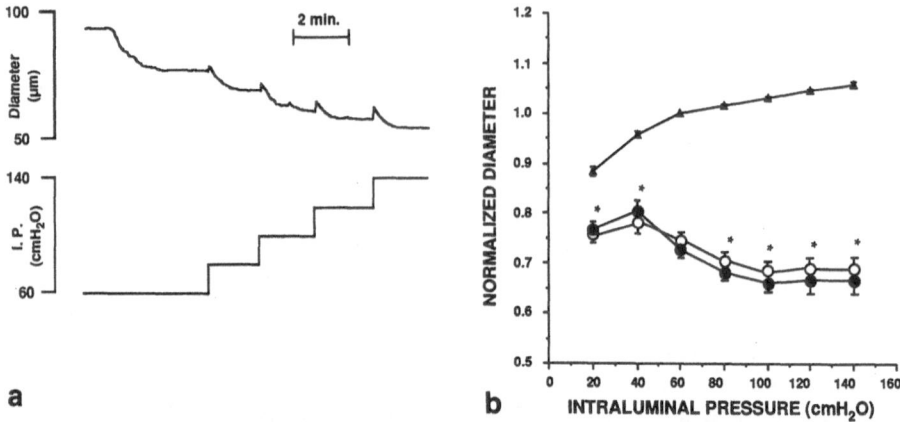

Fig. 1. a Development of spontaneous tone of isolated coronary arteriole at 37 °C and constriction in response to increases in intraluminal pressure (*IP*). **b** Pressure-diameter relations of isolated subepicardial arterioles before (*closed circles*) and after (*open circles*) mechanical denudation of endothelium. Luminal diameters were normalized to diameter at a pressure of 60 cmH$_2$O in the presence of nitroprusside (10^{-4} M, *closed triangles*). *Diameters that were significantly different ($P < 0.05$) from diameters at 60 cmH$_2$O intraluminal pressure. There is no statistical difference in myogenic responses before and after mechanical denudation. Average luminal diameter (*d*) at 60 cmH$_2$O is shown (*d* = 69.1 ± 3.6 μm; *n* = 14). *Vertical bars* denote mean ± SEM. (Data are from [13], with permission of the American Physiological Society)

diameter of 91 to 75 μm, and subsequent constrictions in response to a stepwise-increase in intraluminal pressure from 60 to 140 cmH$_2$O. At each pressure increment, the arteriole initially distended but then increased tone and constricted to smaller than the original diameter (Fig. 1a). The pressure-diameter relations (myogenic responses) from 14 subepicardial arterioles with intact endothelium are summarized in Fig. 1b. In PSS-albumin solution at 60 cmH$_2$O intraluminal pressure, these vessels developed spontaneous tone to about 73 ± 1% of their passive diameter (in nitroprusside solution). All vessels produced a significant decrease in diameter when intraluminal pressure was elevated above 60 cmH$_2$O. Over the lower pressure range (< 60 cmH$_2$O), dilation was observed in 11 of 14 arterioles. In contrast, in the presence of nitroprusside (10^{-4} M), all vessels dilated and behaved passively during changes in intraluminal pressure (i.e., distended at high pressures and collapsed at low pressures) (Fig. 1b). After mechanical removal of endothelium, the active pressure-diameter relations (myogenic responses) were not significantly altered (Fig. 1b).

The response of an isolated coronary arteriole (59 μm) to flow produced by increases in ΔP is shown in Fig. 2a. Note that during constant mean intraluminal pressure (60 cmH$_2$O), as evident from the servo-null intraluminal pressure measurement, a graded vasodilation was observed when ΔP, and thus flow, was

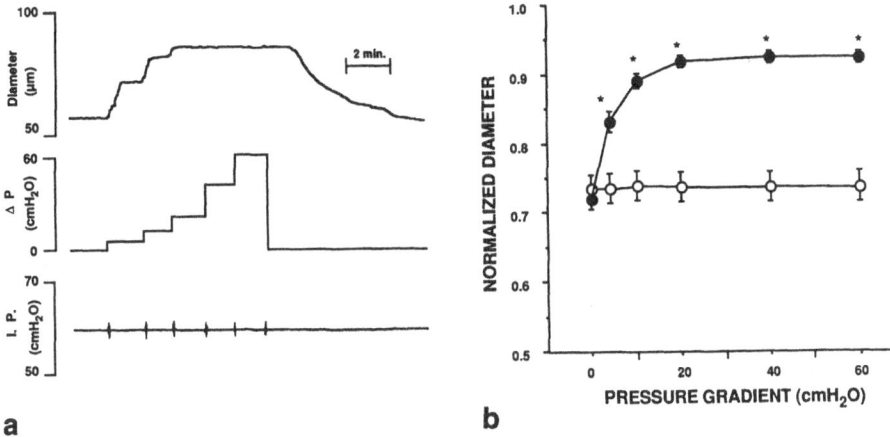

a   **b**

Fig. 2. a Flow-induced vasodilation was observed during production of a pressure difference (ΔP) by raising and lowering two reservoirs. Note that mean intraluminal pressure
(*IP*) was not altered during flow. **b** Flow-induced dilation of 14 vessels at 60 cmH$_2$O
intraluminal pressure before (*closed circles*) and after (*open circles*) mechanical removal of
endothelium. Luminal diameters were normalized to diameter at a luminal pressure of
60 cmH$_2$O in the presence of nitroprusside (10^{-4} M). *Diameters that were significantly
different ($P < 0.05$) from those at zero ΔP. Denudation completely abolished flow-induced
vasodilation. Average luminal diameter (*d*) at 60 cmH$_2$O is shown (*d* = 64.2 $\pm$ 2.1 μm; *n*
= 14). *Vertical bars* denote mean $\pm$ SEM. (Data are from [13], with permission of the
American Physiological Society)

increased in a stepwise manner. It is worth noting that there was a 2–5 s delay
for the onset of vasodilation to flow. When flow was stopped ($\Delta P = 0$), the
diameter returned to control within 6 min (Fig. 2a). The average results from 14
subepicardial arterioles are summarized in Fig. 2b. These vessels dilated from
73% of maximal diameter at $\Delta P = 0$ to 93% of maximal diameter when ΔP was
increased to 20 cmH$_2$O (Fig. 2b). No further dilation was observed at higher
flows ($\Delta P = 40$ and 60 cmH$_2$O). After mechanical denudation of the endothelium,
the flow-induced vasodilatory response was completely abolished (Fig. 2b).

Figure 3 shows examples of the interaction of pressure- and flow-induced
responses in isolated coronary arterioles. The effect of flow on pressure-induced
constriction and dilation is shown in Fig. 3, a and b, respectively. The pressure
changes shown here caused submaximal constriction or dilation. When intraluminal pressure was increased from 60 to 80 cmH$_2$O, the arteriole initially
distended but then increased tone and constricted to smaller than its original
diameter (Fig. 3a). This pressure-induced myogenic constriction was reversed by
flow (when ΔP was increased from 0 to 4 cmH$_2$O). When flow was stopped (ΔP
= 0 cmH$_2$O), vasoconstriction occurred and the diameter gradually returned to
its pre-flow value (Fig. 3a). Figure 3b shows that a myogenic dilation (intraluminal pressure decreased from 60 to 40 cmH$_2$O) was potentiated by an increase

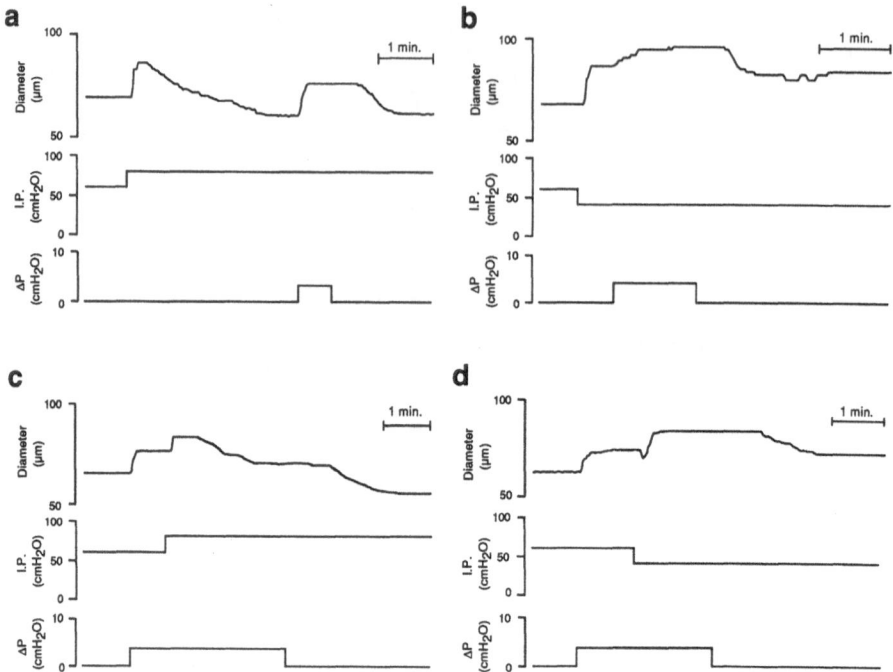

Fig. 3a–d. Interaction of pressure-induced myogenic responses and flow-induced dilations in isolated subepicardial arterioles. **a** Myogenic constriction was inhibited by an increase in flow. **b** Myogenic dilation was additively increased by flow. **c** Flow-induced dilation was attenuated by elevating intraluminal pressure (*IP*). **d** Flow-induced dilation was additively increased by lowering IP. (Data are from [26], with permission of the American Physiological Society)

in flow. When flow was stopped, the vessel returned to the diameter that occurred during the myogenic dilation. Figure 3c,d show the effects of pressure-induced responses on vessels with flow. Under control conditions (intraluminal pressure at 60 cmH$_2$O and $\Delta P - 0$ cmH$_2$O), vasodilation occurred when flow was initiated ($\Delta P = 4$ cmH$_2$O), but the increase in diameter was attenuated by elevating luminal pressure (intraluminal pressure $= 80$ cmH$_2$O) (Fig. 3c). In contrast, myogenic relaxation (an intraluminal pressure decrease from 60 to 40 cmH$_2$O) potentiated the vasodilatory effect of flow (Fig. 3d).

The effect of myogenic tone on flow-induced dilation is summarized in Fig. 4. Flow-induced responses were studied at low, intermediate, and high levels of myogenic tone by setting intraluminal pressure at 20, 60, and 100 cmH$_2$O, respectively. Under zero flow conditions ($\Delta P = 0$ cmH$_2$O), vessels developed myogenic tone, which was directly related to the level of intraluminal pressure. A graded vasodilation was observed when ΔP, and thus flow, was increased in a stepwise manner. Generally, flow-induced dilation reached its plateau when ΔP

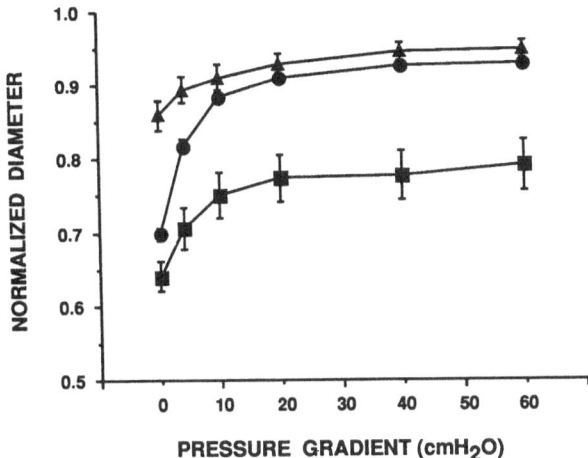

Fig. 4. Flow-induced dilation at different levels of myogenic tone. The magnitude of flow-induced dilation was attenuated at highest myogenic tone (i.e., luminal pressure at 100 cmH$_2$O). All diameters were significantly different ($P < 0.05$) from those at zero pressure gradient. *Squares*, 100 cmH$_2$O; *circles*, 60 cmH$_2$O; *triangles*, 20 cmH$_2$O. (Data are from [26], with permission of the American Physiological Society)

= 20 cmH$_2$O. The magnitude of the flow-induced dilation was greatest at intermediate tone (IP = 60 cmH$_2$O) but was significantly attenuated ($P < 0.05$) at higher levels of tone (IP = 100 cmH$_2$O). At lower pressure, IP = 20 cmH$_2$O, there was very little vasodilatory reserve; therefore, flow-induced vasodilation was minimized.

The effect of flow on pressure-induced myogenic responses is summarized in Fig. 5. Under zero flow conditions, myogenic constrictions and dilations were observed when intraluminal pressure was increased (> 60 cmH$_2$O) and decreased (< 60 cmH$_2$O), respectively. When flow was established (ΔP = 4 cmH$_2$O), the vessels exhibited lower myogenic tone. As a result, the pressure-diameter relationship was shifted upward and the magnitude of myogenic responsiveness was attenuated (Fig. 5). With nitroprusside, all vessels behaved passively during luminal pressure changes.

Discussion

The major findings of this study are: (1) coronary arterioles (40–100 μm in diameter) exhibit pressure-induced myogenic responses which are endothelium-independent, (2) flow-induced vasodilation occurs in coronary arterioles, and this response depends upon an intact endothelium, and (3) myogenic and flow-induced responses interact with each other to regulate coronary arteriolar tone. We speculate that under physiological conditions the interactions between

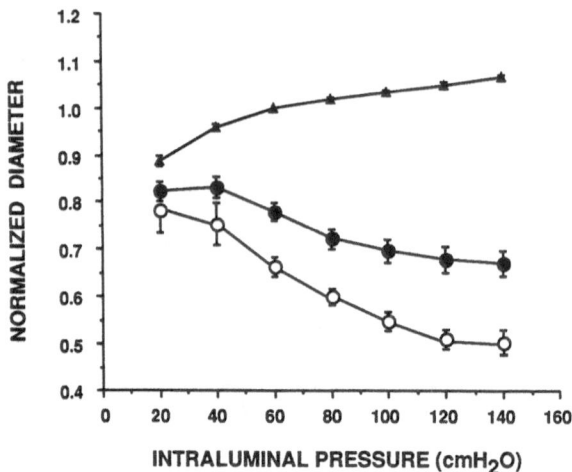

Fig. 5. Pressure-diameter relations (myogenic responses) of coronary arterioles in the presence (*closed circles*) and absence (*open circles*) of luminal flow. Myogenic responsiveness was attenuated in the presence of flow. All vessels dilated maximally to nitroprusside (10^{-4} M) and behaved passively (*triangles*) to pressure changes. All diameters were significantly different ($P < 0.05$) from their control diameter (at $60\,cmH_2O$ intraluminal pressure). (Data are from [26], with permission of the American Physiological Society)

pressure- and flow-induced responses in coronary resistance arterioles may be important determinants of myocardial blood flow regulation.

Myogenic Responses in the Coronary Microcirculation

In many organ systems, namely kidney, skeletal muscle, intestine, and brain, myogenic responses play an important role in autoregulation of blood flow [5]. However, it remained unclear for many years whether myogenic responses were present in the coronary microcirculation, mainly due to the difficulties of studying purely pressure-dependent responses in the intact heart without confounding alterations in myocardial oxygen consumption [6]. For example, an increase in coronary pressure designed to elicit myogenic coronary constriction in vivo may be accompanied by an increase in aortic pressure which can enhance myocardial oxygen consumption and therefore counteract the myogenic constriction due to initiation of metabolic vasodilatory mechanisms. Furthermore, a change in coronary pressure is generally accompanied by a change in flow, the major stimulus for autoregulatory adjustments of coronary vasomotor tone. To overcome these problems, we employed a direct approach, the isolated-arteriole technique, to study the reactivity of coronary arterioles to luminal pressure changes without involvement of neuro-humoral and metabolic control mechanisms.

This study clearly shows that isolated coronary arterioles, 40–100 μm in diameter, developed spontaneous tone and actively responded to luminal pressure changes (Fig. 1). Myogenic constriction was demonstrated during graded increases in pressure from 60 cmH₂O to 140 cmH₂O, and vasodilation occurred during graded decreases in pressure from 60 cmH₂O to 20 cmH₂O. Thus, active alterations in vascular tone due to myogenic responses overcame passive changes in arteriolar diameter. This is first direct evidence to show myogenic regulation of coronary arteriolar tone. Since these vessels were of a size that contributes significantly to coronary vascular resistance ($> 50\%$ of total resistance) in vivo, the myogenic responses demonstrated in this study may be involved in the control of coronary blood flow. Specifically, myogenic dilation in response to a fall in arteriolar pressure during either coronary occlusion or metabolic vasodilation may reduce coronary vascular resistance.

The mechanisms of the myogenic response have been a subject of considerable interest. Mechanical denudation of rabbit ear arteries [15] and small pulmonary arteries [16] was reported not to affect the myogenic responses. Our results showing endothelial independence of coronary myogenic responses (Fig. 1b) extend this concept to the microcirculation. However, in cerebral arteries, the stretch-induced and pressure-dependent vascular constriction required an intact endothelium [17, 18]. The disparity between the findings of these studies is probably due to differences in the selectivity of endothelial denudation. For example, we have previously shown that high intraluminal pressure, excessive mechanical abrasion, and chemical treatment are potentially damaging to microvascular smooth muscle and may thus compromise the myogenic responsiveness [14]. There is now clear evidence that the myogenic responsiveness of arterioles is generally unaffected by careful endothelial removal [19], and is therefore an intrinsic property of vascular smooth muscle.

Flow-induced Responses in the Coronary Microcirculation

In the last decade, dilation in response to an increase in flow has been demonstrated in large coronary arteries from a variety of species, including humans, in vivo [7, 8, 20, 21]. Flow-induced dilation, in an in vitro study, was found to be caused by a local mechanism and not mediated by ascending dilation from the microvasculature [22]. Adrenergic, cholinergic, and ganglionic blockades were found not to influence the dilation [9, 22]. The present study shows that flow-induced dilation also occurs in true coronary arterioles, the major determinants of vascular resistance and blood flow control in the coronary vasculature. In addition, the flow-induced response in coronary arterioles was abolished by selective mechanical endothelial removal (Fig. 2b), in agreement with the findings in large conduit arteries [8, 9, 23–25]. In conjunction with the endothelium dependence of flow-induced response, we previously demonstrated that an L-arginine analogue, N^G-monomethyl-L-arginine (L-NMMA), but not a prostanoid synthesis inhibitor, indomethacin, abolished flow-induced dilation in isolated coronary arterioles [26]. The inhibitory response of L-NMMA was reversed by administration of L-arginine, a physiological precursor for synthesis of nitric oxide, to the vessels [26]. This indicates that a nitrovasodilator may be released from endothelial cells in response to flow. The crucial role for a transferable substance released from endothelium was confirmed by us using a double-microvessel bioassay technique [26], in which an arteriole denuded of endothelium was perfused in series with an arteriole with endothelium intact. The denuded arteriole dilated only when it was perfused downstream from the intact arteriole, indicating that it was dilated by a substance transferred in the flow from the intact arteriole. These studies ruled out the possibility that flow-induced dilation is mediated by electrical conduction mechanisms between endothelial cells and the underlying smooth muscle. However, the signal transduction mechanism and precise cellular events leading to flow-induced dilation are not clear at the present time but probably involve the activation of potassium channels on the endothelial cell membrane by a change in the shear stress imposed on the endothelial cell, leading to hyperpolarization and inducing calcium entry into the cell [27]. It has been shown that an increase in endothelial calcium is necessary for the synthesis and release of nitric oxide [28]. A recent study by one of us (LK), demonstrates that a rise in endothelial calcium was associated with vasodilation to flow in isolated skeletal muscle arterioles [29]. This observation provides convincing support for the possible involvement of calcium as a second messenger in the endothelium-dependent vasodilatory response to flow.

Interaction Between Pressure- and Flow-induced Responses

Based on current evidence, flow-induced dilation has been suggested to be involved in: (1) changing resistance during metabolic hyperemia [8, 13], (2) augmenting collateral flow when a feed artery is occluded [10], (3) modulating neurogenic vasomotor tone [30], and (4) coordinating the hemodynamic properties of a vascular network [31]. The present study suggests that flow is able to modulate myogenic responses in the coronary microcirculation (Figs. 3, 4, and 5). It has been shown that pharmacological inhibition of endothelial function in response to flow could unmask an intrinsic vascular constriction during sudden increases in flow and pressure [32]. This suggests that a competition between flow- and pressure-induced responses occurs in vivo. This idea is directly supported by our data. Within physiological ranges of pressure and flow, myogenic constriction and flow-induced dilation opposed each other (Fig. 3a and c), whereas myogenic dilation and flow-induced dilation were additive (Fig. 3b and d). Functional competition between myogenic constriction and flow-induced dilation was apparent also as diminished flow-induced dilation with increasing myogenic tone (Fig. 4) and diminished myogenic constriction in the presence of flow (Fig. 5). This competitive interaction has also been observed by Hilton [33] in intact animals that flow-induced dilation was attenuated or only occasionally observed when arterial pressure was high (190–220 mmHg). The critical value of systemic blood pressure for appearance of the flow-induced response was usually 90–100 mmHg [33], suggesting that the magnitude of the flow-induced dilation was greatest at resting levels of myogenic tone. These observations are consistent with our in vitro findings that the greatest flow-induced response occurred when the vessels were pressurized to 60 cmH$_2$O (Fig. 4), which is estimated to be the normal luminal pressure for arterioles of this size in the beating heart [3]. Lower levels of tone (intraluminal pressure = 20 cmH$_2$O) also reduced the magnitude of the flow-induced dilation, but this was likely due to the fact that the arterioles were unable to dilate substantially at this pressure. Hilton's observations [33] also indicated that flow-induced responses were smaller or absent in animals with lower systemic arterial pressure. Thus, there is evidence that this interaction occurs under in vivo conditions. In addition, our results also suggest that myogenic constriction is chronically opposed by flow-induced dilation (Fig. 5). This interaction may play an important role in the balance of vascular tone and moment-by-moment (regulation of blood flow during local pressure changes.

Physiological Considerations

Speculatively, the interaction between flow-induced and myogenic responses may be beneficial in the maintenance of myocardial function and perfusion. For instance, during intense sympatho-adrenal excitation, perfusion pressure is elevated due to an increase in peripheral resistance, and coronary blood flow is augmented because of enhanced myocardial oxygen demands. Myogenic constriction would then compete with flow-induced dilation in resistance arterioles. This would minimize the increase in tissue perfusion and the local pressure rise in the exchange vessels, possibly protecting the myocardium from edema. Interaction between pressure and flow could also influence the coronary vasomotor adjustments during an increase in metabolic demand of the myocardium even in the absence of gross changes in arterial pressure. Because small arterioles preferentially dilate in response to an increase in myocardial oxygen consumption [2], both an increase in blood flow and lower upstream local pressure would occur secondary to a decrease in downstream resistance. The additive effects of myogenic and flow-induced dilation in upstream arterioles would further the hyperemia, perhaps to meet the increased metabolic activity. Recently, we found that coronary venules (80–120 µm in diameter) also dilate to flow stimuli by releasing a nitrovasodilator from the endothelium [34]. This venular dilation may contribute to the adjustment of postcapillary resistance to maintain a maximum flow perfusion and a proper fluid movement across the capillary wall during increases in metabolic demand of the myocardium such as physical exercise.

Our results also suggest that under resting conditions the interaction of pressure- and flow-induced responses may play an important role in fine adjustments and moment-by-moment control of blood flow. It is conceivable that during pathological conditions, e.g., hypertension, ischemia-reperfusion, and atherosclerosis, the increased vascular tone and/or functional impairment of the endothelium would attenuate the effect of flow on vascular resistance in the coronary microcirculation and may result in inadequate oxygen supply during intense metabolic demands.

References

1. Berne RM, Rubio R (1979) Coronary circulation, In: Berne RM (eds) Handbook of physiology, section 2: The cardiovascular system—the heart. American Physiological Society, Bethesda, pp 873–952
2. Kanatsuka H, Lamping KG, Eastham CL, Dellsperger KC, Marcus ML (1989) Comparison of the effects of increased myocardial oxygen consumption and adenosine on the coronary microvascular resistance. Circ Res 65:1296–1305
3. Chilian WM, Eastham CL, Marcus ML (1986) Microvascular distribution of coronary vascular resistance in beating left ventricle. Am J Physiol 251:H779–H788
4. Feigl EO (1983) Coronary physiology. Physiol Rev 63:1–205
5. Johnson PC (1981) The myogenic response. In: Bohr DF, Somlyo AP, Sparks HV Jr (eds) Handbook of physiology, section 2: The cardiovascular system. Vascular smooth muscle. American Physiological Society, Bethesda, pp 409–442
6. McHale PA, Dubé GP, Greenfield JC Jr (1987) Evidence for myogenic vasomotor activity in the coronary circulation. Prog Cardiovasc Dis 30:139–146
7. Holtz J, Giesler M, Bassenge E (1983) Two dilatory mechanisms of anti-anginal drugs on epicardial coronary arteries in vivo: Indirect, flow-dependent, endothelium-mediated dilation and direct smooth muscle relaxation. Z Kardiol 72 [Suppl 3]:98–106
8. Lamping KG, Dole WP (1988) Flow-mediated dilation attenuates constriction of large coronary arteries to serotonin. Am J Physiol 255:H1317–H1324
9. Hull SS Jr, Kaiser L, Jaffe MD, Sparks HV Jr (1986) Endothelium-dependent flow-induced dilation of canine femoral and saphenous arteries. Blood Vessels 23:183–198
10. Smiesko V, Lang DJ, Johnson PC (1989) Dilator response of rat mesenteric arcading arterioles to increased blood flow velocity. Am J Physiol 257:H1958–H1965
11. Fujii K, Heistad DD, Faraci FM (1991) Flow-mediated dilatation of the basilar artery in vivo. Circ Res 69:697–705
12. Kuo L, Davis MJ, Chilian WM (1988) Myogenic activity in isolated subepicardial and subendocardial coronary arterioles. Am J Physiol 255:H1558–H1562
13. Kuo L, Davis MJ, Chilian WM (1990) Endothelium-dependent, flow-induced dilation of isolated coronary arterioles. Am J Physiol 259:H1063–H1070
14. Kuo L, Chilian WM, Davis MJ (1990) Coronary arteriolar myogenic response is independent of endothelium. Circ Res 66:860–866
15. Hwa JJ, Bevan JA (1986) Stretch-dependent (myogenic) tone in rabbit ear resistance arteries. Am J Physiol 250:H87–H95
16. Kulik TJ, Evans JN, Gamble WJ (1988) Stretch-induced contraction in pulmonary arteries. Am J Physiol 255 (Heart Circ Physiol 24):H1391–H1398
17. Harder DR (1987) Pressure-induced myogenic activation of cat cerebral arteries is dependent on intact endothelium. Circ Res 60:102–107
18. Katusic ZS, Shepherd JT, Vanhoutte PM (1987) Endothelium-dependent contraction to stretch in canine basilar arteries. Am J Physiol 252:H671–H673
19. Kuo L, Davis MJ, Chilian WM (1992) Endothelial modulation of arteriolar tone. News Physiol Sci 7:5–9
20. Gerova M, Gero J, Barta E, Dolezel S, Smiesko V, Levicky V (1981) Neurogenic and myogenic control of conduit coronary a.: A possible interference. Basic Res Cardiol 76:503–507
21. Drexler H, Zeiher AM, Wollschlager H, Meinertz T, Just H, Bonzel T (1989) Flow-dependent coronary artery dilatation in humans. Circulation 80:466–474

22. Lie M, Sejersted OM, Kiil F (1970) Local regulation of vascular cross section during changes in femoral arterial blood flow in dogs. Circ Res 27:727–737

23. Pohl U, Holtz J, Busse R, Bassenge E (1986) Crucial role of endothelium in the vasodilator response to increased flow in vivo. Hypertension 8:37–44

24. Rubanyi GM, Romero JC, Vanhoutte PM (1986) Flow-induced release of endothelium-derived relaxing factor. Am J Physiol 250:H1145–H1149

25. Tesfamariam B, Halpern W (1987) Modulation of adrenergic responses in pressurized resistance arteries by flow. Am J Physiol 253:H1112–H1119

26. Kuo L, Chilian WM, Davis MJ (1991) Interaction of pressure- and flow-induced responses in porcine coronary resistance vessels. Am J Physiol 261:H1706–H1715

27. Davies PF (1989) How do vascular endothelial cells respond to flow? News Physiol Sci 4:22–25

28. Peach MJ, Singer HA, Izzo NJ, Loeb AL (1987) Role of calcium in endothelium-dependent relaxation of arterial smooth muscle. Am J Cardiol 59:35A–43A

29. Falcone JC, Kuo L, Meininger GA (1993) Endothelial cell calcium increases during flow-induced dilation in isolated arterioles. Am J Physiol 264 (Heart Circ Physiol 33): H653–H659

30. Tesfamariam B, Cohen RA (1988) Inhibition of adrenergic vasoconstriction by endothelial cell shear stress. Circ Res 63:720–725

31. Griffith TM, Edwards DH, Davies RL, Harrison TJ, Evans KT (1987) EDRF coordinates the behaviour of vascular resistance vessels. Nature 329:442–445

32. Griffith TM, Edwards DH (1990) Myogenic autoregulation of flow may be inversely related to endothelium-derived relaxing factor activity. Am J Physiol 258: H1171–H1180

33. Hilton SM (1959) A peripheral arterial conducting mechanism underlying dilatation of the femoral artery and concerned in functional vasodilatation in skeletal muscle. J Physiol (Lond) 149:93–111

34. Kuo L, Arko F, Chilian WM, Davis MJ (1993) Coronary venular responses to flow and pressure. Circ Res 72:607–615

Effects of Coronary Collateral Circulation and Outflow Pressure Elevation on Coronary Pressure-Flow Relationships in The Beating and Non-Beating States

Tomiyoshi Saito, Minoru Mitsugi, Shuichi Saitoh, Masahiko Sato, and Yukio Maruyama[1]

Summary. Many factors are known to influence pressure intercept and curvilinearity of coronary pressure-flow (P-F) relationships. In this paper, we focused on two of them, interarterial collateral circulation and outflow pressure elevation. We reviewed previous reports and reexamined their effects from the viewpoint of the beating/non-beating state. The presence of interarterial perfusion pressure gradients between the left anterior descending artery and the remaining vessels resulted in an increase in zero flow pressure (Pf = 0) and a more linear slope of the P-F relationship to the same degree in both the beating and non-beating states, through interarterial collateral circulation. This indicates that collateral circulation is not necessarily inhibited by cardiac contraction. Coronary outflow pressure elevation due to coronary sinus occlusion caused an increase in Pf = 0 without altering the slope of the P-F relationship, and the slope in the non-beating state shifted to the right more than in the beating state when coronary sinus pressure was elevated to the same level of 30 mmHg.

Key words: Coronary pressure-flow relationship—Collateral circulation—Outflow pressure elevation—Beating/non-beating state

Introduction

Many factors are known to affect zero flow pressure (Pf = 0), and the curvilinearity of the coronary pressure-flow (P-F) relationship in the maximally dilated coronary vascular bed in which the coronary P-F relationship is not influenced by myocardial metabolic changes. This is reviewed by Hoffman and Spaan [1]. In this paper, we focused on two of them; that is, coronary arterial collateral circulation and outflow pressure elevation.

The effects of collateral circulation on coronary P-F relationships have been reported to increase Pf = 0 in dogs [2, 3]. In these studies, perfusion pressures of the left circumflex artery (LCX) varied from 100 mmHg to pressure at zero coronary flow in the beating [2, 3] and non-beating states [2]. The pressures in the remaining vessels were maintained at a constant to create interarterial pressure gradients (PG) as perfusion pressure of LCX was reduced. Although

---

[1] First Department of Internal Medicine, Fukushima Medical College, Fukushima, 960-12 Japan

they compared Pf = 0 of P-F relationships obtained with PG to those without PG, other investigators performed P-F measurements on a single vessel in the non-beating state and reproduced a relatively higher pressure intercept of about 15 mmHg, suggesting the same effect as the presence of PG on P-F relationships. This is because the inflow pressure on the remaining vessels was also maintained at a constant at 100 mmHg [4–6]. Therefore, the effect of collateral circulation, that is, increase in pressure intercept of the P-F relationship, seems to be likely. However, there were no reports in which the effects of collateral circulation were compared in the same heart between the beating and non-beating states.

It has been reported that elevated coronary outflow pressure brings increased Pf = 0 without changing the slope of the coronary P-F relationship in both the beating [7–9] and non-beating states [10–13]. These observations suggest that no change has occurred in coronary vascular resistance following the elevation of coronary outflow pressure, but other investigators showed that raising coronary venous pressures decreased steady-state coronary vascular resistance, probably by distending the vessels [14]. Thus, the effects of outflow pressure elevation as well as their mechanisms is still controversial.

Cardiac contraction or heart beat is known to reduce coronary arterial inflow during systole due to a throttling effect, increase in intramyocardial pressure especially in the subendocardium, and other mechanisms [1, 15–18]. In the beating state, interarterial collateral circulation may also be influenced by cardiac contraction [19] and the effect of outflow pressure elevation may be different from that in the non-beating state. In the present study, we reexamined the effects of collateral circulation and outflow pressure elevation from the viewpoint of the beating/non-beating state. This is because there have been no systematic reports tested how P-F relationships were altered in the same heart in those situations.

Methods

An isolated canine beating heart perfused with a support dog was used. Mongrel dogs used for heart donors were anesthetized with sodium pentobarbital. Other dogs were used as support animals. Bilateral thoracotomy was performed in the donor dog under positive pressure ventilation and the heart was excised without discontinuation of coronary perfusion from the heparinized support dog under Langendorff technique. A Gregg's glass cannula, which had been used for Langendorff perfusion, was advanced into the left main coronary artery of the excised heart, and the proximal portion of the left descending artery (LAD) was ligated to perfuse only in the left circumflex artery (LCX) territory. Other cannulas (ID, 1.3 mm) were inserted via the epicardial side into the just distal portion of the site of ligation on the LAD and the right coronary artery (RCA), respectively. Thus, LAD, LCX and RCA were independently perfused with arterial blood from the support dog. A thin latex balloon was inserted into the left ventricle (LV) to obtain an isovolumic contraction, and LV end-diastolic pressure (LVEDP) was maintained at approximately 5 mmHg throughout the experiment. A balloon catheter 7.5 F in size (Retroperfusion Systems, Inc., Costa Mesa, Calif.) that could be optionally inflated by hand was inserted into the coronary sinus (CS) of donor heart to obtain various degrees of CS pressure elevation.

The pressures of LAD, LCX, and CS were measured with pressure transducers (Nihon Kohden AP-641G, Tokyo, Japan), LV pressure was measured with a microtip manometer, 4 F in size (Camino 420, San Diego, Calif.) and LAD perfusion flow was measured with a cannulating type of electromagnetic flowmeter (Nihon Kohden MFV-3200).

Mean perfusion pressure and mean perfusion flow of LAD were measured under adenosine infusion (1 mg/kg). After atrioventricular block was accomplished by injection of formalin into the atrioventricular junction, the heart was paced at $100-120\,min^{-1}$ on the right ventricular free wall. The non-beating state was brought about by cessation of the pacing.

Coronary P-F relationships were determined under the following four conditions for evaluating effects of collateral circulation. 1) Beating state without perfusion pressure gradients (PG); that is, perfusion pressures of LAD, LCX, and RCA were simultaneously reduced slowly and quasi-linearly by changing common air reservoir pressure from 70–80 mmHg to pressure at zero coronary flow. 2) Beating state with PG; that is, only perfusion pressure of LAD was gradually reduced from 70–80 mmHg to Pf = 0 by adjusting a screw clamp on the LAD circuit, whereas perfusion pressures of LCX and RCA were maintained constantly at an initial value of 70–80 mmHg as mentioned above. 3) Non-beating state without PG. 4) Non-beating state with PG.

Another set of P-F relationships was also obtained under following four conditions with PG for evaluating effects of outflow pressure elevation. 1) Control condition (deflation of balloon in the catheter) in the beating state. 2) CS partial

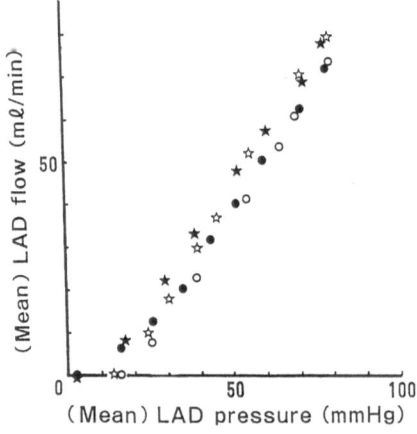

Fig. 1. Coronary pressure-flow relationships obtained under four conditions, i.e., in the beating state without perfusion pressure gradients (*PG*) (*closed circles*), in the beating state with PG (*open circles*), in the non-beating state without PG (*closed stars*), and in the non-beating state with PG (*open stars*) in one experiment. See text for details. *LAD*, Left descending artery

occlusion (CSPO); that is, CS pressure was elevated to 30 mmHg due to the partial balloon inflation in the CS, in the beating state. 3) Control condition in the non-beating state. 4) CSPO in the non-beating state.

Results

Figure 1 shows the first set of P-F relationships under the four different conditions in one experiment. The lowest Pf = 0 was that obtained in the non-beating state without PG (3 mmHg). In addition, the slope of the P-F relationship in the non-beating state without PG was almost linear. In the beating state without PG, the same value of Pf = 0 (3 mmHg) was obtained, and the slope of the P-F relationship was also almost linear but slightly convex toward the pressure axis in the lower range of perfusion pressure, in other words, the slope shifted to the right slightly in the higher range of perfusion pressure from that in the non-beating state without PG. On the other hand, Pf = 0 in the non-beating state with PG was higher (13 mmHg), and the slope of the P-F relationship was also almost linear. Pf = 0 in the beating state with PG (15 mmHg) was similar to or rather larger than that in the non-beating state with PG, but the slope was slightly convex toward pressure axis and shifted to the right in the same fashion as that without PG.

Figure 2 shows another set of P-F relationships in another experiment. In the control condition, in the beating and non-beating states, the same relationships, i.e., similar value of Pf = 0s (12 mmHg in the beating state vs 11 mmHg in the non-beating state) and rightward shift of slope in the beating state from that in the non-beating state at higher perfusion pressures, was observed as described in Fig. 1. When CS pressures were increased to the same level of 30 mmHg, Pf = 0s increased to the almost same values (18 mmHg in the beating state vs 19 mmHg in the non-beating state), and the slopes of P-F relationships in the beating

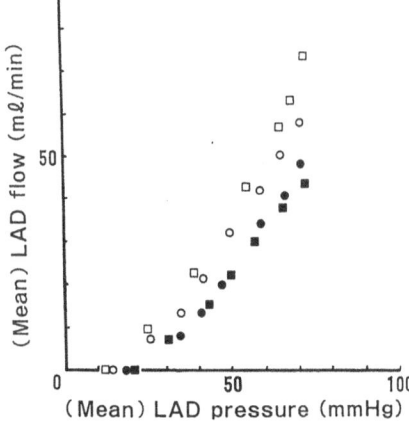

Fig. 2. Coronary pressure-flow relationships obtained under four conditions, i.e., control condition in the beating state (*open circles*), coronary sinus partial occlusion (*CSPO*) in the beating state (*closed circles*), control condition in the non-beating state (*open squares*), and CSPO in the non-beating state (*closed squares*). See text for details

and non-beating states were almost identical. In other words, the slope of P-F relationship in the non-beating state shifted to the right to a greater degree compared with that in the beating state.

Discussion

An isolated and maximally vasodilated heart preparation was used in this study, because this model is well known to avoid or minimize several confounding factors and to obtain reproducible results, although with many variations from the normal physiological situation. The left ventricular filling pressure was kept constant at a low level, thereby left ventricular oxygen requirements were low [20], reducing the likelihood of ischemia and left ventricular dysfunction when coronary pressure was transiently reduced. The coronary arterial capacitive effect was avoided by reducing perfusion pressure gradually [11, 12, 21] at about 1 mmHg/s in this study.

The Effects of Collateral Circulation on P-F Relationships

In this study, interarterial pressure gradients reduced the curvature at low LAD pressure, and increased the pressure axis intercepts from 3 to 13 mmHg and 3 to 15 mmHg in the non-beating and beating states, respectively. Two groups have reported similar observation previously in the beating [2, 3] and non-beating states [2], i.e., the presence of interarterial pressure gradients caused reduction in curvature and increase in pressure intercept. Messina et al. [2], Scheel et al. [3], and ourselves believe that the most likely cause for the effects of the interarterial pressure gradients was the occurrence of collateral flow from LCX and RCA to LAD territories when pressure of LAD was selectively lowered below that in the other arteries.

The slope of P-F relationships in the beating and non-beating states without PG was almost linear in the higher range of perfusion pressure, but was slightly convex toward the pressure axis in the lower range. However, P-F relationship slopes in the beating and non-beating states with PG were more linear even in the lower perfusion pressure range, and were almost identical to those without PG in the higher perfusion pressure range. As a result, P-F relationships both without and with PG diverged in the lower range of perfusion pressure. This observation is also similar to that by Messina et al. [2] who have reported the divergence of the slopes both without and with PG at low perfusion pressure in the beating and non-beating states in different dogs.

Russell et al. [19] reported an inhibitory effect of cardiac contraction on coronary flow using a radiomicrosphere technique. Collateral flow has been noted to be modified by interarterial pressure difference, development of collateral channels, neuro-humorally and metabolically induced blood vessel tone, contractile force, and so on. In the present study, effects of contractile force on LAD P-F relationships through collateral channels were investigated, and this was compared with non-beating P-F relationships in maximally vasodilated state and the similar interarterial pressure gradient. The obtained P-F relationships in the beating and non-beating states without and with PG were similar, indicating that collateral flow is not greatly altered by the contractile force, or since an increase in Pf $= 0$ with PG in the beating state is rather larger than that of non-beating one, collateral circulation might be promoted in the beating condition, in contrast to the previous report [19].

In 1975, Downey and Kirk [18] applied the waterfall concept, in which coronary flow ceases in the different layers of the myocardium at different pressure, to explain curvilinearity and a pressure intercept for the P-F relationship in the beating state. When collateral circulation is present at lower perfusion pressure, the fall crest in the subepicardial layers, especially showing lower values of Pf $= 0$, might be increased due to an increase in extravascular pressure, or intramyocardial pressure elevation according to the degree of pressure gradients between LAD and other major coronary arteries. As a result, the curvilinearity of P-F relationship with PG may become more linear in the lower perfusion pressure range.

The Effects of Outflow Pressure Elevation on P-F Relationships

In this study, we confirmed the effects of outflow pressure elevation due to CS occlusion, and observed a new result that the slope of the P-F relationship in the non-beating state shifted to the right more than in the beating state when CS pressure was similarly elevated.

Several researchers have reported about coronary P-F relationships with elevation of downstream pressure by various methods [7–13]. Their reports indicated similar results either in the beating state [7–9] or in the non-beating state [10–13]; that is, Pf $= 0$ increased without altering the slope of the P-F relationship. On the other hand, there have been controversial reports about the presence of the effects of outflow pressure elevation due to CS occlusion [1, 14], in which raising coronary venous pressure decreased steady-state coronary vascular resistance, probably by distending the vessels. According to the result, the slope of the P-F relationship with CS pressure elevation should be steepened. However, the slope shifted to the right almost in parallel in this study as well as previous reports [7–10].

Bellamy et al. [10] speculated that the increased Pf $= 0$ was induced by an increased extravascular pressure, or intramyocardial pressure elevation due to the CS occlusion. In previous studies [7–12] and the present one, the slope of the P-F relationship was not affected significantly and therefore increased extravascular pressure within the range of the increased CS pressure seems to be insufficient for increasing arterial resistance. However, as for arterial resistance during coronary sinus pressure elevation, a different result has been also reported, as described above. On the contrary, Pantley et al. [8] took notice of CS and extra-CS pathways and supposed that the increase in Pf $= 0$ was probably due to an intrinsically higher zero-flow pressure in the extra-CS drainage pathways, that is, CS occlusion diverted all venous return into the extra-CS pathway which is characterised by intrinsically higher Pf = 0 than the CS pathway. When taking this into account all together, this explanation regading an increased Pf = 0 is likely.

The previous results [10] for the slope and Pf = 0 following CS complete occlusion were similar, i.e., no change in the slope and increased Pf = 0 (31 ± 8 to 52 ± 10 mmHg), in which CS pressure increased from 5 ± 4 to 38 ± 11 mmHg. On the contrary, in our study, Pf = 0 increased by only about two fifth (from 12 mmHg to 18 mmHg) of the increase in CS pressure (from 10 mmHg to 30 mmHg) in the beating state, and Pf = 0 increased similarly from 11 mmHg to 19 mmHg in CS pressure elevation from 6 mmHg to 28 mmHg in the non-beating state. The main reason why CS pressure exceeded Pf = 0 may be existence of an extra-CS pathway as Satoh et al. [11] pointed out, in which they studied the effect of downstream pressure elevation by right atrial and ventricular pressure elevation. In addition, the reason why the magnitude of an increase in Pf = 0 due to outflow pressure elevation to the same level in the beating state tended to be less than that in the non-beating state, may be as follows; coronary venous blood is drained predominantly in systole due to the squeezing effect by myocardial contraction [22–24]. In the beating state, the squeezing effect in the venous site is still expected through an extra-CS pathway even under CS complete occlusion, and results in smaller amounts of intravascular blood volume, which itself may promote coronary inflow [26] or lead likely to less intramyocardial pressure, compared with the beating state.

This concept may also explain our new result that the slope of the P-F relationship in the non-beating state shifted to the right to a greater degree compared with that in the beating state when CS pressure was elevated. Even after CS pressure elevation, as previously mentioned, the blood volume in coronary vasculature in systole may be reduced due to the squeezing effect during systole and diastolic inflow can be facilitated despite prominent systolic CS pressure elevation. In the non-beating state, however, increased venous pooling is kept constant due to the non-squeezing effect, and this may lead to the restriction of coronary inflow. Therefore, it is quite probable that the P-F relationship between the beating and non-beating states did not differ in the outflow pressure elevation, even though intramyocardial pressure in systole increases due to systolic compressive force. Our result also suggests that the effects of outflow pressure elevation on P-F relationship, which have been obtained during long diastole, cannot be directly applied to that in the beating condition and if adopted, the outflow pressure elevation on the P-F relationship might be overestimated.

Since Satoh et al. [11] and Farhi et al. [12] directly observed waterfall behavior to elucidate effects of outflow pressure elevation by right atrial and ventricular pressure elevation, a similar mechanism was likely operating in our results. However, Scheel et al. [9] studied effect of CS pressure on coronary P-F characteristics, and concluded that CS pressure has a direct influence on coronary perfusion without utilizing waterfall concept. Unfortunately our study was not designed to clarify the mechanism of the effects of outflow pressure elevation; further experiments will be needed.

In conclusion, we did show only in the representative examples how coronary P-F relationships without and with collateral circulation were modified by cardiac contraction and also how coronary P-F relationships following outflow pressure elevation was influenced by the same factor, i.e., cardiac contraction. Our present experiments clarified the role of the contraction on coronary P-F relationship in the presence of downstream pressure elevation or without/with collateral circulation, indicating that cardiac contraction does not necessarily inhibit coronary inflow in those situations. That is, P-F relationships with PG were not affected in the beating condition, indicating that collateral flow was not changed, or rather promoted by the cardiac contraction, and following the coronary sinus pressure elevation rightward shift of the P-F relationship was greater in the non-beating state than that in the beating one, indicating that effect of back pressure elevation is greater in the non-beating state and coronary inflow is largely reduced at the same perfusion pressure.

References

1. Hoffman JIE, Spaan JAE (1990) Pressure-flow relations in coronary circulation. Physiol Rev 70(2):331–390
2. Messina LM, Hanley FL, Uhlig PN, Baer RW, Grattan MT, Hoffman JIE (1985) Effects of pressure gradients between branches of the left coronary artery on the pressure axis intercept and the shape of steady state circumflex pressure-flow relations in dogs. Circ Res 56:11–19
3. Scheel KW, Mass H, Williams SE (1989) Collateral influence on pressure-flow characteristics of coronary circulation. Am J Physiol 257:H717–H725
4. Klocke FJ, Weinstein IR, Klocke JF, Ellis AK, Kraus DR, Mates RE, Canty JM, Anbar RD, Romanowski RR, Wallmeyer KW, Eclit MP (1981) Zero-flow pressures and pressure-flow relationships during single long diastole in the canine coronary bed before and during maximum vasodilation. J Clin Invest 68:970–980
5. Dole WP, Bishop VS (1982) Influence of autoregulation and capacitance on diastolic coronary artery pressure-flow relationships in the dog. Circ Res 51:261–270
6. Eng C, Jentzer JH, Kirk ES (1982) The effects of the coronary capacitance on the interpretation of diastolic pressure-flow relationships. Circ Res 50:334–341
7. Rouleau JR, White M (1984) Effects of coronary sinus pressure elevation on coronary blood flow distribution in dogs with normal preload. Can J Physiol Pharmacol 63:787–797
8. Pantley GA, Bristow JD, Ladley HD, Anselone CG (1988) Effect of coronary sinus occlusion on coronary flow, resistance, and zero flow pressure during maximum vasodilation in swine. Cardiovasc Res 22:79–86
9. Scheel KW, Williams SE, Parker JB (1990) Coronary sinus pressure has a direct effect on gradient for coronary perfusion. Am J Physiol 258:H1739–H1744
10. Bellamy RF, Lowensohn HS, Ehrlich W, Baer RW (1980) Effect of coronary sinus occlusion on coronary pressure-flow relations. Am J Physiol 239:H57–H64

11. Satoh S, Watanabe J, Keitoku M, Itoh N, Maruyama Y, Takishima T (1988) Influences of pressure surrounding the heart and intracardiac pressure on the diastolic coronary pressure-flow relation in excised canine heart. Circ Res 63:788–797

12. Farhi ER, Klocke FJ, Mates RE, Kumar K, Judd RM, Canty JM, Satoh S, Sekovski B (1991) Tone-dependent waterfall behavior during venous pressure elevation in isolated canine hearts. Circ Res 68:392–401

13. Uhlig PN, Baer RW, Vlahakes GJ, Hanley FL, Messina LM, Hoffman JIE (1984) Arterial and venous coronary pressure-flow relations in anesthetized dogs. Circ Res 55:238–248

14. Hanley FL, Messina LM, Grattan MT, Hoffman JIE (1984) The effect of coronary inflow pressure on coronary vascular resistance in the isolated dog heart. Circ Res 54:760–772

15. Sabiston DC Jr, Gregg DF (1957) Effect of cardiac contraction on coronary blood flow. Circulation 15:14–20

16. Moir TW (1972) Subendocardial distribution of coronary blood flow and the effect of antianginal drugs. Circ Res 30:621–627

17. Downey JM, Downey HF, Kirk ES (1974) Effects of myocardial strains on coronary blood flow. Circ Res 34:286–292

18. Downey J, Kirk ES (1975) Inhibition of coronary blood flow by a vascular waterfall mechanism. Circ Res 36:753–760

19. Russell RE, Chagrasulis RW, Downey JM (1977) Inhibitory effect of cardiac contraction on coronary collateral flow. Am J Physiol 233:H541–H546

20. McKeever WP, Gregg DE, Canney PC (1958) Oxygen uptake of the non-working left ventricle. Cire Res 6:612–623

21. Aversano T, Klocke FJ, Canty JM Jr (1984) Preload-induced alterations in capacitance-free diastole pressure-flow relationships. Am J Physiol 246:H410–H417

22. Porter WT (1898) The influence of the heartbeat on the flow of blood through the walls of the heart. Am J Physiol 1:145–163

23. Wiggers CJ (1954) The interplay of coronary vascular resistance and myocardial compression in regulating coronary flow. Circ Res 2:271–279

24. Kajiya F, Tsujioka K, Goto M, Wada Y, Chen XL, Nakai M, Tadaoka S, Hiramatsu O, Ogasawara Y, Mito K, Tomonaga G (1986) Functional characteristics of intramyocardial capacitance vessels during diastole in the dog. Circ Res 58:476–485

25. Goto M, Tsujioka K, Ogasawara Y, Wada Y, Tadaoka S, Hiramatsu O, Yanaka M, Kajiya F (1990) Effect of blood filling in intramyocardial vessels on coronary arterial inflow. Am J Physiol 258:H1042–H1048

Abstracts

Effect of Cardiac Contraction on Regional Myocardial Blood Flow

Masami Goto, Arthur E. Flynn, Joseph W. Doucette, Catharina M.A. Jansen, Marjolein M. Stork, Dwain L. Coggins, Derek D. Muehrcke, Waleed K. Husseini, and Julien I. E. Hoffman[1]

The effect of myocardial contraction on the regional myocardial flow and its mechanism are still not clear, although these have been the most fundamental questions in the study of coronary blood flow. To evaluate the effect of myocardial contraction on myocardial flow in the different layers and the roles of intramyocardial forces and systolic ventricular pressures in the blood flow distribution, we measured myocardial flow in rabbit hearts during stable systolic contraction with different left ventricular pressures (60 mmHg; $n = 5$, and 0 mmHg; $n = 5$) and during stable diastolic arrest ($n = 5$). We also measured the number and size of the intramyocardial vessels after perfusion fixation (systolic arrest; $n = 5$, and diastolic arrest; $n = 5$). In 25 rabbits, hearts were excised and perfused from the aortic root. Systolic arrest was achieved by perfusion of a low Ca^{2+} Tyrode's solution containing Ba^{2+} (2.0 mM). Diastolic arrest was achieved by intraventricular injection of pentobarbital (700–1000 mg) and was maintained by perfusion with St. Thomas' cardioplegic solution. At perfusion pressure of 100 mmHg, subendocardial flow was lower than subepicardial flow in systolic hearts, regardless of left ventricular pressure, whereas in diastolic hearts, subendocardial flow was higher than subepicardial flow. In systolic hearts with low left ventricular pressure, subendocardial-to-subepicardial flow ratio for a physiologic range of perfusion pressures was lower than in systolic hearts with high left ventricular pressure. Small arteriolar and capillary densities showed no difference between subendocardium and subepicardium. In systolic hearts, diameters of subendocardial terminal arterioles (4.6 ± 1.3 µm) and capillaries (4.0 ± 1.3 µm) were smaller than those in the subepicardium (8.8 ± 7.7 µm and 7.1 ± 1.6 µm, respectively, $P < 0.0001$), whereas in diastolic hearts, diameters of subendocardial terminal arterioles (10.1 ± 2.0 µm) and capillaries (7.6 ± 1.8 µm) were larger than those in the subepicardium (9.5 ± 1.5 µm and 6.7 ± 1.0 µm, respectively, $P < 0.01$). We concluded that cardiac contraction predominantly affects subendocardial vessels, and impedes subendocardial flow more than subepicardial flow, regardless of left ventricular pressure.

---

[1] Cardiovascular Research Institute, University of California, San Francisco, USA

Variations of Blood Pooling in Coronary Vascular Beds

Jun Watanabe, Katsuyuki Hangai, Shoichi Satoh[1], Yukio Maruyama[2], and Tamotsu Takishima[1]

The myocardium is highly vascular, consisting of 10%–15% of blood at diastolic volume. Thereby coronary blood volume (CBV) affects both coronary hemodynamics and myocardial properties. The aims of this study were to assess: (a) variations of CBV in beating hearts with heart rate (HR) changes and (b) the effects of coronary venous pressure (VP) on diastolic myocardial properties. (a) To assess CBV, we measured myocardial volume in LV-isovolumically beating and vasodilated dog hearts mounted in a newly-developed pressure type plethysmography system. HR varied from 60 to 180 bpm when perfusion pressure (PP) was maintained constant at approximately either 70 or 40 mmHg. Mean CBV decreased linearly with reduced RR intervals at both PP, the decrease being significantly larger at a PP of 70, i.e., $CBV = 3.5 \times RR - 1.8$ (PP = 70) vs $CBV = 2.1 \times RR - 1.0$ (PP = 40), $P < 0.005$. These results suggest that HR and PP play important interdependent roles in determining CBV. (b) To assess the effect of CBV on diastolic myocardial distensibility, we studied excised, LV isovolumic dog hearts in which VP and right ventricular pressure (RVP) were manipulated separately. LV wall volume was determined by subepicardial segment length at end-diastole. Both VP and RVP were increased, from 0 to 30 mmHg, over a range of LV volumes. Left ventricular end-diastolic pressure (LVEDP) (mmHg) data are shown in the Table.

LV volume (ml)	22 ± 2	31 ± 3	40 ± 3		21 ± 3	31 ± 3	41 ± 4
VP(mmHg) 0	5.2 ± 0.3	10.4 ± 0.3*	11.2 ± 1.5*#	RVP 0	5.2 ± 0.2	9.8 ± 0.3	19.0 ± 0.5*
15	8.3 ± 0.9	14.2 ± 0.7*	24.2 ± 1.0*#	15	5.8 ± 0.2	10.5 ± 0.4	20.4 ± 0.5*
30	11.2 ± 1.5	18.2 ± 1.2*	28.8 ± 1.2*#	30	6.6 ± 0.2	11.6 ± 0.6	21.4 ± 0.8*

(Mean ± SEM, *$P < 0.05$ vs VP = 0; #$P < 0.05$ vs RVP study)

In both groups, the LV end-diastolic pressure-volume relation shifted upward in an almost parallel fashion, but the shift was much greater in the VP study. The increase in LV wall dimension was significant and much greater when VP increased. In conclusion, increased coronary venous pressure reduces LV diastolic distensibility with increasing LV all volume.

---

[1] First Department of Medicine, Tohoku University School of Medicine, Sendai, 980 Japan
[2] First Department of Medicine, Fukushima Medical College, Fukushima, 960 Japan

C. Models of Coronary Circulation

Theoretical Relationship among Myocardial Tissue and Cavity Pressures, Contractility, and Vascular Volume Change

R.S. Chadwick[1] and Cheng Dong[2]

Introduction

The intramyocardial pump model introduced by Spaan et al. [1] to explain the reduction of coronary inflow during cardiac contraction has been recently questioned by Krams et al. [2–4]. The latter authors demonstrated that the systolic coronary inflow impediment is only weakly dependent on left ventricular cavity pressure, but is affected more by cardiac contractility and perfusion pressure. Since the pump model, in its simplest form, assumes a direct proportionality between intramyocardial tissue pressure (IMP) and left ventricular cavity pressure (with one zero if the other is zero) it would appear that the pump model fails to explain the results of Krams et al. A recent letter by Kresh [5] discusses the controversy further. In view of the interest and importance of this subject, it seemed helpful to develop a relatively simple model from which various factors affecting IMP can studied. In particular, the relationship between IMP and left ventricular pressure is clarified, and simple formulas are given that can be used in conjunction with a pump model.

Mathematical Description of Model

In Fig. 1a we consider an annular ring of myocardial tissue represented as a continuum of circumferential muscle fibers, an isotropic collagen matrix, and small distensible vessels. Let the outer radius of the ring be R_O and the inner radius be R_i in the reference configuration, which is assumed here to be free of residual stress. The cavity pressure P_{lv}, acting on the endocardial surface at $r = R_i$, is taken to be known and controlled independently of the contraction of the muscle fibers. The capillary pressure P_C is dependent on the radial coordinate, r. The stress acting on the epicardial surface at $r = R_O$ is assumed to be zero.

[1] Theoretical Biomechanics Group, Biomedical Engineering and Instrumentation Program, NCRR, National Institutes of Health, Bethesda, MD 20892, USA
[2] Bioengineering Program, College of Engineering, The Pennsylvania State University, University Park, PA 16802, USA

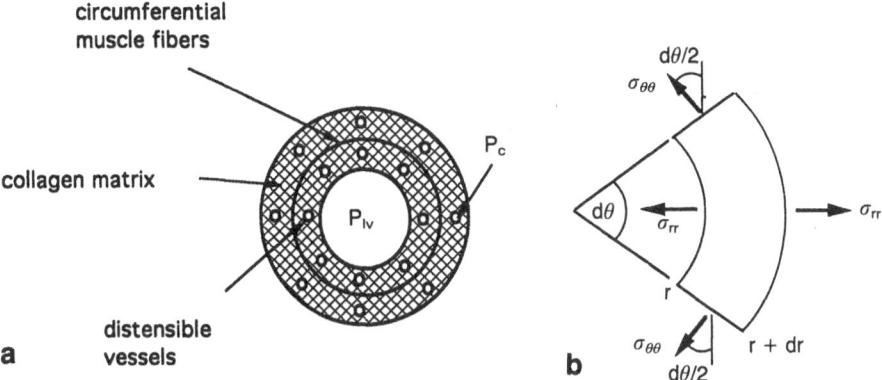

Fig. 1. a Continuum model for intramyocardial tissue pressure. **b** Infinitesimal tissue element showing stress components. P_{lv}, cavity pressure; P_c, capillary pressure; $\sigma_{\theta\theta}$, circumferential stress; σ_{rr}, radial stress

Mechanical Equilibrium

Our aim is to determine the mechanical equilibrium state of the annular ring of tissue at end systole. The resulting spatial distribution of intramyocardial tissue pressure P is of particular interest. With that end in mind, an infinitesimal annular sector of tissue is isolated, to indicate the stresses acting on each face of the element. As shown in Fig. 1b, the radial stress σ_{rr} acts normal to the circular arcs, and the circumferential stress $\sigma_{\theta\theta}$ acts normal to the radial segments. When the deformation is axisymmetric (independent of the angular coordinate θ) there are no shear stresses present. The condition of mechanical equilibrium requires that there must be a balance of forces (stress times segment length times unit depth) in both the horizontal and vertical directions, as well as no net moment. Clearly, the latter two requirements are satisfied by symmetry. Horizontal equilibrium requires

$$\sigma_{rr}(r + dr)[(r + dr)d\theta] - \sigma_{rr}(r)[r\,d\theta] - 2\sigma_{\theta\theta}\sin(d\theta/2)[dr] = 0. \quad (1)$$

Note that $[(r + dr)d\theta]$ is the length of the outer arc, $[rd\theta]$ is the length of the inner arc, and $[dr]$ is the length of the slanted sides of the element. Since the element is infinitesimal, both dr and $d\theta$ approach zero. Consequently, $\sin(d\theta/2) \cong d\theta/2$, and $\sigma_{rr}(r + dr) \cong \sigma_{rr}(r) + dr\,d\sigma_{rr}/dr$. This latter approximation is based on an hypothesis which is central to the subject of continuum mechanics, namely, that the stress varies smoothly between neighboring locations. At the microstructural scale, this approximation cannot be expected to hold. Also, on a macroscopic scale, discontinuous behavior might exist at the interface of muscle fiber bundles. We shall ignore this latter possibility in our development. When these approximations are used and the term involving $(dr)^2$ is neglected compared to the others, the following equation of equilibrium is obtained:

$$\frac{d\sigma_{rr}}{dr} + \frac{1}{r}(\sigma_{rr} - \sigma_{\theta\theta}) = 0 \tag{2}$$

Constitutive Relations

Equation 2 is valid for any continuous material. We now must introduce specialized relations for the stresses that attempt to describe the essential features of myocardial tissue. Consider first the circumferential stress $\sigma_{\theta\theta}$ in Fig. 1b. The face on which it acts is subject to the pull of activated muscle fibers, exerting the stress, T, and the push of the intramyocardial tissue pressure, P. In addition, the collagen matrix exerts a pulling stress $2\mu U/r$, where μ is the effective shear modulus of the matrix, $U(r)$ is the outward radial tissue displacement field, and U/r is the circumferential strain. The displacement U is normally negative unless the cavity pressure is unusually high. So for the normal situation the collagen exerts a pushing stress in the circumferential direction. The systolic fiber stress, T is expressed as the sum of two terms: $T_O + E^*U/r$, where T_O is the active stress present independent of fiber circumferential strain, and E^* is the systolic fiber modulus. Both parameters, T_O and E^*, can be considered to be contractility parameters, and their values would be expected to change with inotropic intervention. The active modulus E^* is the local fiber analog of the global end-systolic ventricular elastance, commonly called E_{max} Following this description we then write

$$\sigma_{\theta\theta} = -P + T_O + (2\mu + E^*)\frac{U}{r}. \tag{3}$$

Similar arguments apply to the radial stress σ_{rr} except for two differences: since the fibers have no radial component in the present model, the parameters E^* and T_O do not enter, and the radial strain is dU/dr instead of U/r. Therefore

$$\sigma_{rr} = -P + 2\mu\frac{dU}{dr}. \tag{4}$$

One further remark should be made concerning the term "effective", which applies to the parameters T_O, μ, and E^*. Each of these quantities should include a weighting fraction that accounts for the fraction of cross sectional tissue area occupied by the corresponding individual tissue component being described. Substitution of Eqs. 3 and 4 into 2 yields a relationship between the intramyocardial tissue pressure gradient and the radial displacement.

$$\frac{dP}{dr} = 2\mu\frac{d}{dr}\left\{\frac{1}{r}\frac{d}{dr}(rU)\right\} - \frac{1}{r}\left(T_O + E^*\frac{U}{r}\right) \tag{5}$$

Conservation of Myocardial Tissue Volume

Equation 5 contains two unknowns, P and U. Therefore another relationship that describes some property of the tissue is required, e.g., the relative change in tissue volume during contraction. A reasonable assumption is that the extravascular space is incompressible, so that the volume change of an element of myocardial tissue corresponds to an equal volume change of the vascular space. Consider again the tissue volume element shown in Fig. 1b. For small displacements the volume change is approximately $U(r + dr)[(r + dr)d\theta] - U(r)[rd\theta]$ $\cong (r\, dU/dr + U)drd\theta$. The change in volume of the total number of vessels contained in the tissue element is $(P_C - P)D_C V_C$, where D_C is the capillary distensibility and V_C is the total volume of capillaries contained in the element. Equating these two expressions and dividing by the volume of the tissue element, $rdrd\theta$, yields the desired relation

$$\frac{1}{r}\frac{d}{dr}(rU) = -(P - P_C)D_C\phi_C = -\delta V/V_O \tag{6}$$

where ϕ_C denotes the volume fraction of the capillary space.

The volume conservation relation, Eq. 6, introduces the capillary pressure P_C into the problem. In principle the spatial distribution $P_C(r)$ can be also calculated by coupling a hydrodynamic model of flow through deformable vessels to Eqs. 5 and 6. This would complicate the analysis considerably, and is beyond the scope of the present analysis. Instead, we will make use of the experimental finding by Judd and Levy [6] that each volume element of tissue has approximately the same decrease in volume at end systole, i.e., $\delta V/V_O$ is independent of the radial coordinate.

Determination of the Distribution of Intramyocardial Tissue Pressure

For constant $\delta V/V_O$, Eq. 6 is easily integrated to give the displacement field

$$U(r) = -\frac{1}{2}\frac{\delta V}{V_O}r + \frac{C}{r} \tag{7}$$

where C is a constant of integration to be found. Note that for constant $\delta V/V_O$, Eq. 5 simplifies considerably, since the first term on the right-hand side is identically zero. Substitution of Eq. 7 into Eq. 5 yields

$$\frac{dP}{dr} = -\left(T_O - \frac{1}{2}E^*\frac{\delta V}{V_O}\right)\frac{1}{r} - E^*\frac{C}{r^3}$$

which can be integrated to give

$$P(r) = -\left(T_O - \frac{1}{2}E^*\frac{\delta V}{V_O}\right)\log r + \frac{1}{2}E^*\frac{C}{r^2} + D \tag{8}$$

where D is another constant of integration. The two unknown constants C and D can be found from the two boundary conditions: $\sigma_{rr}(R_O) = 0$, and $\sigma_{rr}(R_i) = -P_{lv}$, where σ_{rr} is given by Eq. 4. These two boundary conditions lead to a linear algebraic system for the unknowns C and D. The final result for the radial distribution of intramyocardial tissue pressure is

$$P(\rho) = -\mu\frac{\delta V}{V_O} + \left(T_O - \frac{1}{2}E^*\frac{\delta V}{V_O}\right)\log\left(\frac{\alpha}{\rho}\right) +$$

$$\frac{\left[P_{lv} - \left(T_O - \frac{1}{2}E^*\frac{\delta V}{V_O}\right)\log\alpha\right]}{\alpha^2 - 1}\left[\frac{E^*\alpha^2}{(E^* + 4\mu)\rho^2} - 1\right] \qquad (9)$$

where we have written $\rho = r/R_i$, and $\alpha = R_O/R_i$.

Volume-Averaged Intramyocardial Tissue Pressure

The volume-averaged intramyocardial tissue pressure $\langle P \rangle$ is a convenient index of the systolic reduction of coronary flow since, in the present model, increases in $\langle P \rangle$ are accompanied by equal increases in $\langle P_C \rangle$, which is the mean back pressure operating against coronary inflow. $\langle P \rangle$ can be calculated from

$$\langle P \rangle = \frac{2}{\alpha^2 - 1}\int_1^\alpha P(\rho)\rho\,d\rho \qquad (10)$$

where $P(\rho)$ is given by Eq. 9. It is of interest to examine the dependence of $\langle P \rangle$ on P_{lv}. It turns out that the relationship is linear, i.e.,

$$\langle P \rangle = AP_{lv} + B \qquad (11)$$

where

$$B = \frac{1}{2}\left(T_O - \frac{1}{2}E^*\frac{\delta V}{V_O}\right)\left[1 - 4\frac{E^*}{(E^* + 4\mu)}\frac{\alpha^2(\log\alpha)^2}{(\alpha^2 - 1)^2}\right] - \mu\frac{\delta V}{V_O} \qquad (12)$$

$$A = \frac{1}{\alpha^2 - 1}\left[2\frac{E^*}{(E^* + 4\mu)}\frac{\alpha^2\log\alpha}{(\alpha^2 - 1)} - 1\right]. \qquad (13)$$

Theoretical Results and Comparison with Experiments

Nominal Values of Parameters

Unless stated otherwise in the figure captions, the following parameter values are used in the calculations: $\alpha = R_O/R_i = 1.75$; $T_O = 5 \times 10^5\,\mathrm{dyne/cm^2} \cong 375$ mmHg; $\mu = 2 \times 10^5\,\mathrm{dyne/cm^2}$; $E^* = 33 \times 10^5\,\mathrm{dyne/cm^2} \cong 2475\,\mathrm{mmHg}$; $\delta V/V_O = 0.025$. The value for α is typical of mammalian hearts. The value of T_O is similar to that measured by ter Keurs et al. [7] for rat trabeculae. A similar value for μ has been used in an analysis of a biaxial tension test on passive canine myocardium by Ohayon and Chadwick [8]. The value of $\delta V/V_O$ is similar to that measured in the rat by Judd and Levy [6]. The value of E^* was chosen to match an end-systolic elastance value, $E_{max} = 55\,\mathrm{mmHg/ml}$, which was measured by Krams et al. [3] in the cat. The systolic elastance of the present model is related to E^* by

$$E_{max} = \frac{(\alpha^2 - 1)(E^* + 4\mu)}{4\alpha^2 V_{cav}} \qquad (14)$$

where V_{cav} denotes the reference left ventricular cavity volume.

Radial Distribution of Intramyocardial Tissue Pressure

Figure 2 shows the effects of cavity pressure and the relative volume change of the vascular space on the distribution of tissue pressure from the endocardium to the epicardium, as computed from Eq. 9. Notice that a significant tissue pressure is developed even for the case of $P_{lv} = 0$. The distribution appears more linear as the cavity pressure increases, as can be seen in the case of $P_{lv} = 100$ mmHg. Intramyocardial tissue pressure is uniformly reduced by an increasing systolic volume decrement of the vascular space.

Influences of Cavity Pressure, Contractility, and Vascular Volume on the Volume-Averaged Intramyocardial Tissue Pressure

Equations 11–13 are used to compute $\langle P \rangle$ vs cavity pressure in Figs. 3–5. Figure 3 shows the effect of changes in the contractility parameter E^*, the end-systolic muscle fiber modulus of elasticity. Figure 4 shows the effect of changes in the contractility parameter T_O, the end-systolic active muscle fiber stress. Figure 5 shows the effect of changes in $\delta V/V_O$, the relative decrement in vascular volume. Each of these plots indicates a relatively weak increase of $\langle P \rangle$ on the cavity pressure which is consistent with the experiments of Krams et al. [2–4] who found a similar small increase of systolic coronary flow impediment on cavity pressure. Decreasing T_O decreases $\langle P \rangle$ (Fig. 4), but decreasing E^* *increases* $\langle P \rangle$ (Fig. 3). This latter result is contrary to the hypothesis of Krams et al. [2],

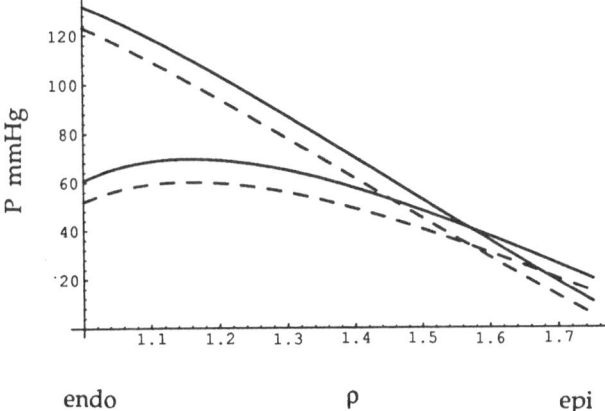

Fig. 2. Transmural distribution of intramyocardial tissue pressure. *Solid lines*, $\delta V/V_O = 0$; *dotted lines*, $\delta V/V_O = 0.025$. Cavity pressure $P_{lv} = 100$ mmHg (*upper curves on left*); $P_{lv} = 0$ mmHg (*lower curves on left*)

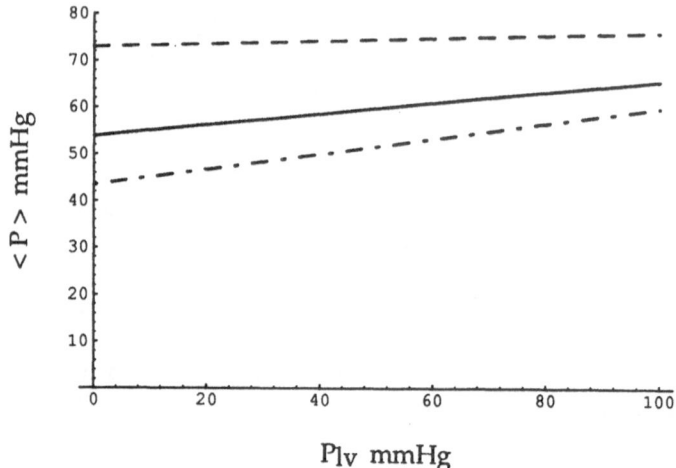

Fig. 3. Volume average intramyocardial tissue pressure $\langle P \rangle$ vs cavity pressure P_{lv}. *Dot-dash*, fiber elastic modulus $E^* = 2475$ mmHg; *Solid*, $E^* = 1768$ mmHg; *dashed*, $E^* = 1061$ mmHg

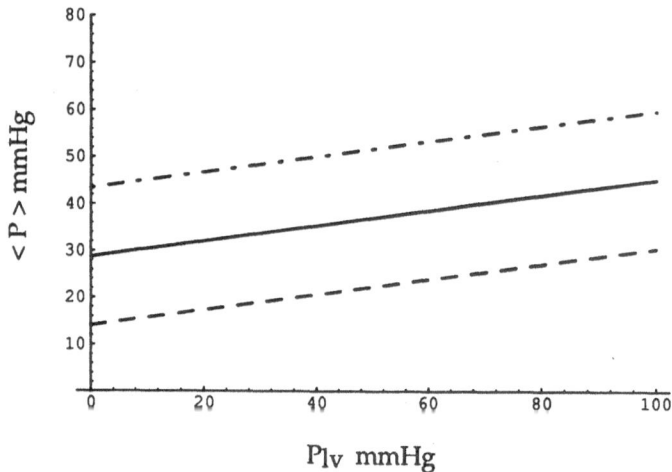

Fig. 4. Volume average intramyocardial tissue pressure $\langle P \rangle$ vs cavity pressure P_{lv}. *Dot-dashed*. Fiber isometric tension $T_O = 375$ mmHg; *Solid*, $T_O = 268$ mmHg; *dashed*, $T_O = 161$ mmHg

which claims that decreasing E_{max} *decreases* the systolic coronary flow impediment. Their hypothesis is based on experiments using negative inotropic agents. Our calculations suggest that the parameter T_O must have also decreased in their experiments, but this is difficult to quantify from their data. A decrease in T_O would result in a downward shift in the end-systolic pressure-volume line. Figure

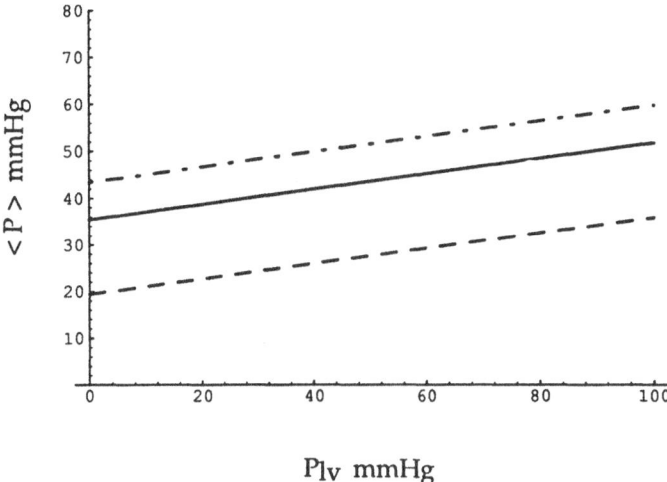

Fig. 5. Volume average intramyocardial tissue pressure $\langle P \rangle$ vs cavity pressure P_{lv}. *Dot-dashed*, Systolic vascular volume decrement $\delta V/V_O = 0.025$; *Solid*, $\delta V/V_O = 0.050$; *dashed* $\delta V/V_O = 0.100$

5 shows an increase of $\langle P \rangle$ with decreasing relative vascular volume decrement $\delta V/V_O$. One expects $\delta V/V_O$ to decrease with increasing perfusion pressure, therefore the results shown in Fig. 5 are consistent with the observation of Krams et al. [4] that systolic coronary flow impediment increases with increased perfusion pressure.

Summary and Conclusions

A simple model of myocardial tissue represented as a continuum of circumferential muscle fibers, an isotropic collagen matrix, and small distensible vessels was developed to examine the effect of various factors on the generation of intramyocardial tissue pressure (IMP). Model calculations show that a significant IMP is developed by contracting cardiac muscle even when the left ventricle is unloaded. The IMP is maximal in the subendocardium for an unloaded ventricle; as cavity pressure increases the transmural distribution of IMP decreases monotonically from the endocardium to the epicardium. Volume-averaged IMP is relatively insensitive to left ventricular cavity pressure. Volume-averaged IMP *increases* with: *increasing* T_O (an active fiber isometric stress contractility parameter; *decreasing* E^* (an active fiber elastic modulus contractility parameter); *decreasing* systolic vascular volume decrement $\delta V/V_O$; *increasing* radius ratio R_O/R_i; and *increasing* collagen matrix stiffness.

The mechanism for the generation of IMP in the unloaded ventricle in our model can be understood as follows. The active tension T_O induces circumferential shortening of the muscle fibers. This fiber shortening compresses the circumferential active spring element enough to put the subendocardial layers into compression. Because of fiber curvature, the compressive circumferential stress must be balanced by a positive gradient of IMP. This effect is amplified by stretching the radial component of the collagen matrix concomitant with wall thickening, and reduced by a decreased volume of the vascular compartment. This theory suggests an experiment to demonstrate that a decrease in systolic coronary flow impediment following perfusion of negative inotropic agents is accompanied by a parallel downward shift component to changes in the end-systolic pressure-volume line.

Acknowledgments. The authors are grateful for many useful discussions with Peter Basser, BEIP, NIH.

References

1. Spaan JA, Bruels NPW, Laird JD (1981) Diastolic-systolic coronary flow differences are caused by intramyocardial pump action in the anesthetized dog. Circ Res 49: 584–593
2. Krams R, Sipkema P, Westerhof N (1989) Varying elastance concept may explain coronary systolic flow impediment. Am J Physiol 257 (Heart Circ Physiol 26): H1471–H1479
3. Krams R, Sipkema P, Zegers J, Westerhof N (1989) Contractility is the main determinant of coronary systolic flow impediment. Am J Physiol 257 (Heart Circ Physiol 26):H1936–H1944
4. Krams R, Sipkema P, Westerhof N (1990) Coronary oscillatory flow amplitude is more affected by perfusion pressure than ventricular pressure. Am J Physiol 257 (Heart Circ Physiol 27):H1889–H1898
5. Kresh JY (1990) Myocardial modulation of coronary circulation [letter]. Am J Physiol 259 (6 Pt 2):H1934–H1937
6. Judd RM, Levy BI (1991) Effects of barium-induced cardiac contraction on large-and small-vessel intramyocardial blood volume. Circ Res 68:217–226
7. ter Keurs HEDJ, Rijnsburger WH, van Heuningen R, Nagelsmit MJ (1980) Tension development and sarcomere length in rat cardiac trabeculae. Circ Res 46:703–714
8. Ohayon J, Chadwick RS (1988) Effect of collagen microstructure on the mechanics of the left ventricle. Biophys J 54:1077–1088

Concepts and Controversies in Modelling the Coronary Circulation

Rafael Beyar, Dan Manor, Daniel Zinemanas, and Samuel Sideman[1]

Summary. Some major concepts and controversies associated with coronary flow dynamics and the interaction between left ventricular myocardial contraction and the coronary vasculature are highlighted. The controversies rise from the interpretations of the forces acting on the intramyocardial vessels and the mechanisms postulated to affect the coronary circulation. These include the waterfall concept and the pressure-dependent resistance and compliance models which are, in essence, refinements of the intramyocardial pump concept. These are reviewed in the light of recent models of the coronary circulation. Coronary ischemia, which is typically associated with the development of collateral flow, is also presented. In addition, the important role of the collagen mesh in the generation of intramyocardial pressure (IMP) at a wide range of loading conditions and the contribution of myocardial fluid transport to IMP distribution and the coronary vasculature are included in a comprehensive model relating the interactions of the above factors.

Key words: Circulation—Models—Flow—Contraction—Intramyocardial transport

The Phasic Coronary Flow Signal

The phasic coronary flow signal is a "mirror" reflecting the dynamics of the myocardial contraction phenomenon and the associated ventricular pressure and aortic pressure dynamics. The exact mechanisms by which these parameters interact are still controversial. However, helpful insights can be gained by studying the phasic characteristics of the normal phasic flow in relation to ventricular and aortic pressures. An example of the coronary flow signal measured in an open chest anesthetized dog with a transonic flowmeter is given in Fig. 1. As also shown by Kouwenhoven et al. [1], the coronary flow has a typical early-systolic flow decrease, with a recognizable early negative peak, which occurs simultaneously with the buildup of left ventricular (LV) pressure. Following this stage, there is a mid-systolic flow increase, which is followed by a late-systolic flow decrease. A rapid rise in the coronary flow to diastolic levels occurs as the ventricular pressure relaxes. The diastolic flow then declines simultaneously with the decline in the aortic pressure. In general, this flow pattern is suggestive of a mechanism by which myocardial contraction, as manifested by the LV cavity

---

[1] Heart System Research Center, The Julius Silver Institute, Department of Biomedical Engineering, Technion-Israel Institute of Technology, Haifa, 32000 Israel

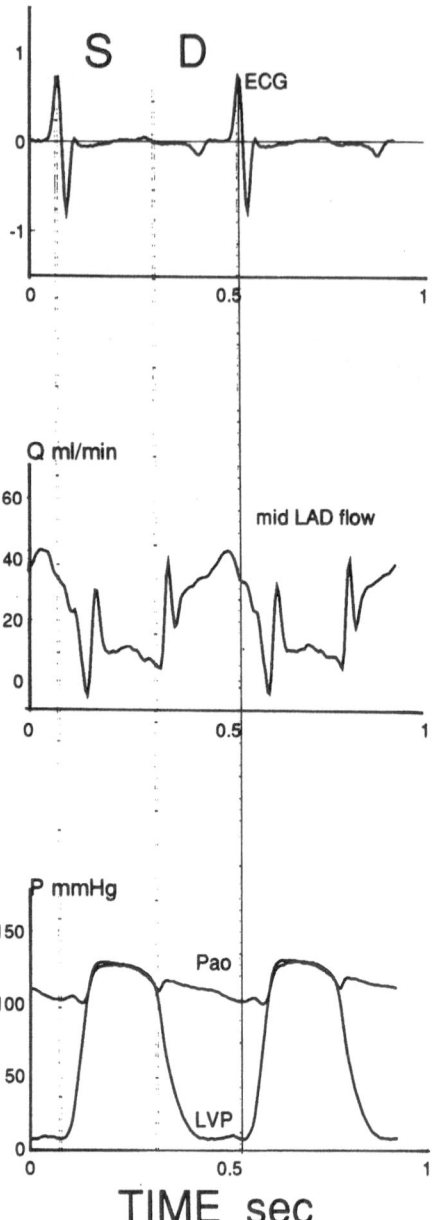

Fig. 1. An example of the phasic flow signal measured in the mid-left anterior descending (*LAD*) segment in relationship to the electrocardiogram (*ECG*), aortic (*Pao*) and left ventricular pressure (*LVP*). Systole (*S*) and diastole (*D*) are delineated by the R wave of the ECG and the closure of the aortic valve

pressure, somehow interferes with the blood flow and impedes the coronary flow during systole.

Coronary Compression: The Intramyocardial Pressure (IMP) and the Extravascular Compressive Pressure (ECP)

Measurement of the IMP

The IMP is commonly defined as the pressure in the interstitial fluid phase of the myocardium. This concept has been used in mathematical models as the parameter which locally compresses the coronary vasculature, and is responsible for the systolic coronary flow impediment [2–6]. Past attempts to characterize the transmural IMP distribution have used different pressure transducers inserted into the myocardium [7–8]. Most have reported that the IMP varies from high values at the endocardium to low values at the epicardium during systole [2–4, 7, 9]. However, there is less agreement on the actual magnitude of these pressures; some studies using pressure transducers report the endocardial pressure to be higher than LV cavity pressures [7], while others report the IMP at the endocardium to be lower, or at most equal, to the LV cavity pressure [9]. The main criticism of the experimental intramural pressure recordings relates to the local myocardial distortion affecting the micro-environment around the transducers, and possibly introducing relatively large artifactual signals.

Diastolic IMP transmural gradients are even more confusing. Measurements with pressure transducers have yielded epicardial pressures which are larger than endocardial pressure during diastole, which is the opposite of the pressure gradient encountered during systole [7]. The reason for that gradient is unclear; it is possibly due to the interstitial fluid transport. This point will be discussed below.

Mathematical Models and Predictions of the IMP

The reported mathematical models concerning IMP yield conflicting results. Models which assume nested shells, with parallel muscle fibers interacting between them by hydrostatic pressure, yield IMPs which decrease from the LV cavity pressure at the endocardium to zero pressure at the epicardium [3, 9, 10]. These models cannot predict endocardial pressures higher than the LV pressures, due to the given endocardial and epicardial boundary conditions [3, 10, 11]. Obviously, these models cannot describe the observation that the epicardial tissue pressure is higher than endocardial tissue pressure during diastole.

The Use of LVP as a Mediator of the Coronary Compression

Consistent with the above models, and in view of the apparent effect of LV pressure (LVP) on the coronary flow during systole, it has been commonly assumed that the coronary compression may be equated, or at least correlated, with the LVP. Thus, many models of the coronary circulation have used the assumption that the coronary compression is related in some way to the LVP wave [4–6].

Dissociation Between Coronary Compression and LVP

The empty beating heart [12] provides most striking evidence that the coronary compression reflected by oscillations in coronary flow can be independent of the LVP; it does not disappear when the LV cavity pressure is eliminated. The fact that the beating heart with zero LV cavity pressure generates considerable coronary compression is, perhaps, the most impressive evidence of the dissociation between LVP and the generation of coronary compression usually attributed to the IMP. Krams et al. [13] have shown that coronary compression is unchanged when the LVP pressure is reduced in the feline heart by preload (volume) manipulation, using coronary flow under constant perfusion pressure as an index of coronary compression. On the other hand, when LVP changes are caused by changes in myocardial contractility, LV pressure and the coronary

a

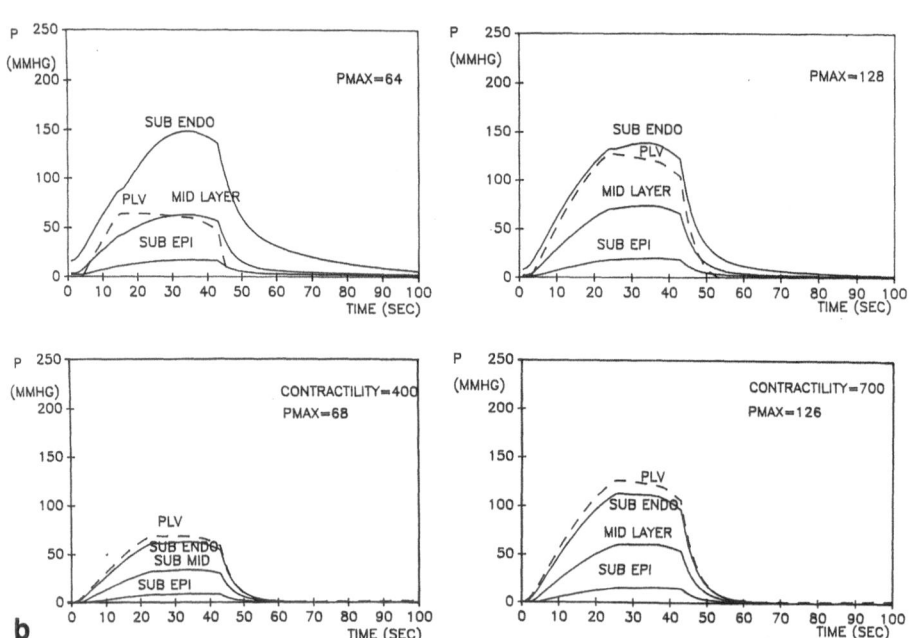

b

Fig. 2. Intramyocardial pressure (*IMP*) distribution (subendocardium [*SUB ENDO*], midwall [MID LAYER] and subepicardium [*SUB EPI*]) predicted by a model that includes radial collagen attachments between muscle fibers. *Top*: Two pressure levels obtained by changing LV volume. *Bottom*: Two LVP levels obtained by changing muscle contractile function

oscillatory flow amplitude are tightly related. These observations have suggested the concept that the coronary compression is determined by the myocardial elastance [14].

Effect of the Collagen Network on the IMP Distribution

The simple straightforward "Laplacian" models, which include muscle fibers embedded in fluid, have been modified in a new generation of models to include the contribution of stiffness elements, typically collagen fibers, in planes other than in the plane of the muscle fibers. These collagen elements interconnect with the muscle fibers in all directions and are responsible for perpendicular stiffness. However, direct mechanical coupling between the cells, as well as other elements, such as vessels, cellular components, etc., may also have a perpendicular component and can contribute to the transverse stiffness. These "stiffness models" predict that the IMP, i.e., the interstitial pressure, exceeds the LV cavity

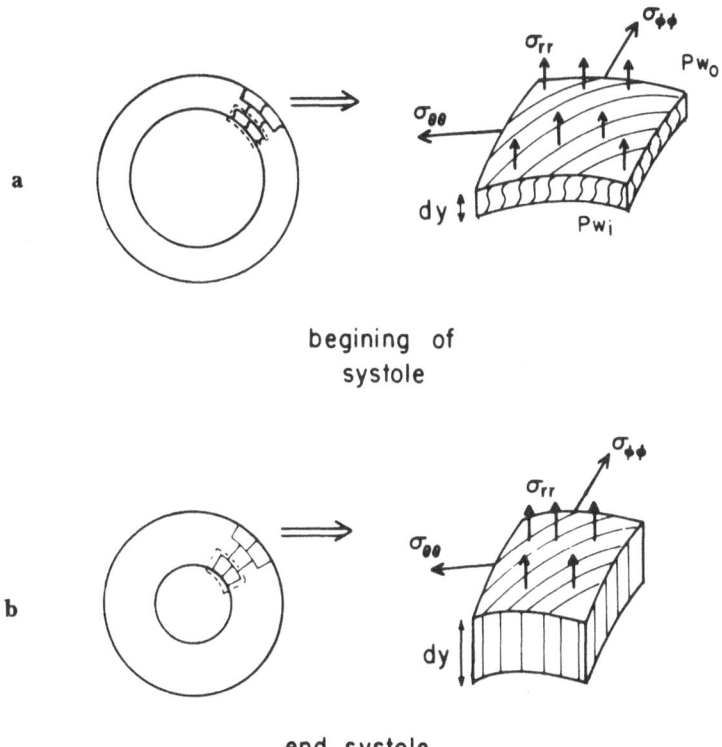

begining of
systole

end systole

Fig. 3a,b. Myocardial contraction **a** from beginning of systole to **b** end-systole generates wall thickening and stretching of the radial collagen fibers in the thin muscle layer (*thickness dy*), associated with the generation of transverse fiber stresses σ_{rr} and modification of the pressure in the interstitial fluid, P_w. The stresses in the circumferential and longitudinal directions are also shown

pressure [15–17]. Utilizing our multilayer model [17], including the transverse stiffness, shows that the endocardial IMPs clearly exceed the LVP. Furthermore, when the LVP is reduced by preload manipulation (i.e., changes in the end-diastolic volume), the IMP exceeds LVP by an increasing amount, so that the peak IMP shows minimal reduction with a reduction in LVP (Fig. 2). When the LVP is reduced by modifying the contractility, the IMP reduces proportionally. Thus, the transverse stiffness elements explain the compression of the empty beating heart and the independence of IMP of the LVP under some conditions.

Figure 3 presents the mechanism associated with the transverse stiffness. As the LV contracts, it thickens, and the radial elements interconnecting the muscle fibers are stretched. The stresses generated in the collagen fibers affect the pressure in the fluid media, increasing it proportionally to the magnitude of the radial fiber stress.

Fluid Transport Modulation of the IMP

While models are becoming more complicated, reality is even more complex. An important factor, which has so far not been considered in the analysis of cardiac mechanics, is the intramyocardial fluid transport which both affects and is affected by the myocardial contraction. Figure 4 presents the myocardial mesh, which is composed of muscle fibers, collagen fibers and blood vessels. The collagen fibers connect between muscle cells, and direct linkage of the collagen to arterioles also exists [18]. The capillaries "leak" into the interstitial space and lymphatic flow is responsible for drainage. In addition, the pressure in the intramyocardial vasculature can, through coronary and myocardial compliance, modify the tissue pressure. Therefore, the goal is an integrated model that accounts for all these interactions.

Preliminary studies by our group have suggested that the calculated transmural distribution of the IMP is greatly affected when the fluid transport is accounted for in an integrated model. This model also predicts that the IMP is higher at the epicardium than at the endocardium during diastole. In addition, the model correctly predicts that the IMP is higher then the LVP at the endocardium and that, as observed by Krams et al. [13], the IMP is relatively independent of the LVP at different preloads.

The Extravascular Compressive Pressure (ECP) versus the IMP

The vessels that are embedded in the myocardium are subjected to an external compressive pressure which is highest in systole. As indicated above, this compressive pressure external to the vessels is actually the hydrostatic pressure in the interstitial space, which is commonly recognized as the IMP. However, it is conceivable that direct contact between muscle fibers, collagen fibers, and blood vessels modifies this IMP. The result is a net "effective" pressure on the vessels, defined here as the ECP. The magnitude of these effects is, as yet, unestablished and the role of these direct connections remains speculative. However, we can

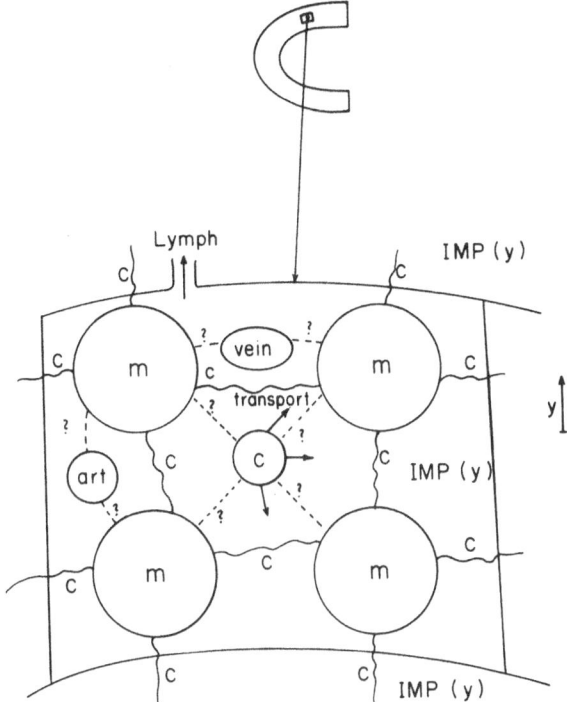

Fig. 4. Important parameters in coronary circulation-LV myocardium modeling. Muscle fibers (m), across the LV wall at location y, are interconnected by collagen, (c). Capillaries (cp), arteries (art), and veins are shown. Possible direct connections by collagen between muscle fibers, capillaries, and other vessels are also shown in *dashed lines*. Transport of fluid from the capillaries to the interstitium is indicated as well

estimate the ECP from the phasic coronary flow signal at constant perfusion pressure conditions. This was done by using an analog model of the coronary circulation. Using this approach, we can calculate the ECP and compare it to the measured LVP and IMP values. Interestingly, the ECP is closely related to the measured IMP signal and weakly related to the LVP.

Mechanisms of Modulating the Coronary Flow by Myocardial Contraction

The Waterfall and Intramyocardial Pump Models

The waterfall concept was derived from experiments in collapsible tubes. According to this concept, the tube partially collapses at the point where the external pressure surrounding the tube equals, or exceeds, the intraluminal pressure. The waterfall concept has been successfully applied to the coronary circulation by Downey and Kirk [6], who predicted microsphere distribution. It has been extensively applied to the coronary circulation thereafter [4, 6, 19–22]. However, it has no predictive value for venous outflow and could not explain retrograde coronary arterial flow. These shortcoming were overcome by the intramyocardial pump concept introduced by Spaan et al. [23].

Pressure-Dependent Resistance and Compliance

The pressure-dependent resistance and compliance model [5, 8, 17, 24] is an attractive one that modifies the original intramyocardial pump concept and takes into account the compression of the intramyocardial compliance. This model assumes that the cross-sectional area of a vessel is a function of the local transmural pressure, i.e., the local intraluminal minus the external pressure. The vessel resistance and compliance are derived from the transmural pressure versus the cross-sectional area relationship of the artery, and are therefore modified dynamically throughout the cycle [25]. Obviously, the pressure-dependent model requires an input function that compresses the intramyocardial vessels, i.e., the IMP. Most of the reported models to date have assumed this pressure to be a priori known and have used it to calculate the flow dynamics. These models represent unidirectional interactions, i.e., the flow dynamics are affected by the IMP input value. However, the inverse effect of the coronary hemodynamics on the IMP has not been accounted for.

Transmyocardial Flow Distribution

The transmural flow distribution in the LV wall is also of great interest [4–6, 26]. Our recent model [25] yields the flow distribution in a three-layered LV myocardium based on the pressure dependence of the resistance and compliance. Figure 5 shows the cross-sectional areas of the larger and smaller ($<150\,\mu$) intramyocardial arteries as well as those in the veins. Note that the largest area reduction occurs in the larger intramyocardial veins, and then in the larger intramyocardial arteries. The smaller vessels are characterized by small area changes. Kajiya et al. [27] have developed an 'endocardial microscope" capable of observing the area changes in small subendocardial vessels. They report that the arteriolar diameter at the subendocardium ($177\,\mu$) changed by 26.9%, while the venular diameter ($107\,\mu$) decreased by 18.8%; this is in good agreement with the prediction of our model for the small subendocardial intramyocardial arteries and veins.

The Importance of the Venous Outflow

The simple waterfall model does not account for the hemodynamics distal to the location of the waterfall. Typically, waterfall models account for systolic flow impediments in the arterial side. It is well recognized that the intramyocardial blood is squeezed during systole antegradely into the venous system [28], as well

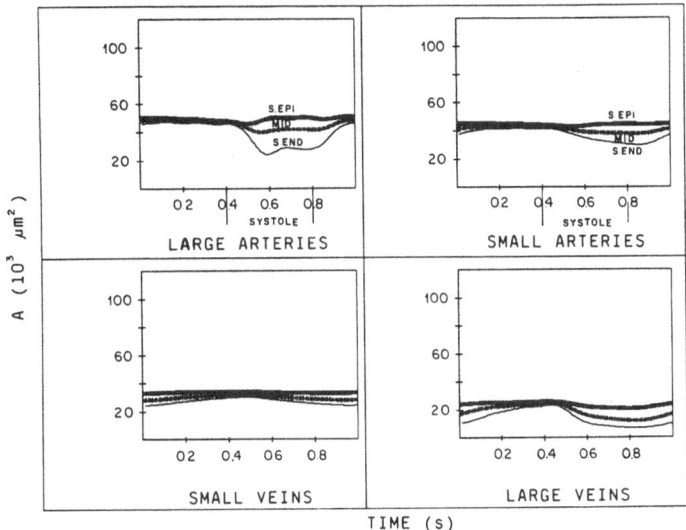

Fig. 5. Pressure-dependent model. The changes in cross-sectional area are given for three layers across the myocardium. *Upper left*— larger intramyocardial arteries (*LA*; $> 150\,\mu$) *upper right*—small arterioles (*SA*; $< 150\,\mu$); *lower left*—small venules (*SV*; $< 150\mu$); *lower right*—large veins (*LV*; $> 150\,\mu$) Reproduced with permission from [25]

as in the retrograde direction [23]. The classic waterfall model cannot explain this flow distribution, and must be linked to a venous compliance to accommodate this phenomenon [19]. In contrast, the systolic venous outflow is clearly modulated by the pressure-dependent resistance and compliance model. As seen in Fig. 5, the systolic venous outflow represents the discharge of the large veins' compliance, accompanied by a significant decrease in their cross-sectional area, with a corresponding large increase in their systolic resistance. Although the venous resistance and compliance have a minor effect on the arterial inflow pattern, the model clearly shows that they are quite important in modifying the dynamics of venous outflow.

Coronary Ischemia

Regional ischemia introduces a number of parameters that must be accounted for:

1. *Regional hemodynamics associated with coronary stenosis.* The simplest approach to represent a stenosis is to introduce an appropriate resistance. However, hemodynamic analysis of flow through stenosed vessels has shown that the flow is a function of a resistive viscous term, which is proportional to the flow rate, and an inertial term which is proportional to the square of the flow rate [29]. This may further be complicated by the possibility that a stenosis can have some compliance and can therefore change its cross-sectional area with a change in the input pressure throughout the cycle.

2. *The development of coronary collateral circulation.* This is an important factor in stenosis modelling. Coronary collaterals are conveniently described [19, 30, 31] as simple resistors. However, there is a possibility that the collaterals, by being intramyocardial in part, may depend on the surrounding external pressures. Irrespective of that possibility, it is clear that collateral resistance is important in modulating the flow in an ischemic region.

3. *Autoregulation.* This parameter needs to be accounted for as long as the stenosis does not reach critical values. Maximum vasodilation is achieved as critical stenosis develops and coronary flow reserve is exhausted.

4. *Myocardial function.* The decreased coronary perfusion pressure may affect the IMP, and possibly the contractile function, even before ischemia is developed. Once ischemia develops, local function is compromised and the local IMP may be modified.

5. *Transmural flow.* This flow is of high interest in modelling ischemia, since ischemia has different effects on the various layers. Subendocardial flow is the most vulnerable. The pattern of transmural coronary flow during ischemia may suggest the mechanism for that vulnerability. The transmural flow distribution during ischemia is given in Fig. 6. Note that arteriolar flow is reversed in systole in the endocardial layers, while flow in the epicardial layers is positive. This suggests that ischemia may affect alternating flow between the layers [17]. Such alternating flows have been suggested by experimentalists and may explain the hidden flow phenomenon in the myocardium [32], i.e., the observation that there is a discrepancy between the flow rate measured by microspheres and that determined by Xenon washout. Clearly, an alternating to-and-fro flow may wash out metabolites, but cannot be detected by microspheres.

The Elastance Model

The elastance model was proposed to explain the magnitude of the coronary compression in terms of global ventricular and myocardial function concepts. It has been suggested that the coronary compression can be estimated by measuring the oscillatory flow amplitude under constant coronary perfusion pressure conditions. A myocardial elastance concept, derived from the ventricular elastance model [33] has been proposed to explain coronary compression [13, 14], as detailed above in the Section on Dissociation Between Coronary Compression and LVP. The observation that the oscillatory flow amplitude is a function of the coronary perfusion pressure [14] has been interpreted as a myocardial elastance effect; the augmented intramyocardial blood volume, caused by increased coronary perfusion pressure, produces large flow oscillations.

While the elastance-based model is attractive for some situations, it fails in others. For instance, local dysfunction by intracoronary Lidocaine completely eliminates the local regional contractility, yet does not hinder the oscillatory flow [34], as shown in Fig. 7. Under such conditions, coronary compression is sensitive to the LVP. These phenomena suggest that both the LVP and the

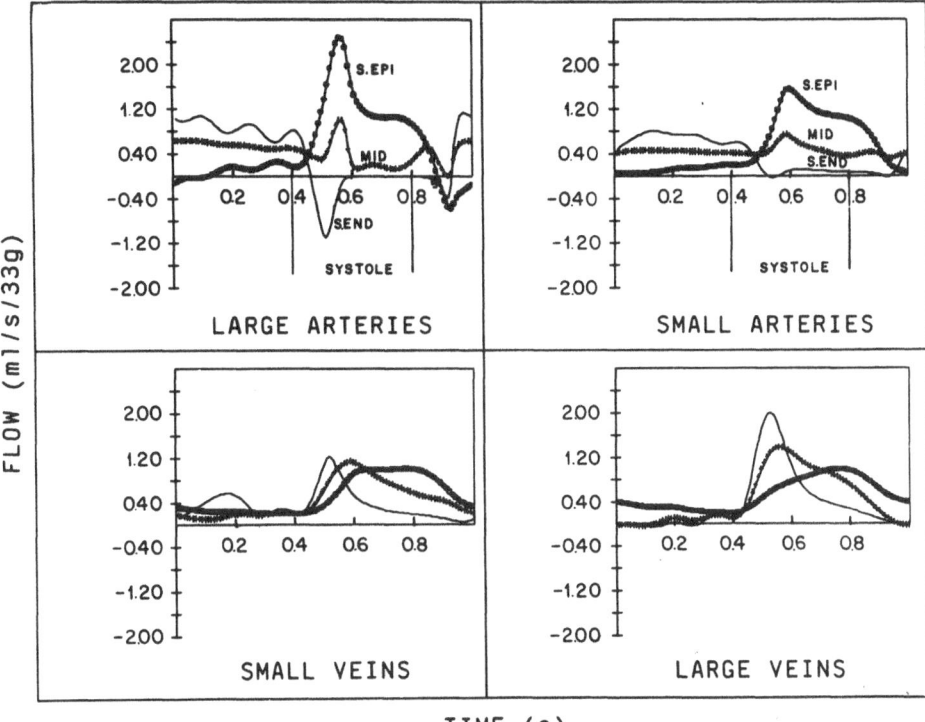

FLOW (ml/s/33g)

Fig. 6. Pressure-dependent model: Transmural flow during ischemia. Similar notation to Fig. 5. Note reverse flow from the subendocardial layers and positive flow at the subepicardial vessels during systole. See text for further explanation. Reproduced with permission from [25]

myocardial elastance play a role in coronary compression [35], and that their relative roles may change in different conditions.

The Integrated Coronary Pressure-Flow Model

This is a new model that explains mechanical and hemodynamic interactions, including those of the intramural interstitial fluid and mass transfer over a wide range of conditions. The constructed model is based on physiological facts and engineering principles, and accounts for the complex interactions of the coronary circulation; it has the following features:

i. The cross-sectional area of the coronary vessels is subjected to changes determined by the transmural pressure [17, 24, 31], and these changes determine the dynamic changes in the resistance and compliance of the coronary circulation. As in our earlier study [25], three layers of the myocardium (epi-, mid-, and endocardium) are considered; each layer consists of four compartments (larger arteries, small arteries, small veins, and larger

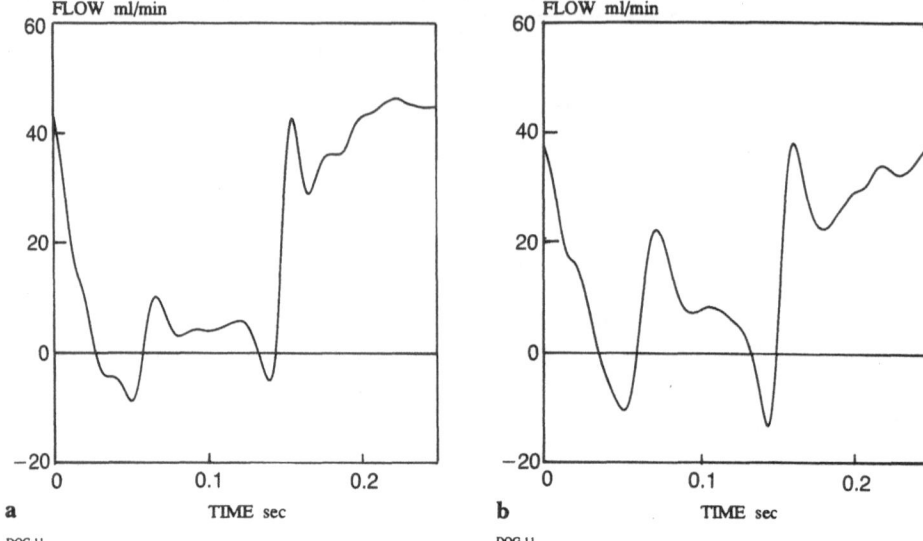

Fig. 7a,b. Oscillatory flow **a** before and **b** during intracoronary Lidocaine administration

veins); each compartment is subjected to dynamic resistance and capacitance behavior. The intramyocardial circulation is coupled to the aortic pressure by the epicardial arterial system, which is also characterized by a resistance and a capacitance.

ii. Ischemic flow is characterized by combining two myocardial regions with a collateral resistance.

iii. The pressure in the interstitial fluid, i.e., the hydrostatic pressure external to the arteries, is used both for calculations of instantaneous vascular transmural pressures and, hence, for the calculations of vessel cross-sectional area and for fluid and mass transport across the capillaries. The pressure in the interstitial fluid is a complex function which is determined by the myocardial contraction, the development of stress in the radial collagen fibers interconnecting the muscle fibers, the ventricular load determined by the ventricular pressure, and the fluid transport across the capillaries, as well as the lymphatic drainage of the heart.

iv. Fluid transport across the capillaries and the lymphatic drainage of the heart are most important in determining the transmural distribution of the diastolic and systolic IMP. The interactions between the intramyocardial coronary compliance, the fluid transport, and the local myocardial distensibility, which is affected by the degree of activation and deformation of the myocardial collagen mesh, have to be taken into account in studying the interactions between cardiac mechanics, coronary flow, and mass transport. Such a model, currently being developed by us, will, we hope, help in the understanding of the complex puzzle of coronary-myocardial-ventricular interactions.

An effect of mass transport on cardiac mechanics is predicted by such a model. When the coronary perfusion pressure is changed over a wide range of pressure, and maximum vasodilatation and sufficient time for equilibration of mass transport are assumed, the IMP is affected by the increase in the perfusion pressure. This effect is mostly contributed by increased water content in the interstitial fluid.

Conclusions

While questions exist and the secrets of the cardiac system are still bewildering, some recent basic studies on the various causes and effects of the cardiac system and the coronary circulation may open the door for better understanding of the interactions between myocardial mechanics and coronary flow dynamics. The remaining controversies and the unresolved questions serve as catalysts in the continuous attempt to decipher one of the most intriguing secrets of life.

Acknowledgment. This study was supported by grants from the Fund for Promotion of Research in the Technion, the Wolf award, the Anna George Ury Fund (Chicago, USA), Mr. Yochai Schneider (Las Vegas, USA), the Michael and Adelaide Kennedy-Leigh Fund (London, UK), and the numerous beautiful ladies of the Women's Division of the American Society for the Technion, NY, USA.

References

1. Kouwenhoven E, Vergrossen I, Han Y, Spaan JAE (1992) Retrograde coronary flow is limited by time-varying elastance. Am J Physiol 263 (Heart and Circ Physiol 32):H484–H490
2. Hoffman JIE, Spaan JAE (1990) Pressure flow relation in the coronary circulation. Physiol Rev 70:331–390
3. Beyar R, Sideman S (1984) A computer study of left ventricular performance based on fiber structure, sarcomere dynamics, and transmural electrical propagation velocity. Circ Res 55:358–375
4. Beyar R, Sideman S (1987) Time-dependent coronary blood flow distribution in the left ventricular wall. Am J Physiol 252 (Heart Circ Physiol 21):H417–H433
5. Bruinsma P, Arts T, Dankelman J, Spaan JAE (1988) Model of the coronary circulation based on pressure dependence of the coronary resistance and compliance. Basic Res Cardiol 83:510–524
6. Downey JM, Kirk ES (1985) Inhibition of coronary blood flow by vascular waterfall mechanism. Circ Res 36:753–760
7. Stein PD, Marzilli M, Sabbah HN, Tennyson L (1980) Systolic and diastolic pressure gradients within the left ventricular wall. Am J Physiol 238 (Heart Circ Physiol 7):H625–H630

8. Kresh JY, Fox M, Brockman SK, Noordergraaf (1990) A Model-based analysis of transmural vessel impedance and myocardial circulation dynamics. Am J Physiol 258:H262–H276

9. Heineman FW, Grayson J (1985) Transmural distribution of intramyocardial pressure measured by micropipette technique. Am J Physiol 249:H1216–H1223

10. Feit TS (1979) Diastolic pressure-volume relations and distribution of pressure and fiber extension across the wall of a model left ventricle. Biophys J 28:143–166

11. Arts T, Reneman TS, Veenstra PC (1979) A model of the mechanics of the left ventricle. Ann Biomed Eng 7:299–318

12. Baird RJ, Goldbach MM, De La Rocha (1970) Intramyocardial pressure: The persistence of its transmural gradients in the empty heart and its relationship to myocardial oxygen consumption. J Thorac Cardiovasc Surg 59:810–823

13. Krams K, Sipkema P, Zegers J, Westerhof N (1989) Contractility is the main determinant of coronary systolic flow impediment. Am J Physiol 257 (Heart Circ Physiol 26):H1936–H1944

14. Krams K, Sipkema P, Westerhof N (1989) Coronary oscillatory flow amplitude is more affected by perfusion pressure than ventricular pressure. Am J Physiol 258 (Heart Circ Physiol 27):H1936–H1944

15. Chadwick RS, Tedgui A, Michel JB, Ohayon J, Levy BI (1988) A theoretical model for myocardial blood flow. In: Brun P, Chadwick RS, Levy BI (eds) Cardiovascular dynamics and models, Proceedings of NIH-INSERM Workshops, vol 183. Paris, INSERM, pp 77–90

16. Nevo E, Lanir Y (1989) Structural finite deformation model of the left ventricle during diastole and systole. Trans Am Soc Mech Eng 111:342–349

17. Beyar R, Kamminker R, Manor D, Ben Ari R, Sideman S (1991) On the mechanism of transmural myocardial compression and perfusion. In: Sideman S, Beyar R, Kléber A (eds) Cardiac electrophysiology, circulation and transport. Kluwer, Boston, pp 245–258

18. Caulfield JB, Borg TK (1979) The collagen network of the heart. J Lab Invest 40:364–372

19. Beyar R, Guerci A, Halperin H, Tsitlik J, Weisfeldt M (1989) Intermittent coronary sinus occlusion following coronary arterial ligation results in venous retroperfusion. Circ Res 65:695–707

20. Bellamy RF (1978) Diastolic coronary pressure flow relationship in the dog. Circ Res 43:92–101

21. Klocke FJ, Mates RE, Canty JM, Ellis AK (1985) Coronary pressure flow relationships: Controversial issues and probable implications. Circ Res 56:310–323

22. Uhlig PN, Baer RW, Vlahakes GJ, Hanley FL, Messina LM, Hoffman JIE (1984) Arterial and venous pressure flow relations in anesthetized dogs. Evidence for a vascular waterfall in epicardial coronary veins. Circ Res 55:238–248

23. Spaan JAE, Breuls NPW, Laird JD (1981) Diastolic-systolic coronary flow differences are caused by intramyocardial pump action in the anesthetized dog. Circ Res 49:584–593

24. Chadwick RS, Tedgui A, Michel JB, Ohayon J, Levy BI (1990) Phasic regional myocardial inflow and outflow: Comparison of theory and experiments. Am J Physiol 258:H1687–H1698

25. Beyar R, Caminker R, Manor D, Sideman S (1993) Coronary flow patterns in normal and ischemic hearts: Transmyocardial and artery to vein distribution. Ann Biomed Eng (in press)

26. Lee J, Chambers D, Akizuki S, Downey J (1984) The role of vascular capacitance in the coronary arteries. Circ Res 55:751–762

27. Kajiya F, Goto M, Yada T, Ogasawara Y, Kimura A, Hiramasatu O, Tsujioka K (1992) How does myocardial contraction affect intramyocardial microcirculation (abstract)2. Heart and Vessels [Suppl 8]:129

28. Armour JA, Klassen GA (1983) Pressures and flows in the epicardial coronary veins of the dog heart: Responses to positive inotropism. Can J Physiol Pharmacol 62:38–48

29. Gould KL (1985) Quantification of coronary artery stenosis in vivo. Circ Res 57: 341–353

30. Scheel KW, Mass H, Williams SE (1989) Collateral influence on pressure-flow characteristics of coronary circulation. Am J Physiol 257 (Heart Circ Physiol 26): H717–H725

31. Manor D, Beyar R, Sideman S (1991) On the pressure-flow relationship of the coronary collaterals: A model study. In: Proceedings of Computers in Cardiology, 23–26 Sept 1991, Venice, Italy. IEEE Computer Society Press, California, pp 713–716

32. Yoshida S, Akizuki S, Gowski D, Downey JM (1985) Discrepancy between microspheres and diffusible tracer estimates of perfusion to ischemic myocardium. Am J Physiol 248 (Heart Circ Physiol 17):H255–H264

33. Suga H, Sagawa K, Shoukas AA (1973) Load independence of the instantaneous pressure-volume ratio of the canine left ventricle and effects of epinephrine and heart rate on the ratio. Circ Res 32:314–322

34. Doucette JW, Goto M, Flynn AE, Husseini WK, Hoffman JIE (1990) Effect of left ventricular pressure and myocardial contraction on coronary flow (abstract). Circulation 82 [Suppl III]:379

35. Kresh JY (1989) Myocardial modulation of coronary circulation (letter). Am J Physiol 26:H1934–H1935

Simple vs Complex Models of the Coronary Circulation—The Tradeoffs

Robert E. Mates [1]

Summary. Mathematical models of biological systems are useful in testing hypotheses, synthesizing a variety of data, and evaluating the effect of interventions on model parameters. A number of models of the coronary circulation have been published in recent years, motivated by increased interest in the dynamics of coronary blood flow. The number of circuit elements ranges from 4–60. This paper discusses the relative advantages of simple and complex models. Simple models permit parameter identification using optimization techniques and thus can be used to measure the effect of interventions on model parameters. They are not able to describe more detailed features of coronary flow such as transmural flow variations, and unique anatomic interpretation of model parameters is not always possible. More complex models provide a more precise anatomic interpretation of parameters and a more detailed description of flow. In these models it is generally necessary to estimate model parameters from a variety of sources, and different combinations of model parameters may provide an equally good fit to the data. The best choice of model depends on the objective of a particular study.

Key words: Modeling—Pressure/flow relations—Input impedance—Parameters

Introduction

One of the major contributions of bioengineers is the use of mathematical models to explain biological phenomena. In fact, many of the basic laws of mechanics were developed in attempts to better understand the function of the human body. Names such as Poiseuille, Hooke, and Boyle are as familiar in physiology as in engineering. Mathematical models of biological systems are employed for a number of purposes, including the testing of a hypothesis, development of a coherent explanation for a variety of experimental findings, prediction of quantities which cannot be directly measured, and parameter identification in an individual or group for diagnostic purposes.

Perhaps the first model of the coronary circulation was a form of Ohm's Law, which is shown schematically in Fig. 1. Ohm's Law states that mean coronary flow Q is proportional to the difference between arterial and venous pressure $(P_{art}-P_{ven})$ divided by the resistance of the circulation (R). In spite of its limitations, Ohm's Law still forms the basis for the explanation of coronary

---

[1] Departments of Mechanical and Aerospace Engineering and Medicine, State University of New York, Buffalo, NY, USA

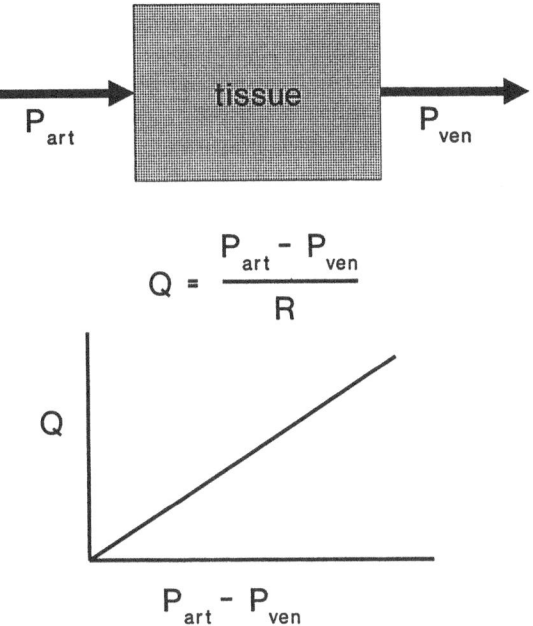

Fig. 1. Classical view of the coronary circulation. Flow (Q) is given as the difference between arterial pressure P_{art} and venous pressure P_{ven} divided by the lumped resistance of the circulation (R)

pressure/flow relationships in most introductory physiology courses. Ohm's Law provides a useful starting point, but it fails to explain many observed phenomena.

While it had long been noted that coronary inflow was impeded during systole, the classic experiments of Bellamy [1] indicated that even diastolic inflow may become zero at arterial pressure well above venous levels. This finding renewed interest in the dynamic behavior of the coronary circulation and has led to much greater insight into coronary physiology.

The initial attempts to explain the Bellamy results centered on the concept of a vascular waterfall, originally applied to systolic flow in the coronary circulation by Downey and Kirk [2]. This hypothesis suggests that a back pressure to coronary flow is produced by intramyocardial forces which cause a partial collapse of the vessels when intravascular pressure falls below the effective intramyocardial pressure. An alternative hypothesis by Spaan et al. [3] was that the apparent zero-flow pressure could be explained on the basis of an intramyocardial compliance which charges during diastole and discharges when the ventricle contracts.

Like most new findings, Bellamy's report was greeted with a great deal of skepticism. Eng et al. [4] were the first to point out that, since the coronary circulation was composed of compliant vessels, phasic capillary flow may persist after inflow has become zero. At the time, there was no method available to measure phasic flow in the microcirculation and hence mathematical models

were the only way to approach this problem. Although rapid progress has been
made in developing means to visualize the microcirculation, there still is no well-
established technique for measuring phasic flow in the intramyocardial circulation. The interest in the dynamics of coronary pressure-flow behavior led to
attempts to explain the phasic nature of coronary inflow and outflow. While
arterial inflow is reduced during systole, venous outflow peaks during ventricular
contraction. A number of mathematical models have been developed in attempts
to explain such behavior [5–13].

Coronary Models

Circulatory models can be classified as either distributed- or lumped-parameter.
The former treat the blood and vessels as continua and describe the mechanics
by partial differential equations. Lumped models treat groups of vessels as a
single entity characterized by a finite number of parameters and generally are

Table 1. Summary of model parameters

Reference	Type[1]	Linearity[2]	Layers	Circuit[3]	Elements[4]
5	i	n	10	a	41
6	i	n	8	e	24
7	i	n	1	f[5]	4
8	io	n	10	b	60
9	i	l	3	c	9
10	i	l	1	d	4
11	i	n	3[6]	c	32
12	i	l	1[7]	e	4
13	io	l	1	f	7

Notes:
1. Type: *i*, inflow; *io*, inflow/outflow
2. *l*, linear (constant elements); *n*, nonlinear (variable elements)
3. Letters refer to Fig. 2. Circuit elements shown are for one layer.
4. Number of circuit elements. Linear models have one parameter per
   circuit element; nonlinear models require more than one parameter per
   element.
5. In [7], the parallel paths containing (C_2, P_2) and (R_2) were employed
   alternately. When inflow pressure P_1 exceeded back pressure P_2, R_2 was
   adjusted to maintain a constant back pressure and no flow passed
   through C_2. When P_2 exceeded P_1, R_2 was made infinite and all flow
   passed through C_2.
6. The model in [11] included 3 layers. Each layer contained multiple
   pathways with differing parameter values.
7. The model in [12] included 10 layers; however, the pressure P_2 was
   constrained to vary linearly with depth in the myocardium and the
   parameters R_1, R_2, and C were the same for all layers. As shown in Fig. 4,
   this reduces the 10-layer model to an equivalent single layer.

described by sets of ordinary differential equations. While the coronary circulation is truly continuous in nature, practical calculations require some simplification of the actual physiology such that even distributed models must be approximated by a finite number of segments. While it would be conceptually possible to develop a model with each vessel treated as a separate element, such a model would require enormous computational time to obtain results. Perhaps more important, it would be necessary to provide geometric and material property data for each vessel which are not presently available. Even if data were available, it is not clear that such a model would provide substantially more information than much simpler models.

There is considerable variability in the complexity of published lumped-parameter models of the coronary circulation. Some models consider outflow as well as inflow while others describe only the relationship between arterial inflow and flow in the microcirculation. Some of the models account for differences in the pressure-flow characteristics in different layers of the myocardium, while others consider only a single layer. All of the dynamic models include capacitive and resistive elements; however, the arrangement of these elements varies. Some of the models account partially for the nonlinear characteristics of the circulation by allowing the values of lumped resistances and compliances to depend on perfusion pressure.

The characteristics of a number of recent models are summarized in Table 1, which includes the type of model (inflow or inflow-outflow), the number of myocardial layers, whether the model is linear (constant parameters) or nonlinear (variable parameters), the circuit employed, and the total number of circuit elements. Linear elements require one parameter per element while nonlinear elements require more than one. Electrical analogs of the various circuit elements are given in Fig. 2. All of the models included in Table 1 provide a good description of phasic coronary inflow, and those which include outflow also provide adequate descriptions of the observed flow patterns. It is immediately clear from this comparison that there is no single model which best describes the pressure-flow characteristics of the coronary circulation. What, then, are the relative merits of the various approaches?

Simple vs Complex Models

Table 2 summarizes the primary advantages and disadvantages of "simple" and "complex" models. Simple models have the obvious advantage of computational simplicity. In addition, the small number of parameters makes it possible to identify parameter values directly from experimental measurements using optimization techniques. This allows the model to be used to examine parameter variations resulting from particular interventions, and potentially to be used for diagnostic purposes by evaluating model parameters in individuals. An example from our laboratory is shown in Fig. 3 [10]. Here, model parameters from an

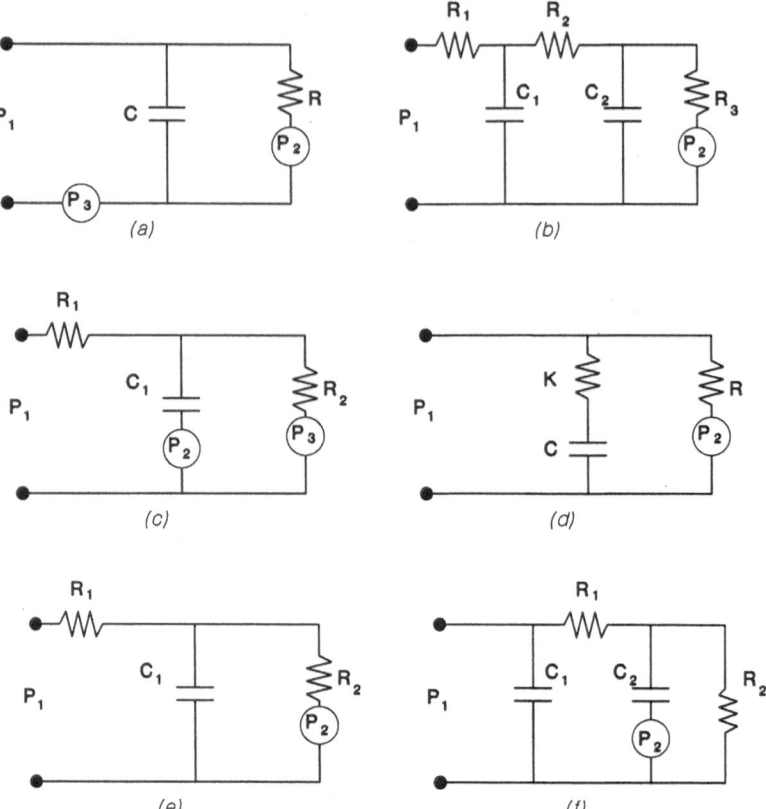

Fig. 2. Circuit analogs used in lumped-parameter models of the coronary circulation. R represents resistance; C, capacitance; P, pressure. The letters a-f refer to the different analogs used in the models listed in Table 1

open-chest paced canine preparation are summarized as a function of heart rate both with vasomotor tone intact and during adenosine vasodilation. Resistance is the parameter R in Fig. 2d. In this model, the pressure P_2 is related to ventricular pressure P_v by the relation $P_2 = .5 P_v + Pd$, where Pd is the value of P_2 when ventricular pressure is zero.

Model parameters were assessed using an optimization algorithm by comparing measured and model-predicted flows in the time domain [10]. Both increases in heart rate and adenosine infusion produced increases in coronary inflow at fixed perfusion pressures. Pacing-induced flow increases with vasomotor tone intact reduced the model resistance with little change in diastolic back-pressure Pd. Adenosine infusion, on the other hand, significantly reduced both resistance and back pressure. These results suggest that there may be differences in the mechanisms responsible for metabolic as opposed to pharmacologic vaso-

Table 2. Tradeoffs

Simple models	Complex models
Advantages	
Parameters can be estimated from a single experiment.	Parameters can be associated with anatomical features of the circulation.
Physiological variations in individual parameters can be evaluated.	Model can predict transmural flow distributions.
Disadvantages	
Parameters have limited physiological significance.	It is difficult to estimate parameter values from a single experimental model.
Model cannot predict transmural flow distribution.	Parameter variations due to physiological interventions are difficult to evaluate.
	Model may not be unique; different parameter sets may give nearly the same predictions.

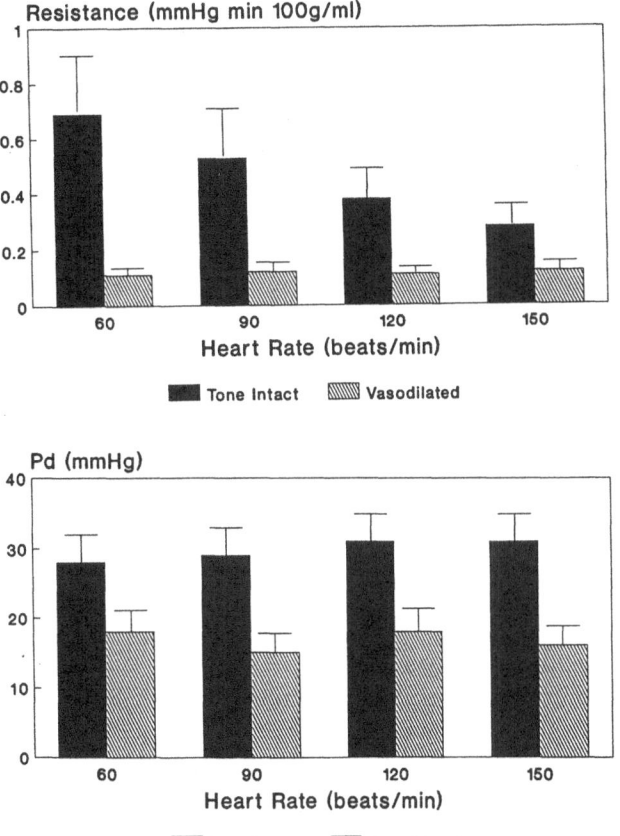

Fig. 3. Resistance and diastolic back pressure (*Pd*) at various heart rates calculated from the model in [10]. The *solid bars* are with tone intact and the *hatched bars* are during adenosine vasodilation. (From [10] with permission)

dilation. It would not have been possible to anticipate these differences without the ability to identify parameters directly from the data.

The primary disadvantage of simpler models is the difficulty of associating model parameters with specific anatomic elements in the circulation. In the model described above, it is not possible to determine unambiguously the location of the primary resistance, the capacitance, or the back pressure. Additionally, simple models are unable to address more detailed questions such as the transmural distribution of flow or differences in phasic flow patterns among different vessels in the microcirculation.

More complex models provide a better anatomic interpretation of model parameters. They are also able to address issues such as transmural flow variations. Their primary disadvantage is the need to assign numerical values to a large number of parameters. Generally these parameters are not all measurable in the same preparation, so it is necessary to choose values from the literature, some of which may have been obtained in isolated vessel preparations and others in the intact circulation. It is therefore not possible to study parameter variations caused by a particular intervention. Complex models are frequently much more sensitive to some parameters than to others, thus it is necessary to perform a detailed sensitivity analysis before drawing conclusions about the appropriateness of the chosen model parameters. Another potential disadvantage is lack of uniqueness in the model description. In our development of the optimization methods mentioned above, we found that in models with more than 4 or 5 parameters, there were local minima in the objective function which produced very similar values for inflow; that is, several sets of parameters gave essentially the same results.

Frequently it is possible to determine equivalence between simpler and more complex models. An example is shown in Fig. 4. By comparing the equations for an n-layer model with those for a single layer model with the same circuit elements, it is easy to shown that if R, C and K are assumed the same for each layer and P_b is assumed to vary linearly across the myocardium, then a single layer model with the indicated parameters is functionally equivalent to the multilayered model. Values for P_b are difficult to determine, and the values obtained depend strongly on the method used. Developers of multilayered models frequently assume a linear distribution of P_b and uniform values for the other parameters. In that situation, the simpler model provides the same results with the added advantages described above.

There is, then, no single model which provides the optimal properties for all purposes. The user of the model must balance the relative advantages of the various possibilities. Most modelers, this author included, are tempted to make their models more complicated in order to simulate more realistically the actual circulation. The danger is that we may believe too strongly in the predictions of a particular model. A useful guide is to include in the model no more parameters than can be estimated with some confidence from available data.

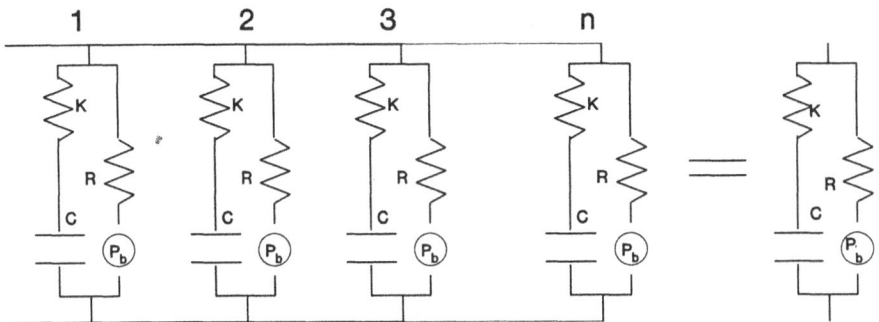

Fig. 4. Equivalence of single and multiple layer models of the coronary circulation R, represents resistance; C, capacitance; K, the viscoelastic constant; Pb, back pressure. If R, C and K are uniform, P_b varies linearly with layer, equivalent single layer values are: $R = R/n$, $K = K/n$, $C = C_n$, and P_b = average of P_b

Acknowledgement. This work was supported by National Heart, Lung and Blood Institute Program Project Grant 2-PO1-HLB-15194

References

1. Bellamy RF (1978) Diastolic coronary artery pressure-flow relations in the dog. Circ Res 43:92–101
2. Downey JM, Kirk ES (1975) Inhibition of coronary flow by a waterfall mechanism. Circ Res 36:753–760
3. Spaan JAE (1981) Coronary diastolic pressure-flow relation and zero flow pressure explained on the basis of intramyocardial compliance. Circ Res 56:293–309
4. Eng C, Jentzer JH, Kirk ES (1982) The effects of coronary capacitance on the interpretation of diastolic pressure-flow relationships. Circ Res 50:334–341
5. Beyar R, Sideman S (1987) Time-dependent coronary blood flow distribution in left ventricular wall. Am J Physiol 252 (Heart Circ Physiol 21):H417–H433
6. Bruinsma T, Arts T, Dankelman J, Spaan JAE (1988) Model of the coronary circulation based on pressure dependence of coronary resistance and compliance. Basic Res Cardiol 83:510–524
7. Burattini R; Sipkema P, van Huis GA, Westerhof N (1985) Identification of canine intramyocardial compliance on the basis of the waterfall model. Ann Biomed Eng 13:385–404
8. Chadwick RS, Tedgui A, Michel JB, Ohayon J, Levy BI (1990) Phasic regional myocardial inflow and outflow: comparison of theory and experiments. Am J Physiol 258 (Heart Circ Physiol 27):H1687–H1698
9. Holenstein R, Nerem RM (1990) Parametric analysis of flow in the intramyocardial circulation. Ann Biomed Eng 18:347–365
10. Judd RM, Mates RE (1991) Coronary input impedance is constant during systole and diastole. Am J Physiol 260 (Heart Circ Physiol 29):H1841–H1851

11. Kresh JY, Fox M, Brockman SK, Noordergraaf A (1990) Model-based analysis of transmural vessel impedance and myocardial circulation dynamics. Am J Physiol 258 (Heart Circ Physiol 27):H262–H276

12. Lee J, Chambers DE, Akizuki S, Downey JM (1984) The role of vascular capacitance in the coronary arteries. Circ Res 55:751–762

13. Spaan JAE, Breuls NPW, Laird JD (1981) Diastolic-systolic coronary flow differences are caused by intramyocardial pump action in the anesthetized dog. Circ Res 49:584–593

D. Regulatory Mechanism of Coronary Circulation and Its Clinical Relevance

Pulsatile Flow, Arterial Chaos, and EDRF Activity

T.M. Griffith[1], D.H. Edwards[2], and I.R. Hutcheson[1]

Summary. We have investigated (i) the role of endothelium-derived relaxing factor (EDRF) in the transduction of mechanical signals from pulsatile flow by the endothelium to the media of the arterial wall in conduit vessels and (ii) the modulatory effect of EDRF on intrinsic oscillations in vasomotor tone in resistance vessels. Cascade bioassay techniques were used to show that EDRF release from conduit arteries is maximal at pulse frequencies in the range of 4–6 Hz, and is inversely proportional to pulse pressure amplitude. Nonlinear techniques were used to characterize irregular histamine-induced pressure oscillations in an isolated rabbit ear resistance artery and to provide evidence that they were generated by deterministic rather than stochastic mechanisms, and are therefore "chaotic". The fractal dimension of the oscillations was generally less than three, thus implying that three (or more) independent control variables were necessary to account for the complexity of their dynamics. EDRF suppressed the oscillations and exerted marked effects on their frequency and amplitude. Their fractal dimension was, however, not altered by graded stimulation or inhibition of EDRF activity. This implies that EDRF is not one of the primary control variables involved in the genesis of vasomotion and that its role is purely modulatory in this preparation.

Key words: Vasomotion—Nonlinear dynamics—Fractal Dimension—Hemoglobin—N^G-nitro-L-arginine methyl ester

Introduction

It is now well-established that both conduit and resistance arteries respond actively to changes in local blood flow through adaptive changes in diameter mediated by release of the endogeneous nitrovasodilator endothelium-derived relaxing factor (EDRF) from the vascular endothelium in response to fluid shear stress [1–5]. EDRF has now been detected extracellularly as the nitric oxide (NO) radical and is synthesized from a terminal guanidino nitrogen atom of the amino acid L-arginine [6, 7].

As the heartbeat is oscillatory in nature, blood flow is intrinsically pulsatile. While the distensibility of the aorta and its major distributive conduit arteries help to convert this to essentially continuous tissue perfusion at the microcircu-

Departments of [1] Diagnostic Radiology and [2] Cardiology, University of Wales College of Medicine, Heath Park, Cardiff, CF4 4XN, UK

latory level there is, nevertheless, marked spatial and temporal heterogeneity of flow [8]. This arises locally as a consequence of both the fractal nature of arterial branching and the spontaneous changes in the diameter of small arteries and arterioles ("vasomotion"). In vivo, such oscillations are generally irregular and can encompass a wide range of frequencies in different arterial generations within the same vascular bed [9–14]. The present report addresses two aspects of the role of EDRF in such dynamic vascular phenomena: (i) the characteristics of pulsatile flow which are important in stimulating EDRF production and, conversely, (ii) the influence of EDRF on oscillations in tone intrinsic to vascular smooth muscle.

To address the former question, we have employed a cascade bioassay system in which the frequency and amplitude of the pressure pulse of the perfusate through an isolated segment of conduit artery could be independently controlled and varied, and the EDRF activity of effluent from the donor continuously monitored [5]. The role of EDRF activity in vasomotion was investigated in an isolated, buffer-perfused, resistance artery from the rabbit ear, with irregular pressure oscillations being induced by histamine. The complexity of this rhythmic behaviour was quantified by a nonlinear analysis able to distinguish between deterministic "chaos" and stochastic "noise" [15]. Such techniques have already provided evidence that fluctuations in vascular caliber and red cell velocities at the microcirculatory level in vivo may be deterministic in origin [16].

Methods

Conduit Arteries

Three-cm-long segments of rat thoracic aorta, rabbit abdominal aorta, or pig coronary artery were perfused with oxygenated Holman's solution at 37 °C at a constant flow rate of $9 \, ml \, min^{-1}$. An air-filled compliance chamber of variable volume was connected to the perfusion circuit via a side-arm to control the level of damping. A ring of endothelium-denuded rabbit aorta, preconstricted by superfusion with phenylephrine ($0.3 \, \mu M$), served as the detector tissue to monitor EDRF release. In all experiments studying the effects of increases in frequency, maximum damping of the perfusion circuit was employed, giving a low constant pulse pressure amplitude of 2 mmHg. The frequency of the pressure pulse in the donor vessel was varied over the range 0.15–12 Hz. In experiments studying the role of the amplitude of the pressure pulse, the compliance was varied to produce excursions of 4, 8, and 16 mmHg at a constant perfusion frequency of 0.1 Hz. Higher amplitudes could not be achieved because of the intrinsic compliance of the system. To confirm the specificity of the responses, experiments were repeated after 45-min perfusion of the EDRF donor with N^G-nitro-L-arginine methyl ester (L-NAME, $100 \, \mu M$), a substituted L-arginine analogue which inhibits NO synthesis [7]. The effects of flow on prostanoid release were excluded by addition of $10 \, \mu M$ indomethacin to the perfusate.

Vasomotion

First branch generation arteries (around 1–1.5 cm long, 150-μm diameter) arising from the rabbit central ear artery were perfused in situ with oxygenated Holman's buffer at 37 °C at a flow rate of 0.5 ml/min, with pressure monitored continuously via a side-arm. Histamine, which does not stimulate EDRF release in the rabbit ear, was used to induce pressure oscillations [17]. In some experiments, the vessels were denuded of endothelium by intermittent air perfusion.

In nonlinear systems governed by three or more control variables, trajectories in "phase space" are at least three-dimensional and can therefore pass over and under each other without intersecting, leading to irregular motion now classified as "chaotic". Such motion is said to be fractal, as its form (and thus complexity) remains the same on both small and large measurement scales, and is therefore "scale-independent" or "self-similar". The dynamics can then be described by a single parameter called the fractal (= fractional) dimension, which provides an estimate of the minimum number of independent contributing variables [8, 15]. In the present study the experimental data, perfusion pressure as a function of time, p(t), were digitized as a series of temporally equidistant points to

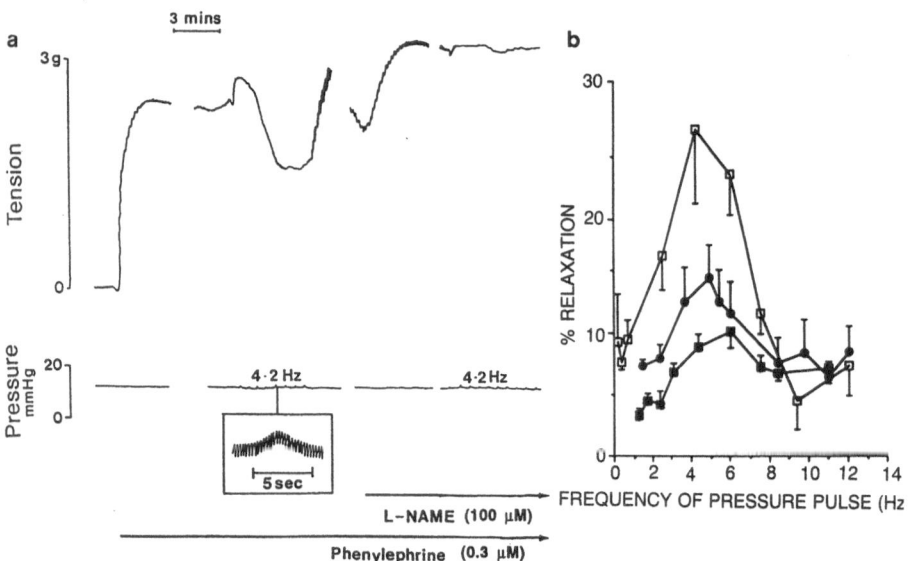

Fig. 1. a Increasing the pulse frequency of the flow through the donor vessel (in this instance rat aorta) from 0,1 to 4.2 Hz caused reversible relaxation of the recipient tissue. This phenomenon was abolished by 100 μM N^G-nitro-L-arginine methylester (*L-NAME*) confirming the involvement of endothelium-derived relaxing factor (*EDRF*). **b** Relaxation of the recipient tissue (expressed as a percentage of initial phenylephrine-induced tone) was highly sensitive to pulse frequency, EDRF release being maximal over the range 4–6 Hz in three different artery types: pig coronary artery (*closed squares*); rat thoracic aorta (*open squares*); and rabbit abdominal aorta (*closed circles*)

generate the vector set $P_i = [p_i(t), p_i(t + \tau), \ldots p_i(t + (m - 1)\tau)]$, where τ is a suitably chosen value of time delay and m is integral and known as the embedding dimension (D_E). This vector was then used to calculate the correlation integral

$$C(l) = \lim_{N \to \infty} \frac{I}{N^2} \times \begin{Bmatrix} \text{number of pairs (i,j) whose distance} \\ \text{from each other, } |Pj - Pi|, \text{ is less than } I \end{Bmatrix}$$

where I is an arbitrarily chosen distance in the n-dimensional phase space generated by the analysis. Grassberger and Procaccia [15] argue that

$$C(I) \propto I^\gamma$$

where γ is a lower bound on the fractal dimension of the time series. The analysis involves calculating the slope of log $C(I)$ as a function of log I for increasing values of the embedding dimension D_E. The slope of this plot theoretically plateaus at a value which gives the so-called correlation dimension (D_c), provided that D_E is sufficiently large and the time series under evaluation is fractal. D_c is a measure of the fractal dimension and thus allows the minimum number of control variables necessary to generate the pressure responses to be estimated. In the case of stochastic "white" noise, no plateau is found.

Results

Frequency Responses

Increasing the pulse frequency of the flow through the donor arteries induced reversible relaxation of the recipient tissue (Fig. 1). This effect peaked at similar frequencies in pig coronary artery and rat and rabbit aorta, and was lost when the donor was pre-incubated with L-NAME or denuded of endothelium, consistent with the involvement of EDRF.

Pressure Responses

Increasing the amplitude of the pressure pulse of the flow through segments of rabbit aorta from the control value of 2 mmHg caused reversible enhancement of recipient vessel tone, reaching a plateau at high amplitudes (Fig. 2). This amplitude-induced constriction was abolished by L-NAME and by endothelial denudation.

Vasomotion

In approximately 70% of all rabbit ear resistance arteries studied, histamine $(1-10\,\mu M)$ induced oscillations in perfusion pressure with an amplitude generally in the range 5–30 mmHg. These were usually irregular, but in some cases were nearly-periodic (Fig. 3a, i). Some traces exhibited characteristic nonlinear features such as period 2 dynamics, the mode of transition from regular to irregular behavior then providing evidence for the existence of a period-doubling

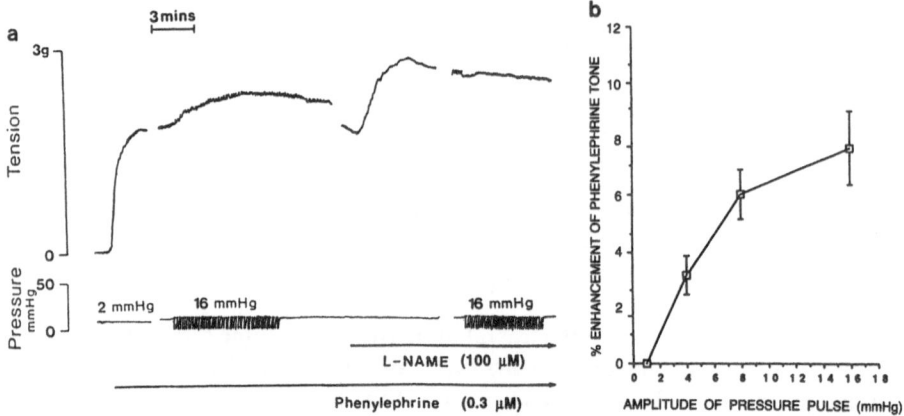

Fig. 2. a Increasing the amplitude of the pressure pulse of the flow through a donor rat aorta from a control value of 2 mmHg to 16 mmHg augmented phenylephrine-induced tone in the recipient tissue. This phenomenon was again abolished by 100 µM L-NAME, confirming the involvement of EDRF. **b** Plot of the percent increase in phenylphrine-induced tone as a function of the amplitude of the pressure pulse, with rabbit aorta as the recipient and rat thoracic aorta (*open squares*) as the donor tissue

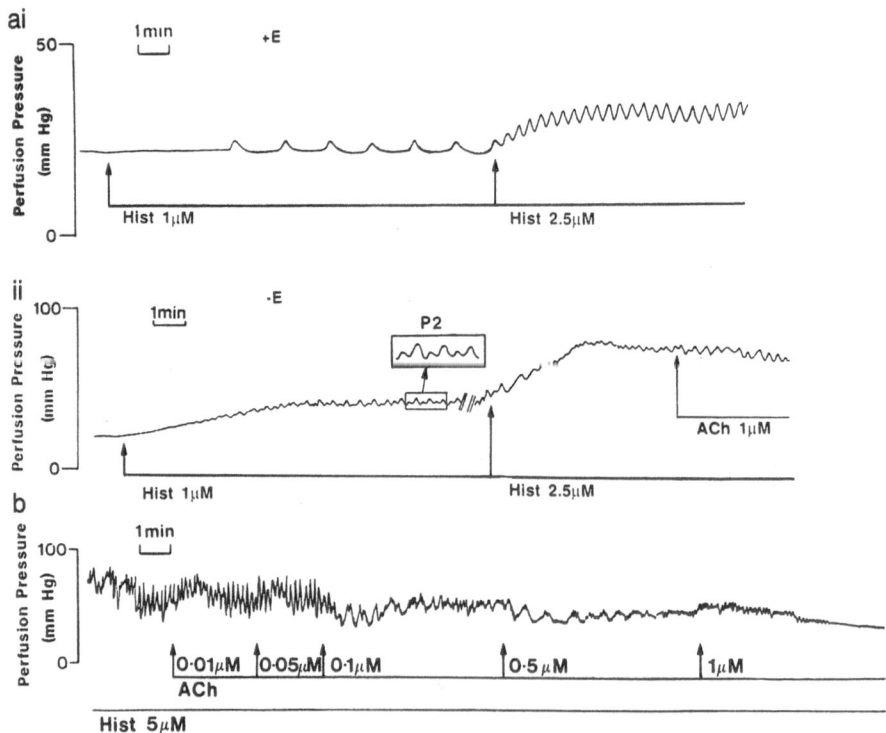

route to chaos (Fig. 3a, ii). Pressure oscillations still occurred after endothelial denudation (Fig. 3a, ii), and in the presence of 50 µM L-NAME or indomethacin (not shown), so that their existence did not have an obligatory requirement for intact endothelium and was EDRF- and prostanoid-independent.

Stimulation of EDRF activity by acetylcholine in endothelium-intact arteries caused a concentration-dependent fall in mean perfusion pressure and the amplitude of the pressure oscillations, to the extent of abolishing them completely at a concentration of 1 µM in almost all instances (Fig. 3b). The inhibitory effects of acetylcholine could be reversed by hemoglobin, which scavenges EDRF activity in these rabbit ear vessels [17], and by 50 µM L-NAME (not shown).

Neither inhibition of basal EDRF activity by 50 µM L-NAME nor stimulation of EDRF activity by 0.01 or 0.10 µM acetylcholine affected the fractal dimension D_c of the pressure oscillations ($n = 7$ in each case, Fig. 4). Furthermore, D_c did not alter when complete inhibition of the effects of histamine by 1 µM acetylcholine was progressively reversed by 0.01, 0.1, and 1 µM hemoglobin ($n = 5$, Fig. 4). Thus, under all experimental conditions, the mean value of D_c remained between 2 and 3. While EDRF activity therefore exerts a major influence on the absolute magnitude of the irregular responses to histamine, it does not modify their intrinsic complexity, as they remained "self-similar".

Discussion

We have examined the role of EDRF activity in different aspects of dynamic vascular physiology. A cascade bioassay was used to dissociate the effects of the frequency and amplitude of pulsatile flow on EDRF release from conduit arteries. The findings confirm that pulsatile flow is a more powerful stimulus for EDRF release than steady flow [3, 4], but indicate that the relationship between EDRF release and the rate of change of shear stress is complex. Thus, EDRF release exhibited high sensitivity to the pulse frequency of perfusion in experiments in which the amplitude of the pressure pulse was kept low. Interestingly, EDRF release was found to be maximal over the same range (4–6 Hz) in arteries from three different species.

Fig. 3. a Effects of endothelial denudation on histamine-induced pressure oscillations. (*i*) With intact endothelium (+*E*) the frequency of nearly-periodic behavior increased by a factor of 4 when the concentration of histamine (*Hist*) was increased from 1 to 2.5 µM. (*ii*) After air perfusion to remove endothelium (−*E*), 1 µM histamine induced stable (time gap, 10-min) oscillations of alternating small and large amplitude (period 2 [*P2*], *inset*), and 2.5 µM histamine induced irregular dynamics. These observations are consistent with a period-doubling route to chaos. Functional loss of endothelium was confirmed by failure to dilate in response to 1 µM acetylcholine (*ACh*). **b** Effects of stimulated EDRF activity on histamine-induced pressure oscillations. These were suppressed in a concentration-dependent fashion by acetylcholine and completely abolished by 1 µM

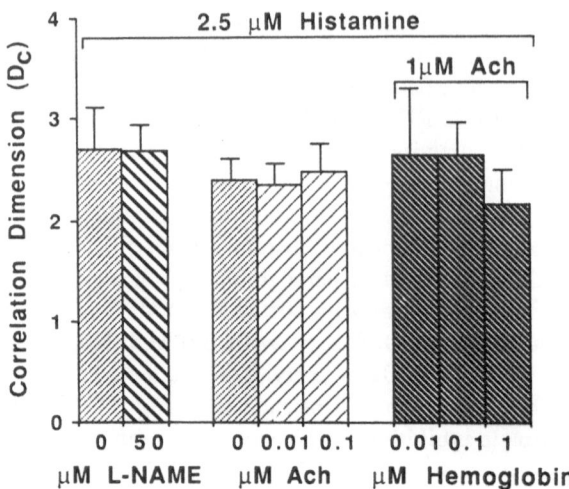

Fig. 4. Histograms demonstrating that EDRF did not significantly affect the fractal dimension (D_c) of the pressure oscillations induced by 2.5 μM histamine. The dynamics thus remained chaotic ($2 < D_c < 3$) and "self-similar" whether the interventions increased (L-NAME, hemoglobin) or decreased (acetylcholine, *ACh*) mean perfusion pressure

Increases in the amplitude of the pressure pulse through isolated donor segments augmented constrictor tone in the detector tissue. This was shown to be a consequence of depressed EDRF synthesis rather than being due to the release of a contracting factor, as the effect was lost after inhibition of EDRF activity by L-NAME. The findings are analogous to those of Rubanyi [18], who reported that an abrupt, sustained elevation of intraluminal pressure (by 30 mmHg) depressed basal, flow-induced, and acetylcholine-stimulated EDRF release from the canine carotid artery. Consistently, it has been reported that NO release from cultured endothelial cells is constant between 80–120 mmHg, increasing at lower and decreasing at higher pressures [19]. We found, however, that increases in pulse pressure amplitude also decreased mean perfusion pressure, implying a concomitant increase in the diameter of the donor vessel, as mean flow was maintained constant experimentally. This effect was endothelium-independent and therefore seems likely to be a consequence of passive distension following an increase in pulse pressure. In our constant-flow system, such a phenomenon would necessarily lead to a fall in the shear stress experienced by the donor endothelium, and thus an associated fall in flow-induced EDRF release. Two distinct mechanisms may therefore contribute to pulse pressure amplitude-induced depression of EDRF release.

The physiological importance of these findings remains to be evaluated. In a damped hydrodynamic circuit, such as the cardiovascular system, the amplitude of periodically forced pressure oscillations will decline as the frequency of the heartbeat increases [20]. As frequency-induced EDRF release is maximal at frequencies above the resting heart rate in the three species studied, its release could therefore be further enhanced by the effects of reduced pulse pressure amplitude under physiological circumstances when there is tachycardia, such as during exercise.

We also investigated the influence of EDRF activity on irregular pressure oscillations induced by an exogenous constrictor agonist, histamine, in isolated rabbit ear resistance arteries, the frequency of the oscillations nevertheless being within the range (0.01–0.20 Hz) reported to occur spontaneously in vivo [9]. The traces illustrated the phenomenon of period-doubling, which is one of the recognized routes for the transition from regular to chaotic dynamics, and which occurs when the steady-state response of a nonlinear system becomes unstable and successively splits into 2, 4, 8 ... 2^n ($n \rightarrow \infty$) possible oscillatory states prior to the onset of irregular, chaotic behavior [21]. The results of a nonlinear analysis were consistent with the hypothesis that the oscillations were generated deterministically, rather than by "random" events, the fractal dimension of the responses generally being >2 but <3. This parameter provides a lower bound on the number of independent control variables involved in generating a chaotic time series, therefore indicating the involvement of at least three control variables. This relatively low value suggests that the overall chaotic behavior of an isolated artery may, in fact, reflect molecular events within vascular smooth muscle at the cellular level.

The fractal dimension (and thus intrinsic complexity) of the pressure oscillations was unaffected by pharmacological stimulation or inhibition of EDRF activity, implying that this is not one of the primary control variables involved in the genesis of the vasomotion. Nitrovasodilators such as EDRF ultimately relax vascular smooth muscle by decreasing $[Ca^{2+}]_i$, through mechanisms which enhance Ca^{2+} extrusion [22], inhibit extracellular Ca^{2+} influx [23], increase intracellular sequestration of Ca^{2+} within sarcoplasmic reticulum [24], and induce membrane hyperpolarization by activating Ca^{2+}-sensitive outward K^+ channels [25, 26]. These events follow activation of the enzyme soluble guanylyl cyclase, with the consequent elevation of cyclic guanosine monophosphate (cyclic GMP) levels and the phosphorylation of a variety of target proteins by cyclic GMP-dependent protein kinase [27]. In view of these multiple sites of action of EDRF, it is perhaps surprising that it can suppress chaotic oscillations in tone without affecting their fractal dimension.

The characteristic sensitivity of nonlinear systems to pertubation and the resulting unpredictability of their responses may explain why interventions such as manipulation of ion fluxes, mechanical stretch, flow rate, and pharmacologically-induced tone, can induce a wide variety of apparently contradictory effects in different artery types [9–13, 28–30]. Small changes in the coefficients (rather than variables) which occur in the continuous-time differential equations which describe the behavior of nonlinear systems can, in fact, convert irregular oscillations to steady-state responses. This could explain observations that rhythmic fluctuations in tone can be induced by EDRF or "exogeneous" nitric oxide in the hamster aorta [28] and canine renal vessels [31], but in other preparations these fluctuations are EDRF- and nitric oxide-independent [30, 32, 33]. Nonlinear analysis may thus provide a powerful method for exploring interactions between the mechanisms which regulate vasomotor tone. "Conventional" approaches such as pharmacological and/or biochemical "isolation" and characterization of the specific components involved may, in comparison, provide only limited insight into their overall functional integration.

Chaos can be generated by periodic "forcing" of nonlinear oscillatory systems, the nature of the resulting dynamics then being dependent on the strength of the coupling [34]. Indeed, it is theoretically conceivable that the arterial wall is "forced" in this way by the pulsatile nature of blood flow itself. In the context of an interconnected microvascular network, forcing could also arise through mechanical coupling between arteries in successive arterial generations, as vasomotion can propagate from "pacemaker" sites at vascular bifurcations into both parent and daughter vessels in vivo [11, 13]. Wave-like contractions could also introduce time delays into nonlinear feedback loops, a mechanism which can per se predispose to high-dimensional chaotic dynamics [8]. Additional factors could therefore contribute to vasomotion in vivo. Nevertheless, nonlinear analysis of variations in microvascular diameter and red cell velocities in rabbit tenuissimus muscle suggest that the fractal dimension of vasomotion/flowmotion in vivo may lie between 3 and 4 and is therefore only slightly higher than that we report in isolated vessels [16].

The physiological "benefits" of vasomotion are unclear. Indeed, it may simply be an inevitable consequence of nonlinearity in the mechanisms controlling smooth muscle tone. Suggested roles [13, 14] include: (i) regulation of vascular resistance in such a way that all tissue elements receive at least intermittent blood flow, thus eliminating regions of permanent anoxia, (ii) minimization of fluid filtration into the extravascular space by reducing distal hydrostatic pressure during periods of low flow, and (iii) promotion of lymphatic drainage through "pumping", as in some tissues lymphatics lie in immediate proximity to arterioles and the presence of lymphatic valves will prevent retrograde flow. Chaos may be central to the ability of biological systems to "learn" and may promote adaptability by permitting a system to escape from an established pattern of behavior which has become disadvantageous [21]. Conversely, the high sensitivity of chaotic trajectories to initial conditions may facilitate dissipation of perturbations, thereby ensuring stability [35]. It remains to be determined to what extent oscillatory vasomotor phenomena in the time domain contribute to the profound spatial heterogeneity of blood flow found in certain vascular beds in vivo [8].

Acknowledgments. The work was supported by the British Heart Foundation. The authors thank M. Stanton and D. Harvey for help with software development, and Mrs R. Maylin for secretarial assistance.

References

1. Holtz J, Forstermann U, Pohl U, Giesler M, Bassenge E (1984) Flow-dependent, endothelium-mediated dilation of epicardial coronary arteries in conscious dogs: Effects of cyclooxygenase inhibition. J Cardiovasc Pharmacol 6:1161–1169
2. Melkymyants AM, Balashov SA, Veselova ES, Khayutin VM (1987) Continuous control of the lumen of feline conduit arteries by blood flow rate. Cardiovasc Res 21:863–870
3. Pohl U, Bussee R, Kuon E, Bassenge E (1986) Pulsatile perfusion stimulates the release of endothelial autocoids. J Appl Cardiol 1:215–235
4. Rubanyi GM, Romero JC, Vanhoutte PM (1986) Flow-induced release of endothelium-derived relaxing factor. Am J Physiol 250:H1145–H1149
5. Hutcheson IR, Griffith TM (1991) Release of endothelium-derived relaxing factor is modulated both by frequency and amplitude of pulsatile flow. Am J Physiol 261:H257–H262
6. Malinski T, Taha Z (1992) Nitric oxide release from a single cell measured in situ by a porphyrinic-based microsensor. Nature 358:676–678
7. Moncada S, Palmer RMJ, Higgs EA (1991) Nitric oxide: Physiology, pathophysiology, and pharmacology. Pharmacol Rev 43:109–142
8. Glenny RW, Robertson HT, Yashashiro S, Bassingthwaighte JB (1991) Applications of fractal analysis to physiology. J Appl Physiol 70:2351–2367
9. Clark ER, Clark EL (1943) Caliber changes in minute blood vessels observed in the living mammal. Am J Anat 73:215–250
10. Johansson B, Bohr DF (1966) Rhythmic activity in smooth muscle from small subcutaneous arteries. Am J Physiol 210:801–806
11. Colantuoni S, Bertuglia S, Intaglietta M (1984) Quantitation of rhythmic diameter changes in arterial microcirculation. Am J Physiol H508–H517
12. Slaaf DW, Tangelder GJ, Teirlinck HC, Reneman RS (1987) Arteriolar vasomotion and arterial pressure reduction in rabbit tenuissimus muscle. Microvasc Res 33:71–80
13. Meyer J-U, Bergstrom P, Intaglietta M (1989) Is vasomotion due to microvascular pacemaker cells? In: Intaglietta M (ed) Vasomotion and flow modulation in the microcirculation. Progress in Applied Microcirculation, vol 15. Karger, Basel, pp 41–48
14. Secomb TW, Intaglietta M, Gross JF (1989) Effects of vasomotion on microcirculatory mass transport. In: Intaglietta M (ed) Vasomotion and flow modulation in the microcirculation. Progress in Applied Microcirculation, vol 15. Karger, Basel, pp 49–61
15. Grassberger P, Procaccia I (1983) Measuring the strangeness of strange attractors. Physica 9D:189–208
16. Yamashiro SM, Slaaf DW, Reneman RS, Tangelder GJ, Bassingthwaighte JB (1990) Fractal analysis of vasomotion. Ann NY Acad Sci 591:410–416
17. Griffith TM, Edwards DH, Davies RL, Harrison TJ, Evans KT (1988) Endothelium-derived relaxing factor (EDRF) and resistance vessels in an intact vascular bed: A microangiographic study of the rabbit isolated ear. Br J Pharmacol 93:654–662
18. Rubanyi GM (1988) Endothelium-dependent pressure-induced contraction of isolated canine carotid arteries. Am J Physiol 255:H783–788
19. Hishikawa A, Nakaki T, Suzuki H, Saruta T, Kato R (1992) Pure transmural pressure inhibits nitric oxide release from cultured human endothelial cells. J Vasc Res 29:36

20. Nichols WM, O'Rourke MF (1990) Measuring Principles of Arterial Waves. In: Nichols WM, O'Rourke MF (eds) McDonald's blood flow in arteries, 3rd edn. Edward Arnold, London, pp 143–195
21. West BJ (1990) Fractal physiology and chaos in medicine. World Scientific, Singapore, pp 124–140, 239–241
22. Popescu LM, Panoiu C, Hinescu M, Nutu O (1985) The mechanisms of cGMP-induced relaxation in vascular smooth muscle. Eur J Pharmacol 107:393–394
23. Collins P, Griffith TM, Henderson AH, Lewis MJ (1986) Endothelium-derived relaxing factor alters calcium fluxes in rabbit aorta: A cyclic guanosine monophosphate-mediated effect. J Physiol (Lond) 381:427–437
24. Twort CHC, Van Breeman C (1988) Cyclic guanosine monophosphate-enhanced sequestration of Ca^{2+} by sarcoplasmic reticulum in vascular smooth muscle. Circ Res 62:961–964
25. Tare M, Parkington HC, Coleman HA, Neild TO, Dusting GJ (1990) Hyperpolarization and relaxation of arterial smooth muscle caused by nitric oxide derived from the endothelium. Nature 346:69–71
26. Thornbury KD, Ward SM, Dalziel HH, Carl A, Westfall DP, Saunders KM (1991) Nitric oxide and nitrocysteine mimic nonadrenergic, noncholinergic hyperpolarization in canine proximal colon. Am J Physiol 261:G553–G557
27. Fiscus RR, Rapoport RM, Murad F (1984) Endothelium-dependent and nitrovasodilator-induced activation of cyclic GMP-dependent protein kinase in rat aorta. J Cyclic Nucleotide Protein Phosphor Res 9:415–425
28. Jackson WF, Mulsch A, Busse R (1991) Rhythmic smooth muscle activity in hamster aortas is mediated by continuous release of NO from the endothelium. Am J Physiol 260:H248–H253
29. Fujii K, Heistad DD, Faraci FM (1990) Ionic mechanisms in spontaneous vasomotion of the rat basilar artery in vivo. J Physiol (Lond) 430:389–398
30. Katusic ZS, Shepherd JT, Vanhoutte PM (1988) Potassium-induced endothelium-dependent rhythmic activity in the canine basilar artery. J Cardiovasc Pharmacol 12:37–41
31. Hester PK, Weiss GB (1984) Effects of nitroprusside and D600 on norepinephrine- and KCl-stimulated Ca^{2+} activation and contraction systems in canine renal vein as compared to canine renal artery. J Cardiovasc Pharmacol 6:762–771
32. Myers JH, Lamb FS, Webb RC (1985) Norepinephrine-induced phasic activity in tail arteries from genetically hypertensive rats. Am J Physiol 248:H419–H423
33. Stein PG, Driska SP (1984) Histamine-induced rhythmic contraction of hog carotid artery smooth muscle. Circ Res 55:180–185
34. Tomita K, Kai T (1979) Chaotic response of a limit cycle. J Stat Phys 21:65–86
35. Hoppensteadt FC (1989) Intermittent chaos, self-organization, and learning from synchronous synaptic activity in model neuronal networks. Proc Natl Acad Sci USA 86:2991–2995

Effects of Endogenous Nitric Oxide on Basal Vasomotor Tone and Stimulated Endothelium-Dependent Responses in the Coronary Arterial Circulation

Frederick R. Cobb[1], *Chang-Chyi Lin*[1], *Richard M.J. Palmer*[2], *Salvador Moncada*[2], *and Alan Chu*[1]

Summary. This study assesses the role of nitric oxide in basal vasomotor tone and stimulated endothelium-dependent dilations in the coronary arteries of chronically-instrumented awake dogs by examining the responses to inhibiting endogenous nitric oxide formation with the specific inhibitor of nitric oxide formation, N^G-monomethyl-L-arginine (L-NMMA). Basal epicardial coronary diameter (piezoelectric crystals), acetylcholine-stimulated endothelium-dependent dilation, flow induced endothelium-dependent dilation of the epicardial arteries, and phasic blood flow (Doppler probes) were recorded before, and after infusion of 5, 15, 50, and 120 mg/kg of L-NMMA. L-NMMA induced a dose-related increase in basal epicardial coronary vasomotor tone. There was an accompanying increase in aortic pressure and a decrease in heart rate. At doses ≥ 50 mg/kg, rest phasic coronary blood flow was also decreased. Left ventricular end-diastolic pressure and contractility were not significantly changed. In contrast, the flow-induced or acetylcholine-stimulated endothelium-dependent responses were attenuated approximately 50% only after infusion of the highest doses of L-NMMA (120 mg/kg). The changes in basal vasomotor tone and acetylcholine-stimulated endothelium-dependent responses returned towards the control states in the presence of L-arginine (660 mg/kg). These data support the view that nitric oxide plays a significant role in modulating basal vasomotion and endothelium-dependent dilation stimulated by acetylcholine or increase in blood flow in epicardial coronary arteries and also influences the regulation of coronary blood flow during physiologic conditions.

Key words: Coronary vasomotion—Awake dogs

Introduction

Following the description by Furchgott and Zawadzki [1] of endothelium-derived relaxing factor (EDRF), intensive efforts have been made to characterize and

---

[1] Department of Medicine, Division of Cardiology, Duke Medical Center and Durham Veterans Administration Medical Center; [2] Wellcome Research Laboratories, Langley Court, Beckenham, Kent BR335, United Kingdom

Supported in part by Grants IROL HL 17670 and HL 42562 from the National Heart, Lung, and Blood Institute, Bethesda, Maryland and from the Research Service, Veterans Administration Medical Center. Alan Chu was supported by a Career Development Award by the Veterans Administration

171

identify the nature of this biologic mediator [2–5]. Increasing evidence suggests that nitric oxide or a closely related compound represents at least one type (if not the only type) of EDRF because both nitric oxide and EDRF have the same biologic and pharmacologic properties [3, 6–8] and nitric oxide is released in sufficient quantities to explain the biological actions of EDRF [3]. Other investigators have provided data that suggest nitric oxide may not account for all the actions of EDRF, and that, depending on the vascular bed and the activator, there may be more than one type of EDRF [9–13].

Nitric oxide is synthesized by endothelial cells from the terminal guanidino-nitrogen atom of the amino acid L-arginine [14] and N^G-monomethyl-L-arginine (L-NMMA) has been shown to inhibit its formation in a concentration-dependent and enantiomorphic specific manner [15]. In addition, L-NMMA increases basal tone in the rings of the rabbit aorta [15, 16], the guinea pig pulmonary artery [17], and Langendorff perfused rabbit-heart preparations [18]. Acetylcholine-induced relaxation is also attenuated by L-NMMA in these preparations [16–18]. Furthermore, L-NMMA increases mean arterial pressure in anesthetized rabbits and this is associated with reduced release of nitric oxide from the perfused aorta of treated animals [19]. A similar rise in blood pressure induced by L-MNNA has been reported in guinea pigs [20]. More recently, the rise in blood pressure induced by L-NMMA in conscious rats has been shown to be accompanied by a substantial fall in regional blood flow in the renal, mesenteric hindquarters, and internal carotid vascular beds [21]. The relevance of these findings to man has recently been highlighted by the observation that infusion of L-NMMA into the brachial artery causes vasoconstriction and inhibition of the vasodilation induced by acetylcholine [22]. All these effects of L-NMMA are reversed by an excess of L-arginine.

This study was designed to examine the role of endogenous nitric oxide in coronary vasomotion in chronically-instrumented awake dogs by evaluating the effects of a wide range of inhibition of endogenous nitric oxide formation by L-NMMA on basal coronary vasomotion and endothelium-dependent stimulated vasodilation by acetylcholine and increases in blood flow.

Methods

Mongrel dogs, 30–35 kg (n = 4), were subjected to left thoracotomy under general anesthesia with intravenous thiamylal sodium (60–80 mg/kg). Heparin-filled polyvinyl catheters were inserted into the ascending aorta via the left internal thoracic artery, in the left atrium via the atrial appendage, and in the left ventricular chamber via the apex. A proximal segment (0.5–1 cm) of the left circumflex artery just distal to the atrial appendage was minimally dissected. Miniature 7 MHz piezoelectric crystals (1.5 × 2.5 mm, 15–20 mg), attached to a Dacron backing, were sutured to the adventitia on opposite surfaces of the vessel segment with 6-0 prolene (Ethicon, Somerville, NJ). Oscilloscope monitoring and on-line sonomicrometry (Sonomicrometer 120-2; Triton Technology, San Diego,

CA) were used to verify proper crystal alignment. A pulse Doppler flow probe (10 MHz, cuff type) was implanted distal to the crystals. An inflatable balloon occluder was also placed distal to the flow probe. All arterial branches between the crystals and the occluder were carefully ligated. The catheters and electrode wires were tunnelled to a subcutaneous pouch at the base of the neck.

The dogs were allowed to recover for 10–14 days. The catheters and wires were then exteriorized under lidocaine infiltration anesthesia. On a day before subjecting the dogs to the study protocol, each animal was given a bolus injection of nitroglycerin (0.4 mg) to ensure a responsive vasculature (> 5% dilation). On the study day, the dogs were loosely restrained and lying awake on their right side. Aortic pressure, left ventricular end-diastolic pressure, dP/dt, external coronary diameter, coronary flow, and electrocardiograms were continuously recorded. Pressure tracings from the fluid-filled catheters were optimally damped with a Corrector device (Norton Health Care Products, Akron, OH).

After obtaining the baseline coronary and systemic hemodynamic parameters, each dog was subjected to the following two interventions: (a) a bolus injection (left atrial catheter) of acetylcholine (4 µg) during constant phasic coronary flow maintained by partial inflation of the pneumatic occluder, and (b) 20-s transient coronary occlusion followed by complete release of the occluder. The 20-s occlusion below the crystal site was used to cause a transient increase in coronary blood flow, the reactive hyperemic response, which in turn was followed by flow-mediated vasodilation of the epicardial coronary artery. At least 15 min were allowed between interventions and the order was randomized. L-NMMA acetate salt (Wellcome Research Laboratories, Beckenham, UK) (5 mg/kg dissolved in saline at a concentration of 30–50 mg/ml), was infused slowly into the left atrium at a rate of < 100 mg/min to minimize acute hemodynamic changes. At 5 min after the hemodynamic parameters reached a new steady state, the stimuli for the endothelium-dependent responses (20-s transient occlusion, flow-induced dilation, and acetylcholine injection) were repeated. Subsequently, higher doses of L-NMMA (15, 50, 120 mg/kg) were administered in an increasing order. Each

Table 1. Resting hemodynamic measurements after increasing inhibition of nitric oxide formation

	Control	L-NMMA (mg/kg)				L-Arg
		5	15	50	120	
Aortic pressure (mmHg)	94 ± 6	114 ± 4[1]	120 ± 7[1]	121 ± 11[1]	129 ± 15[1]	104 ± 5[2]
Heart rate (HR)	62 ± 2	57 ± 2[1]	46 ± 3[1]	41 ± 3[1]	43 ± 4[1]	62 ± 6[2]
LVEDP (mmHg)	10 ± 1	13 ± 1	14 ± 1	15 ± 1	14 ± 2	11 ± 1
LV dP/dt (mmHg/s)	2483 ± 306	2500 ± 153	2533 ± 167	2467 ± 213	2633 ± 120	2817 ± 117

Results expressed in mean ± SEM. *LV*, left ventricular, *EDP*, end-diastolic pressure.
[1] Significant when compared with control. [2] Significant when compared with measurement at 120 mg/kg N^G-monomethyl-L-arginine (L-NMMA).

administration was followed by a repeat measurement of the basal and the stimulated endothelium-dependent responses. After the stimulated responses after the highest dose of L-NMMA, the nonendothelium-dependent dilator nitroglycerin (0.04 mg) was given as a bolus in the left atrium. After 20–30 min, when the basal coronary diameter had returned to the level before nitroglycerin injection, L-arginine (660 mg/kg dissolved in saline at 100 mg/ml) (Sigma Chemical Co., St. Louis, MO) was infused slowly over 20–30 min into the left atrium. The basal and stimulated endothelium-dependent responses were again repeated. Nitroglycerin (0.4 mg) was given at the end of the study day.

All hemodynamic measurements were compared with measurements obtained before infusion of L-NMMA using the same dog as its own control. Statistical analyses were performed using an analysis of variance for repeated measures.

Results

Table 1 summarizes the baseline hemodynamic changes before and after the infusion of each dose of L-NMMA. L-MNNA (5–120 mg/kg) caused significant dose-related increases in mean aortic pressure from a control of 94 ± 6 to 129 ± 15 mmHg after the highest dose of 120 mg/kg. The largest effects of inhibition of endogenous nitric oxide production on aortic pressure occurred after the lowest dosages of L-NMMA 5–15 mg/kg with minimal further increase despite a > 10-fold increase in L-NMMA dosage. Heart rate decreased from 62 ± 2 to 43 ± 4 beats/min. At > 15 mg/kg, L-NMMA caused intermittent second- or third-degree atrioventricular block. Changes in aortic pressure and heart rate were reversed with L-arginine. Left ventricular end-diastolic pressure tended to increase from 10 ± 1 to 14 ± 2 mmHg but the change was not statistically significant. Left ventricular dP/dt was not significantly altered after infusion of L-NMMA.

The effects of increasing doses of L-NMMA on basal epicardial coronary dimensions in individual dogs are illustrated in Fig. 1. Although the magnitude of the response was variable in individual animals, L-NMMA-induced dose-related vasoconstriction in each animal. The epicardial diameter decreased from 3.67 ± 0.10 to 3.38 ± 0.06 mm after the highest dose of L-NMMA representing an 8% decrease in vessel diameter or 15% decrease in cross-sectional area. As with aortic pressure, the largest effect of inhibition of endogenous nitric oxide production on coronary dimensions occurred after the lower dosages of L-NMMA, with only small changes occurring with dosages > 15 mg/kg. Again, the epicardial vasoconstriction was partially reversed toward control levels with L-arginine.

In contrast to the effect on aortic pressure and basal epicardial dimension, flow-induced endothelium-dependent dilation was not significantly altered after infusions of up to 50 mg/kg L-NMMA, but was significantly reduced after 120 mg/kg L-NMMA from 0.209 ± 0.024 mm (5.71% ± 0.70%) to 0.076 ± 0.024 mm (2.26% ± 0.073%) (Fig. 2). The response, however, was variable in

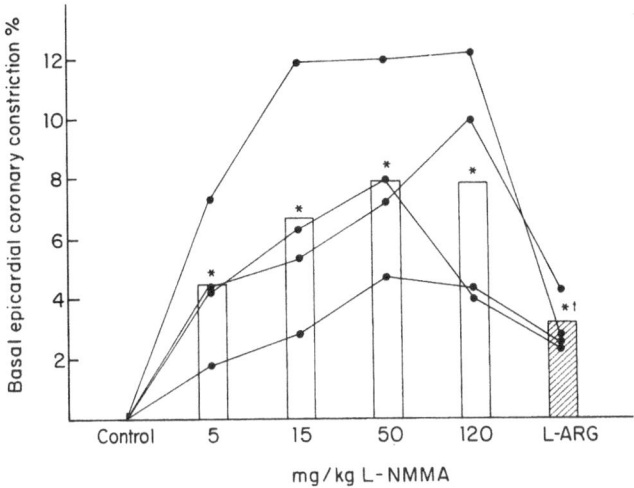

Fig. 1. Effects of increasing doses of N^G-monomethyl-L-arginine (L-NMMA) on basal epicardial coronary vasomotor tone. Individual (*circles*) and mean (*bars*) data are plotted as percent change of control. Coronary dimensions decreased in a dose-related fashion after infusion of L-NMMA; L-arginine (L-ARG) partially reversed the response. *Significant when compared with control. †Significant when compared with 120 mg/kg L-NMMA

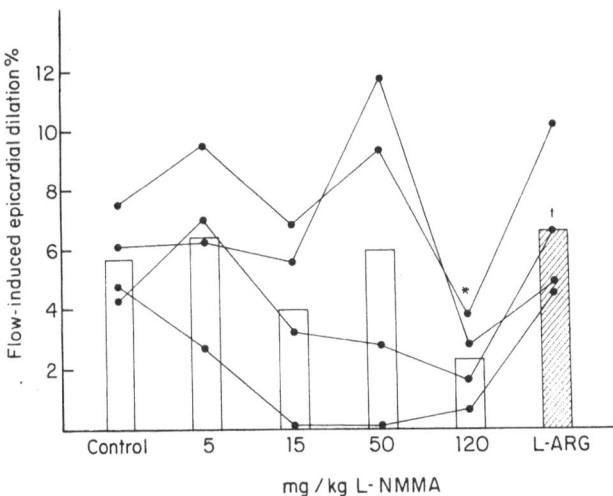

Fig. 2. Effects of increasing doses of N^G-monomethyl-L-arginine (L-NMMA) on flow-induced dilation. Individual (*circles*) and mean (*bars*) data are plotted as percent dilation. Although the effects of L-NMMA were variable at lower doses, the reductions were consistent at the 120 mg/kg dose *when compared with control. L-arginine (L-ARG) reversed the reduction in acetylcholine-induced dilation. †Significant when compared with 120 mg/kg L-NMMA

individual animals; one animal was particularly sensitive to inhibition of L-NMMA. The flow-induced, endothelium-dependent dilation was reversed to control levels after infusion of L-arginine. The maximal epicardial coronary diameter after nitroglycerin was not significantly changed in the presence of L-NMMA.

A similar pattern was seen for the acetylcholine-induced endothelium-dependent dilation which was not significantly inhibited until after infusion of 120 mg/kg of L-NMMA [control, 0.154 ± 0.016 mm (4.20% ± 0.48%); L-NMMA, 0.71 ± 0.008 (2.11% ± 0.21%)] (Fig. 3). The response was also variable in individual animals. The same animal that demonstrated greater sensitivity to L-NMMA inhibition of flow-induced vasodilation also demonstrated greater sensitivity to L-NMMA inhibition of acetylcholine-induced vasodilation. The acetylcholine-induced dilation response was reversed to control levels after infusion of L-arginine.

The effects of L-NMMA on rest phasic coronary blood flow (vasomotion of distal resistance coronary vessels) were modest; maximum decrease was 19%. Although there was a tendency for blood flow to decrease at the 15 mg/kg dose, because of the variable responses, significance was not achieved until the 50 mg/kg dose (Table 2). Peak reactive hyperemic flow-induced after 20-s transient coronary occlusion was, however, not significantly different even after

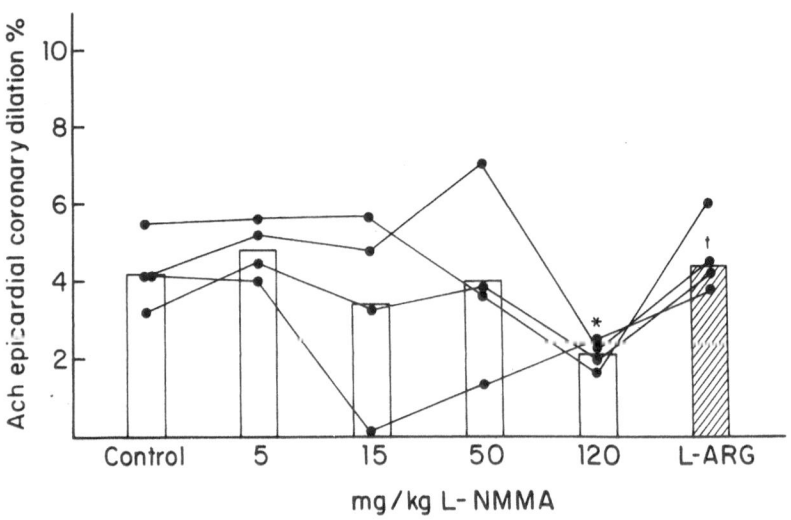

Fig. 3. Effects of increasing doses of N^G-monomethyl-L-arginine (L-NMMA) on acetylcholine-induced dilation. Individual (*circles*) and mean (*bars*) data are plotted as percent dilation. Although the effects of L-NMMA were variable at lower doses, the reductions were consistent at the 120 mg/kg dose *when compared with control. L-arginine (L-ARG) reversed the reduction in acethylcholine-induced dilation. †Significant when compared with 120 mg/kg L-NMMA

Table 2. Effects of increasing inhibition of nitric oxide formation on rest and peak reactive hyperemic flow after a 20-s occlusion

	Control	L-NMMA (mg/kg)				L-Arg
		5	15	50	120	
Rest flow (KHz)	2.18 ± 0.42	2.09 ± 0.34	1.76 ± 0.36	1.78 ± 0.34[1]	1.78 ± 0.30[1]	1.92 ± 0.28
Peak reactive hyperemic flow (KHz)	8.37 ± 0.63	8.68 ± 0.25	7.91 ± 0.56	7.21 ± 0.17	7.32 ± 0.48	8.92 ± 0.61

Results expressed in mean ± SEM. [1] Significant when compared with control.

the infusion of the highest dose of L-NMMA. Therefore, the reduction of flow-induced epicardial coronary dilation seen after L-NMMA resulted from a direct effect of inhibition of nitric oxide formation rather than being secondary to reduced peak flow. L-arginine reversed the resting flow toward control levels.

Discussion

The present study is the first to assess the effects of a wide range of inhibition of endogenous nitric oxide production on basal or rest vasomotion, acetylcholine and flow-stimulated endothelium-dependent vasodilation of coronary epicardial conductance vessels, and systemic hemodynamics in an intact awake physiologic model [23]. In addition, the effects of a wide range of inhibition of endogenous nitric oxide production on basal or rest phasic coronary blood, a measure of resistance vessel vasomotion, was also assessed. L-NMMA has been demonstrated to inhibit endogenous nitric oxide formation from L-arginine in vitro and in anesthetized preparations [15, 16, 18, 19]. Using increasing doses of L-NMMA, the present study demonstrated dose-related decreases in basal epicardial coronary dimensions. The changes in epicardial dimensions were accompanied by increases in aortic pressure and decreases in heart rate. The largest change in epicardial dimensions and aortic pressure occurred with the lower doses of L-NMMA (5 and 15 mg/kg), with only minimal further change in dimensions or pressure despite > 10-fold further increase in L-NMMA. At doses of ≥ 50 mg/kg L-NMMA, rest coronary flow was also decreased; the maximum decrease was 19%. In contrast to the effects on basal vasomotion, the flow-induced or acetylcholine-stimulated, endothelium-dependent dilation responses in the coronary arteries were attenuated only after the highest dose of L-NMMA (120 mg/kg). Thus there was a > 20-fold differential effect of L-NMMA on basal as compared with stimulated epicardial vasomotion. The effects of L-NMMA on rest coronary flow, and flow-induced or acetylcholine-stimulated, endothelium-dependent dilation responses were variable in individual animals; certain animals were affected by lower doses of L-NMMA. It is possible that with larger numbers of animals, a statistically significant effect of L-NMMA on these parameters may be achieved at doses < 120 mg/kg. Left ventricular end-diastolic pressure and dP/dt were not significantly altered after L-NMMA. The changes in basal coronary vasomotion and stimulated endothelium-dependent responses were reversed by a high dose of L-arginine. In an earlier study, in a different group of animals, we described the effects of a low dose of L-NMMA, 5 mg/kg, on basal coronary vasomotion [24]. The present study extends this initial observation to a wide range of inhibition of nitric oxide production and assesses the effects on stimulated as well as basal vasomotion for the first time.

Previous studies by Amezcua et al. [18] demonstrated that L-NMMA, at a dose between 10 and 100 µM, produced a dose-related sustained increase in coronary perfusion pressure (resistance coronary artery vasomotion) in a Langendorff-perfused rabbit heart preparation. In the Langendorff preparation, acetylcholine (0.3 µM) caused a decrease (vasodilation) before and an increase (vasoconstriction) in coronary perfusion pressure after inhibition of nitric oxide formation. In our present study, we observed that as compared to basal coronary vasomotion, the flow-induced and acetylcholine-stimulated coronary endothelium-dependent responses required a 20-fold higher dose of L-NMMA for significant inhibition. Rees et al. [16] also demonstrated similar effects in in vitro rabbit aortic ring preparations after inhibition of nitric oxide formation on endothelium-dependent relaxations induced by acetylcholine, calcium ionophore, or substance P. In that study, the concentration of L-NMMA that produced 50% contraction of the rings was six times lower than that required to achieve 50% inhibition of acetylcholine-induced relaxation [16]. Although the reason for this difference between basal and stimulated responses is not known, several possibilities may be considered. The stimulated responses, as compared to the basal responses, may increase mobilization of the stored substrate within the endothelial cells that required higher concentrations of inhibitor for blockade [16]. An alternative explanation may be the presence of more than one EDRF, unrelated to nitric oxide. Nitric oxide may be the EDRF primarily responsible for modulating basal coronary and peripheral vasomotor tone whereas other EDRFs may contribute to stimulated responses. The fact that a substantial portion ($>50\%$) of the stimulated responses were inhibited at the higher doses of L-NMMA argues against this possibility.

Coronary blood flow decreased when the dose of L-NMMA was 50 mg/kg or greater. Because aortic pressure increased with the lowest dose of L-NMMA and coronary flow was unchanged, distal coronary vascular resistance was elevated even after the initial dose of L-NMMA. Previous studies from our laboratory have demonstrated that cyclic guanosine monophosphate-mediated dilators such as EDRF, nitrates, and atrial natriuretic peptides preferentially dilate proximal conductance arteries as compared to distal regulatory resistance coronary arteries in awake dogs [25–28]. The vasomotor state of the distal resistance vasculature is influenced by the dominant effects of local myocardial metabolic demands [29]. Because nitric oxide has been demonstrated to mediate its vasodilator effects through activation of guanylate cyclase [30–31], it is reasonable to speculate that the increased vasomotor tone of the distal regulatory resistance arteries (as a result of inhibition of nitric oxide formation) may have been over-ridden by the dominant vasodilating stimuli mediated by increased myocardial metabolic demands. The lack of significant change in peak hyperemic flow also indicates that the coronary reserve for increasing blood flow was present and that the reduction in flow-induced dilation of the proximal epicardial arteries after L-NMMA was a direct effect of the inhibition of nitric oxide formation and not secondary to a difference in the flow stimuli. In recent studies, Vallance et al. [22] observed that the infusion of L-NMMA into the brachial artery of healthy volunteers caused a 50% decrease in the basal blood flow and attenuation of the dilator response to acetylcholine, but not to the infusion of nitrates, suggesting that endothelial-derived nitric oxide influences basal and stimulated limb blood flow in man.

The persistent dose-dependent hypertensive effect seen after L-NMMA in the present study is consistent with the observations of Rees et al. [19] who reported a sustained (15–19 min) dose-dependent increase in mean systemic arterial pressure after infusion of L-NMMA (3–100 mg/kg) in anesthetized rabbits. Indomethacin, prazosin, or vagotomy did not alter the hypertensive response to L-NMMA [19], whereas it was reversed by L-arginine, as in our present study. The hypertensive response to L-NMMA probably contributed to the decrease in heart rate seen after the lower doses used in our study (Table 1). Second- and third-degree atrioventricular block was observed after the 15–50 mg/kg dose which may have been influenced by the increase in blood pressure and/or a direct effect of inhibition of nitric oxide production or L-NMMA on the AV node. The effects on pressure, heart rate, and atrioventricular conduction were reversed by L-arginine.

In summary, the results from the present study support the view that endogenous nitric oxide plays a significant physiologic role in modulating basal vasomotor tone in the proximal epicardial coronary arteries and distal regulatory resistance coronary vessels. These data also demonstrate that endogenous nitric oxide plays a role in the endothelium-dependent dilation induced by increases in flow or by acetylcholine in the epicardial coronary vessels. The effects of endogenous nitric oxide on epicardial dimension and aortic pressure were more sensitive to inhibition by L-NMMA with the largest change occurring at the lower dosage, whereas the effect on endothelium-dependent stimulated vasodilation occurred consistently at a > 20-fold higher dose of L-NMMA.

Acknowledgments. We would like to thank Mr. Joe Long for technical assistance, Medical Media Service for illustrations, and Ms. Cathie Collins for her expert secretarial assistance.

References

1. Furchgott RF, Zawadzki JV (1980) The obligatory role of endothelial cells in the relaxation of arterial smooth muscle by acetylcholine. Nature 288:373–376

2. Furchgott RF (1983) Role of endothelium in responses of vascular smooth muscle. Circ Res 53:557–573

3. Moncada S, Radomski MW, Palmer RMJ (1988) Endothelium-derived relaxing factor: Identification as nitric oxide and role in the control of vascular tone and platelet function. Biochem Pharmacol 37:2495–2501

4. Furchgott RF, Vanhoutte PM (1989) Endothelium-derived relaxing and contracting factors. FASEB J 3:2007–2018

5. Ignarro LJ (1989) Biological action and properties of endothelium-derived nitric oxide formed and released from artery and vein. Circ Res 65:1–21

6. Furchgott RF (1987) Studies on relaxation of rabbit aorta by sodium nitrite: The basis for the proposal that the acid-activatable inhibitory factor from bovine retractor penis is inorganic nitrite and the endothelium-derived relaxing factor is nitric oxide. In: Vanhoutte PM (ed) Mechanisms of vasodilation. Raven Press, New York, pp 401–414

7. Ignarro LJ, Byrns RE, Wood KS (1988) Biochemical and pharmacological properties of endothelium-derived relaxing factor and its similarity to nitric oxide radical. In Vanhoutte PM (ed) Mechanisms of vasodilation. Raven Press, New York, pp 427–435

8. Palmer RMJ, Ferrige AG, Moncada S (1987) Nitric oxide release accounts for the biological activity of endothelium-derived relaxing factor. Nature 327:524–526

9. Rubanyi GM, Vanhoutte PM (1987) Nature of endothelium-derived relaxing factor: Are there two relaxing mediators? Circ Res 61 [Suppl II]:II61–67

10. Boulanger C, Hendrickson H, Lorenz RR, Vanhoutte PM (1989) Release of different relaxing factors by cultured porcine endothelial cells. Circ Res 64:1070–1078

11. Shikano K, Ohlstein EH, Berkovitz BA (1987) Differential selectivity of endothelium-derived relaxing factor and nitric oxide in smooth muscle. Br J Pharmacol 92:483–485

12. Kontos HA, Wei EP, Povlishock JT, Christman CW (1984) Oxygen radicals mediate the cerebral arteriolar dilation from arachidonate and bradykinin in cats. Circ Res 55:295–303

13. Myers PR, Guerra R Jr, Harrison DG (1989) Release of NO and EDRF from cultured bovine aortic endothelial cells. Amer J Physiol 256 (Heart Circ Physiol 25): H1030–H1037

14. Palmer RMJ, Ashton DS, Moncada S (1988) Vascular endothelial cells synthesize nitric oxide from L-arginine. Nature 333:664–666

15. Palmer RMJ, Rees DD, Ashton DS, Moncada S (1988) L-arginine is the physiological precursor for the formation of nitric oxide in endothelium-dependent relaxation. Biochem Biophys Res Commun 153:1251–1256

16. Rees DD, Palmer RMJ, Hodson HF, Moncada S (1989) A specific inhibitor nitric oxide formation from L-arginine attenuates endothelium-dependent relaxation. Br J Pharmacol 96:418–424

17. Sakuma I, Stuehr DJ, Gross SS, Nathan C, Levi R (1988) Identification of arginine as a precursor of endothelium-derived relaxing factor. Proc Natl Acad Sci 85:8864–8867

18. Amezcua JL, Palmer RMJ, deSouza BM, Moncada S (1989) Nitric oxide synthesized from L-arginine regulates vascular tone in the coronary circulation of the rabbit. Br J Pharmacol 97:1019–1024

19. Rees DD, Palmer RMJ, Moncada S (1989) Role of endothelium-derived nitric oxide in the regulation of blood pressure. Proc Natl Acad Sci USA 86:3375–3378

20. Aisaka K, Gross SS, Griffith OW, Levi R (1989) N^G-Methyl arginine, an inhibitor of endothelium-derived nitric oxide synthesis is a potent pressor agent in the guinea pig: Does nitric oxide regulate blood pressure in vivo? Biochem Biophys Res Commun 160:881–886

21. Gardiner SM, Compton AM, Bennett T, Palmer RMJ, Moncada S (1990) Control of regional blood flow by endothelium-dependent nitric oxide. Hypertension 15:486–492
22. Vallance P, Collier J, Moncada S (1989) Effects of endothelium-derived nitric oxide on peripheral arteriolar tone in man. Lancet 334:997–1000
23. Chu A, Lin C-C, Chambers DE, Kuehl WD, Palmer RMJ, Moncada S, Cobb FR (1991) Effects of inhibition of nitric oxide formation on basal tone and endothelium-dependent responses of the coronary arteries in awake dogs. J Clin Invest 87: 1964–1968
24. Chu A, Chambers D, Lin C-C, Kuehl W, Cobb FR (1990) Nitric oxide modulates epicardial coronary basal vasomotor tone in awake dogs. Am J Physiol 258 (Heart Circ Physiol 27):H1250–H1254
25. Chu A, Cobb FR (1987) Effects of atrial natriuretic peptide on proximal epicardial coronary arteries and coronary blood flow in conscious dogs. Circ Res 61:485–491
26. Chu A, Murray JJ, Kuehl W, Lin C-C, Russell M, Hagan P-O, Cobb FR (1990) Preferential proximal coronary dilation by activators of guanylate cyclase in awake dogs. Am J Physiol 259 (Heart Circ Physiol 28):H340–H345
27. Chu A, Cobb FR, Hagen P-O, Murray JJ (1989) Effects of a stabilized endothelium-derived relaxing factor on the coronary vasculature in awake dogs. Am J Physiol 257 (Heart Circ Physiol 26):H1895–1899
28. Chu A, Morris KG, Kuehl WD, Cusma J, Navetta F, Cobb FR (1989) Effects of atrial natriuretic peptide on the coronary arterial vasculature in humans. Circulation 80:1627–1635
29. Rubio R, Berne RM (1975) Regulation of coronary blood flow. Prog Cardiovasc Dis 18:105–135
30. Ignarro LJ, Buga GM, Wood KS, Byrns RE, Chadhuri G (1987) Endothelium-derived relaxing factor produced and released from artery and vein is nitric oxide. Proc Natl Acad Sci USA 84:9265–9269
31. Ignarro LJ, Byrns RE, Buga GM, Wood KS, Chadhuri G (1988) Pharmacological evidence that endothelium-derived relaxing factor is nitric oxide: Use of pyrogallol and superoxide dismutase to study endothelium-dependent and nitric oxide-elicited vascular smooth muscle relaxation. J Pharmacol Exp Ther 244:181–189

Mechanical Properties of Coronary Arterioles under Pulsation

Masami Goto, Ed VanBavel, Maurice J.M.M. Giezeman, and Jos A.E. Spaan[1]

Summary. Recently it has become clear that the deep myocardial vessels are pulsating due to cardiac contraction. Although it is known that the coronary arterioles show myogenic responses and play a role in the regulation of blood flow, the influence of pulsation on the mechanical propertes of the arterioles is not clear. We evaluated the mechanical properties of the coronary arterioles under pulsation with basic physiological frequency and different basic tone. Isolated porcine coronary arterioles (I.D. = 100–150 μm) were cannulated with two micropipettes. The luminal cross-sectional area (CSA) was measured using a fluorescence technique under cyclically (1 Hz) changing transmural pressure. After vascular tone was induced (active) by acetylcholine or abolished (passive) by bradykinin, the CSA was measured while the pressure amplitude was changed at a fixed mean level (60 mmHg). At all amplitudes tested, the isolated coronary arterioles in the active condition showed smaller compliance compared to those in the passive condition. The compliance was larger with greater amplitude of the pulsatile transmural pressure than with smaller amplitude of the pulsatile transmural pressure for both the active and passive coronary arterioles. Therefore, the physiological active arterioles pulsating with larger transmural pressure may have benefits in accommodating and discharging blood, and thus may contribute to the compliant blood flow. Raising the amplitude caused the normalized mean CSA in a steady state to increase under active conditions, and to decrease under passive conditions. The vasodilating effects of the pulsation may compensate for the extra compressing effects with increasing cardiac contraction.

Key words: Compliance—Myogenic response

Introduction

Myocardial vessels are compressed by cardiac contraction, because the vessels are within the myocardium [1]. Thus, cardiac contraction causes two major pulsatile mechanical changes to the myocardial vessels, i.e., flow and pressure. The effects of cardiac contraction on the blood flow in the myocardial vessels have been discussed in relation to the phasic nature of the coronary inflow and outflow, which was first speculated on by Scaramucci in 1695 [2]. Scaramucci [2] speculated that cardiac contraction squeezes myocardial vessels and causes

---

[1] Department of Medical Physics, Faculty of Medicine, University of Amsterdam, Meibergdreef 15 1105 AZ Amsterdam Z0 The Netherlands

pulsation in the coronary arterial and venous flows. This speculation was later followed by the controversy about whether contraction had beneficial or deleterious effects on coronary flow [3, 4].

In 1957, Sabiston and Gregg [4] first provided direct evidence for the impeding effect of cardiac contraction on coronary artery inflow, although the magnitude was probably overestimated [5]. Cardiac contraction affects coronary blood flow in two ways: 1) reduction in the time-averaged flow, depending on the depth in the myocardium, and 2) pulsation of the flow. Recently, it has become clear that inflow to the myocardium is almost exclusively limited to diastole, and cardiac contraction even causes retrograde flow in the coronary artery from the deep myocardium [6, 7]. These findings have been reported repeatedly with development of new measuring techniques of the instantaneous blood flow in the coronary vessels, which are applicable to beating hearts. When the instantaneous blood flow pattern is measured in the septal artery using a 20 MHz 80-channel ultrasonic-pulsed Doppler velocimeter, a substantial amount of the blood entering the myocardium during diastole returned to the proximal coronary artery [8]. With coronary artery stenosis, the systolic retrograde flow was enhanced [9], and was augmented further by coronary vasodilation [10]. Therefore, under physiological conditions and with coronary artery disease the deep myocardial vessels, which form the origin of the periodic systolic retrograde coronary arterial blood flow, should be pulsating. Recently it has been demonstrated that the diameter of 100 μm subendocardial arterioles decreases during systole [11]. The direct observation of the subendocardial vessels became possible by introducing a portable needle-probe video-microscope with a charge-coupled-device (CCD) camera. From these findings, it is concluded that the coronary arterioles in the deep myocardial layer face two major pulsatile mechanical stimuli, i.e., flow and pressure.

It is generally considered that flow is the major stimulus to the vascular endothelium, whereas pressure is the major stimulus to the vascular smooth muscle. As for the effect of pulsatile flow to the arterial endothelium, Hutcheson and Griffith [12] showed that endothelium-derived relaxing factor ($EDRF$) release is inversely proportional to the amplitude of the pulsatile flow at constant mean flow. Based on their findings, it is considered that the subendocardial arterioles change their mechanical properties in response to the changes in amplitude of the pulsatile flow. On the other hand, the influence of pulsation of the vessels on the mechanical properties of the arterioles is not clear, although it is known that the coronary arterioles have myogenic responses, like other arterial vessels in most of the organs in the body [13, 14]. In the present study, we analyzed the mechanical properties of the coronary arterioles at various tone levels under pulsation with a basic physiological frequency to obtain some insights into the role of basic mechanical properties of the resistance vessels in the regulation of blood flow.

Methods

The coronary arterioles of anesthetized pigs ($n = 7$, diameter: 100–150 μm) were dissected, and the surrounding tissue was removed carefully. In a vessel chamber filled with MOPS-buffered Ringer solution containing 1% dialyzed albumin (4°C), the isolated arteriole was cannulated with two micropipettes. The cannulated arteriole was filled with MOPS-buffered Ringer solution containing FITC-dextran (40 mg/l, Sigma, St. Louis, MO) and 1% predialyzed albumin. Using a fluorescence technique [15], the luminal cross-sectional area (CSA) of the cannulated vessels was measured: the FITC-dextran present in the lumen of the vessel was excited by a weak light source (lambda 400–480 nm). The total amount of fluorescence light (lambda > 515 nm) from the lumen was measured using a photo-multiplier tube. The amount of this light is proportional to the amount of excited FITC molecules, and therefore to the CSA of the vessel.

The vessels were pressurized from both ends with the same pressure to minimize the possible contribution of flow-induced vascular responses. After vascular tone was induced (active condition) by acetylcholine (10^{-5} M) or abolished (passive condition) by bradykinin (10^{-4} M), the CSA was measured. Transmural pressure was changed rectangularly with a physiological basic frequency of 1 Hz at fixed mean pressure.

Results

Figure 1 shows vascular compliances which were calculated at steady state for the activated and unactivated arterioles from the ratio of the amplitude of the CSA (maximum CSA minus minimum CSA) to the amplitude of the transmural pressure (maximum transmural pressure minus minimum transmural pressure). At all amplitudes of transmural pressure tested, the vascular compliance was different between the active and passive conditions; the compliance was smaller in the active condition than in the passive condition. The vascular compliance increased with increasing amplitude of the pressure pulse in the active and passive conditions. Figure 2 shows the percentage change in the mean CSA under passive and active conditions at different pressure pulse amplitude (ranging from 0–100 mmHg). When vascular tone was abolished, the mean CSA decreased with increasing pressure pulse amplitude. In contrast, preconstricted vessels dilated with increasing pressure pulse amplitude.

Discussion

Mechanical properties of arterial vessels vary depending on the frequency of transmural pressure fluctuation and basic vascular tone. Recently, Giezeman [16] evaluated the static and dynamic properties of isolated small coronary arterioles and reported that there was a considerable decrease in dynamic dis-

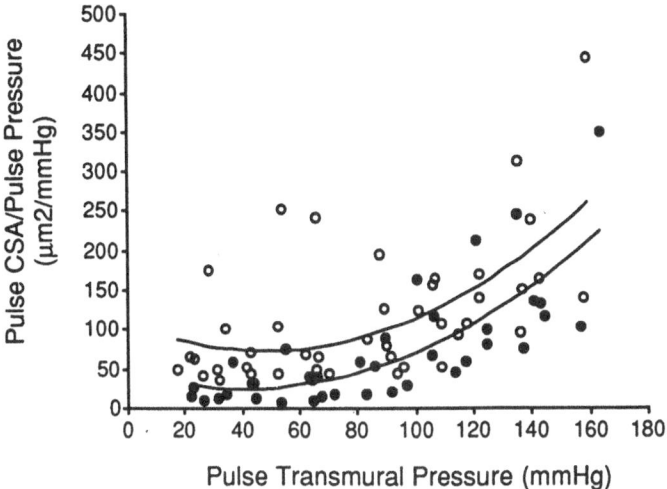

Fig. 1. Isolated coronary arterioles in the active condition (*closed circles*) showed smaller compliance (pulse cross-sectional area/pulse pressure) compared to those in the passive condition (*open circles*) at all amplitudes of pulsatile transmural pressure studied. The compliance was larger with greater amplitude of the pulsatile transmural pressure than with smaller amplitude of the pulsatile transmural pressure for both active and passive coronary arterioles. *CSA*, Cross-sectional area

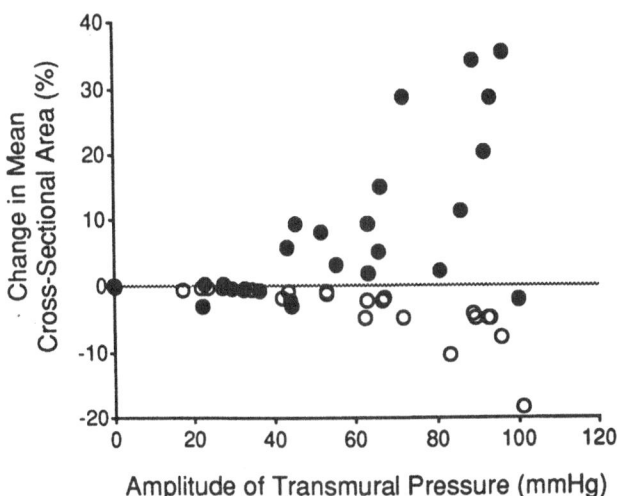

Fig. 2. When vascular tone was abolished (passive; *open circles*) the time averaged vascular cross-sectional area at steady state decreased with increasing the pressure pulse. When the vessels were pre-constricted (active; *closed circles*) the vessels dilated at constant mean transmural pressure with increasing pressure pulsatility

tensibility with respect to static distensibility at frequencies well below the heart rate. Therefore, dynamic properties of blood vessels should not be analyzed using distensibilities obtained from static measurements. In the present study, the compliance of the isolated coronary arterioles were evaluated at a frequency of 1 Hz. At a physiological basic frequency, the isolated coronary arterioles in the active condition showed smaller compliance than those in the passive condition for all amplitudes of pulsatile transmural pressure (Fig. 1). The compliance was larger with greater amplitude of the pulsatile transmural pressure than with smaller amplitude of the pulsatile transmural pressure for both the active and passive coronary arteriole (Fig. 1). Therefore, the physiologically active arterioles pulsating with larger transmural pressure may have benefited in accommodating and discharging blood, and thus may contribute to the compliant blood flow.

Information on the capacitance of the vessels in the myocardium has been limited. Recently, regional blood flow has been compared systematically in superficial and deep myocardial layers in systolically and diastolically arrested hearts. From these investigations, it has been shown that there are significant differences in the effect of cardiac contraction on the myocardial vessels in these layers [17]. Cardiac contraction causes greater diameter changes to the subendocardial microvessels than those in the subepicardium. Diameters of the subendocardial arterioles with relatively larger diameters of about 100 µm have also been measured in the beating hearts using a newly-developed needle-type CCD microscope [11]. Diameters of the subendocardial arterioles significantly decrease during systole. On the other hand, Kanatsuka et al. [18] observed only 1.1% diameter change of the left ventricular epicardial arterioles during the cardiac cycle as measured by an intravital floating microscopic system. It can be concluded that cardiac contraction causes greater pulsation predominantly to the arterioles in the subendocardium. Thus, the subendocardial arterioles may have larger compliance than the subepicardial arterioles.

The response of the coronary arterioles with active tone to the changes in the amplitude of pulsatile transmural pressure is quite different from that of the coronary arterioles without active tone (Fig. 2). When the vessels have active tone, raising amplitude of the pulsatile transmural pressure has a vasodilating effect [19]. The flow impediment by compression, enhanced during vasodilation, is most markable in the inner layers of heart muscle where tissue pressure pulsations are largest [20]. The difference in compression effects between inner and outer heart muscle layers must be equalized by the control action of smooth muscle. From Fig. 2, it is concluded that pressure pulsations themselves form a stimulus for vasodilation and compensate for extra-compression effects. This vasodilating mechanism may support the perfusing subendocardium, which is the most vulnerable area in the myocardium, under a variety of pathophysiological conditions.

References

1. Hoffman JIE, Spaan JAE (1990) Pressure-flow relations in coronary circulation. Physiol Rev 70:331–390
2. Scaramucci J (1695) Theoremata familiaria viros eruditos consulentia de variis physicomedicis lucubrationibus juxta leges mecanicas. Apud Joannem Baptistam Bustum, pp 70–81
3. Wiggers CJ (1954) The interplay of coronary vascular resistance and myocardial compression in regulating coronary flow. Circ Res 2:271–279
4. Sabiston DC, Gregg DE (1957) Effect of cardiac contraction on coronary blood flow. Circulation 15:14–20
5. Katz SA, Feigl EO (1988) Systole has little effect on diastolic coronary blood flow. Circ Res 62:443–451
6. Chilian WM, Marcus ML (1982) Phasic coronary flow velocity in intramural and epicardial coronary arteries. Circ Res 50:775–781
7. Kajiya F, Tomonaga G, Tsujioka K, Ogasawara Y, Nishihara H (1985) Evaluation of local blood flow velocity in proximal and distal coronary arteries by laser Doppler method. J Biomech Eng 107:10–15
8. Kajiya F, Ogasawara Y, Tsujioka K, Nakai M, Goto M, Wada Y, Tadaoka S, Matsuoka S, Mito K, Fujiwara T (1986) Evaluation of human coronary blood flow with an 80-channel pulsed Doppler velocimeter and zero-cross and Fourier transform methods during cardiac surgery. Circulation 74 [Suppl III]:53–60
9. Kimura A, Hiramatsu O, Yamamoto T, Ogasawara Y, Yada T, Goto M, Tsujioka K, Kajiya F (1992) Effect of coronary stenosis on phasic pattern of septal artery in the dog. Am J Physiol 262:H1690–H1698
10. Goto M, Flynn AE, Doucette JW, Kimura A, Hiramatsu O, Yamamoto T, Ogasawara Y, Tsujioka K, Hoffman JIE, Kajiya F (1992) Effect of intracoronary nitroglycerin administration on phasic pattern and transmural distribution of flow during coronary artery stenosis. Circulation 85:2296–2304
11. Kajiya F, Goto M, Yada T, Kimura A, Yamamoto T, Hiramatsu O, Ogasawara Y, Tsujioka K, Yamamori S, Hosaka H (1991) In-vivo evaluation of endocardial blood vessels by a new needle type CCD microscope (abstract). Circulation [Suppl II] 84:271
12. Hutcheson IR, Griffith TM (1991) Release of endothelium-derived relaxing factor is modulated both by frequency and amplitude of pulsatile flow. Am J Physiol 261: H257–H262
13. Kuo L, Davis MJ, Chilian WM (1988) Myogenic activity in isolated subepicardial and subendocardial coronary arterioles. Am J Physiol 255:H1558–H1562
14. Kuo L, Chilian WM, Davis MJ (1990) Coronary arteriolar myogenic response is independent of endothelium. Circ Res 66:860–866
15. VanBavel E, Mooij T, Giezeman MJMM, Spaan JAE (1990) Cannulation and continuous cross-sectional area measurement of small blood vessels. J Pharmacol Methods 24:219–227
16. Giezeman MJMM (1992) Static and dynamic pressure-volume relations of isolated blood vessels. University of Amsterdam, Amsterdam, pp 37–53
17. Goto M, Flynn AE, Doucette JW, Jansen CMA, Stork MM, Coggins DL, Muehrcke DD, Husseini WK, Hoffman JIE (1991) Cardiac contraction affects deep myocardial vessels predominantly. Am J Physiol 261:H1417–H1429

18. Kanatsuka H, Lamping KG, Eastham CL, Dellsperger KC, Marcus ML (1989) Comparison of the effects of increased myocardial oxygen consumption and adenosine on the coronary microvascular resistance. Circ Res 65:1296–1305
19. Goto M, Giezeman MJMM, VanBavel E, Spaan JAE (1992) Increase in amplitude of pulsatile transmural pressure dilates coronary arterioles (abstract). Circulation [Suppl I] 86:508
20. Hoffman JIE (1987) Transmural myocardial perfusion. Prog Cardiovasc Dis 29: 429–464

Regulation of Coronary Circulation During Rigid and Pliable Coronary Stenosis

Mitsuhiro Yokoyama, Yuichi Matsuda, and Hozuka Akita[1]

Summary. We developed a canine model of pliable coronary stenosis that preserves active vasomotion in a stenosed segment and is suitable for investigation of the role of active vasomotor tone of a large coronary artery during a preexisting coronary stenosis in the genesis of myocardial ischemia. Using this model, we studied the effects of vasoactive intestinal peptide (VIP) and substance P (SP), both vasoactive neuropeptides present in hearts and nerves of coronary arteries, on the coronary circulation. Without coronary stenosis, intracoronary infusion of each peptide increased coronary blood flow (CBF) dose-dependently. During pliable coronary stenosis, SP still increased CBF and decreased stenosis resistance (SR) due to dilation of the stenotic segment, but VIP decreased CBF and increased SR due to passive narrowing of the stenotic segment as a result of preferential dilation of small coronary arteries. Following endothelial denudation of the proximal portion of the coronary artery, the SP-induced increment of CBF during coronary stenosis was attenuated markedly, whereas the effects of VIP were not affected. These results suggest that SP and VIP regulate coronary vascular tone at different vessel sizes in a different manner.

Key words: Vasoactive intestinal peptide—Substance P—Coronary stenosis—Coronary circulation—Coronary vasodilation

Introduction

An increase of the vasomoter tone in a large epicardial artery plays an important role in initiation of ischemic episodes including variant angina, exercise-induced angina, unstable angina, myocardial infarction, and sudden death. We developed an experimental model of a pliable coronary stenosis that preserves active vasomotion of the stenosed segment. In this model, coronary stenosis was produced by an inflation of a microballoon within the proximal coronary artery [1–3]. This type of coronary stenosis seems to be relevant to the clinical features of human atherosclerotic coronary lesions that contain some intact smooth muscle and maintain stenosis vasomobility [4–6]. This is in contrast to rigid coronary stenosis that precludes active vasomotion of the stenosed segment. With use of this model of coronary stenosis, in which direct measurements of coronary blood

---

[1] The First Department of Internal Medicine, Kobe University School of Medicine, 7-5-1, Kusunoki-cho, Chuo-ku, Kobe, 650 Japan

flow and pressure gradients across the stenosis could allow calculation of coronary resistance, we have been investigating the effects of a variety of vaso-constricting and vasodilating agents in recent years [1–3, 7–9]. This has proven to be suitable for this purpose.

In this paper we evaluated the effects of vasoactive intestinal peptide (VIP) and substance P (SP), both vasoactive neuropeptides present in nerves of coro-nary arteries and putative mediators of nonadrenergic noncholinergic peptidegic nerves [10–12], on coronary hemodynamics using a pliable coronary stenosis model. In addition, we discuss the regulation of coronary circulation in response to a variety of vasoactive substances during rigid and pliable coronary stenosis.

Methods

Experimental Preparation

Thirty-two mongrel dogs of either sex, weighing 10–15 kg, were pretreated with subcutaneous morphine (1 mg/kg), anesthetized with intravenous alpha-chloralose (100 mg/kg) and ventilated by a mechanical respirator with positive pressure, using room air supplemented with oxygen. Blood gases and acid/base balance were maintained within normal limits. A catheter was passed through the right femoral artery and advanced into the aortic arch for aortic pressure monitoring. A left thoracotomy was performed and the heart was suspended in a pericardial cradle. A stiff catheter, 10-cm-long, was inserted into the left ventricle through the apex to record left ventricular pressure and rate of rise of pressure (dP/dt). The left common carotid artery was exposed and a 0.5–1 cm segment of the left circumflex coronary artery was dissected free. After administration of heparin, 5000 units, the circumflex artery was ligated, promptly cannulated with a thin metal cannula (inner diameter, 2.4 mm) and continuously perfused from the left carotid artery through the perfusion tubing with a minimum inner diameter of 2.4 mm (Fig. 1). Heparin, 2000 units, was supplemented every 30 min. Coronary perfusion pressure at the tip of the cannula was measured, and it was confirmed that mean coronary perfusion pressure and mean aortic pressure were identical in each experiment. Circumflex coronary blood flow was measured with an extracorporeal electromagnetic flow probe (Nihon Koden MF-26, Tokyo, Japan, inner diameter 3 mm). A polyethylene catheter was placed in a small branch of the circumflex coronary artery distal to the occlusion site to record distal coronary pressure. Pressures were measured with Statham trans-ducers (P23Db Gould, Inc., Glen Burnie, Md.). Measurements of heart rate and all systemic and coronary hemodynamic variables were continuously recorded.

Partial coronary obstruction was produced by an inflation of the specially made microballoon occluder, consisting of a minute rubber balloon attached to the tip of polyethylene tubing (outer diameter 0.6 mm). The construction of the microballoon occluder and the characteristics of the experimental model were

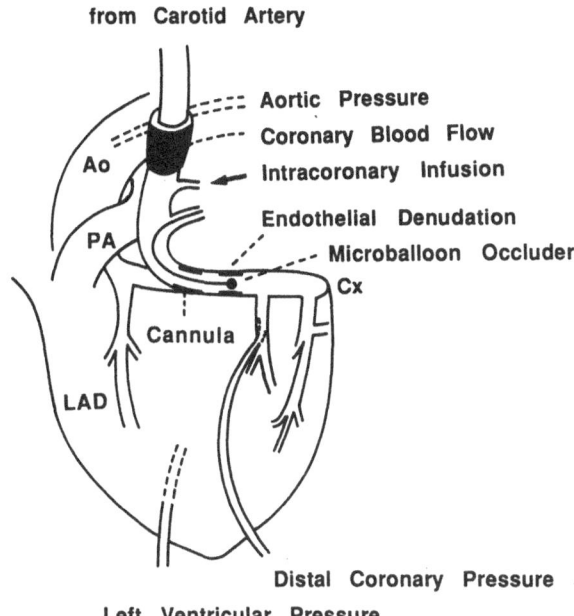

Fig. 1. Experimental preparation. *LAD*, Left anterior descending coronary artery; *Cx*, left circumflex coronary artery; *Ao*, aorta; *PA*, pulmonary artery

reported previously [1–3]. This occluder was inserted through the side arm of the perfusion tubing and advanced into the intact proximal portion of the circumflex coronary artery. There was no major branch between the cannula and the occluder. It was ascertained that placement of the microballoon occluder in the coronary artery affected neither coronary blood flow at rest and its phasic pattern nor the peak reactive hyperemic response before operation of the balloon occluder. The size of the balloon was finely adjusted by expansion with saline solution sufficient to reduce coronary blood flow at rest by approximately 30–40%, and the expansion volume was kept constant. Stenosis resistance was calculated by dividing mean pressure gradient across the stenosis by mean coronary blood flow. Mean pressure gradient was calculated as mean aortic pressure minus mean distal coronary pressure.

Experimental Protocol

The effects of intracoronary injection of VIP and SP were examined in the presence and absence of coronary stenosis. The dogs were divided into two separate treatment groups in which each peptide was administered. Each group was divided into two subgroups, eight dogs without coronary stenosis and eight dogs with coronary stenosis. In the ischemic series, coronary stenosis was produced after control recording at rest. Under both conditions, a 10-min stabilization period was allowed before drug administration, and then an intracoronary injection was performed through the coronary perfusion line at a rate of 0.2 ml/min. It was confirmed that this perfusion rate of physiologic saline solution and a vehicle of the peptides did not affect coronary hemodynamic variables, and that the transit time between injection site and cannula tip was less than a few seconds in the presence and absence of coronary stenosis in the preliminary examinations.

The effects of VIP (0.2, 1 and 5 µg/min) for 1 min and SP (0.001, 0.01, 0.1, and 1 pmol/kg per min) for 40 s were evaluated in each group. The infusion periods were determined by the periods sufficient to obtain the constant coronary effects of each agent in the presence and absence of coronary stenosis. The sequence of the doses of peptides to be examined was selected randomly. There was a 30-min interval between respective drug injections for recovery. Results were uninfluenced by the sequence of doses. At the conclusion of each experiment, dye injection at the same infusion rate and period was performed, and uniform staining of the distal circumflex coronary artery bed was revealed.

Endothelial Denudation of Stenotic Segment

To examine the role of the endothelium in response to intracoronary neuropeptide administration, endothelial denudation was performed around the site where the microballoon occluder was placed in 16 dogs. After the effects of neuropeptide infusion were examined without and with coronary stenosis as control studies, endothelial denudation was performed. A 2F Fogarty embolectomy catheter (Baxter Healthcare Corporation Edwards LIS Division, Santa Ana, CA) was inserted through the side arm of the perfusion tubing and placed in the proximal portion of the circumflex artery. The intima of this portion was mechanically de-endothelialized by balloon inflation and drawing of the catheter. The balloon size

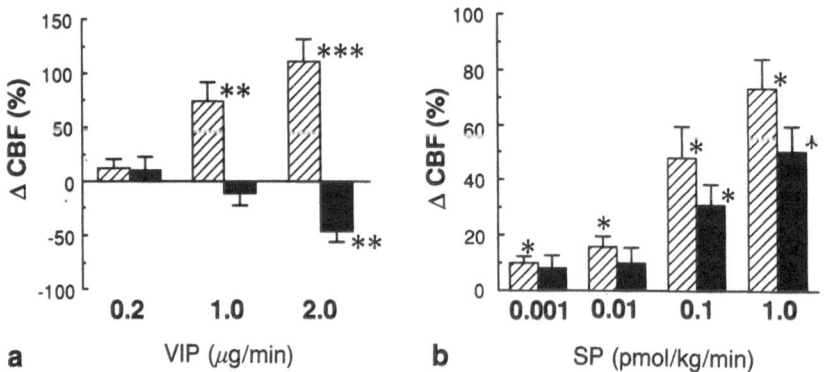

Fig. 2a,b. Percent change in coronary blood flow (*ΔCBF*) in response to various doses of **a** vasoactive intestinal peptide (*VIP*) and **b** substance P (*SP*) with (*solid columns*) and without (*cross-hatched columns*) coronary stenosis. *$P < 0.01$ compared with preinjection; **$P < 0.005$ compared with preinjection; ***$P < 0.001$ compared with preinjection

of the embolectomy catheter was previously adjusted so as not to distend the artery. The microballoon occluder was then placed in the portion of the artery where mechanical rubbing had been applied.

The effects of intracoronary neuropeptide were again examined in the absence and presence of coronary stenosis. To exclude the possibility that endothelial damage might alter the responsiveness of the vascular smooth muscle, the responses to a 40-s infusion of nitroglycerin (1 μg/kg per min) were also compared before and after the procedure.

Statistical Analysis

The mean and standard error of the mean (SEM) were calculated for all variables. Data for a single response in the same dog were analyzed by the t-test for paired comparison. Differences among the different doses of each agent and groups were analyzed by analysis of variance and Tuskey's test.

Results

There were no significant hemodynamic differences in preinjection values in dogs without coronary stenosis.

Effects of VIP

In the absence of coronary stenosis, intracoronary administration of VIP (0.2, 1, and 2 μg/min) increased coronary blood flow in a dose-dependent manner by 111 ± 20.6% at the highest dose without any significant changes in systemic hemodynamic variables (Fig. 2a). Coronary blood flow reached its peak within 20 s after the start of the infusion and sustained the peak during the infusion period.

Application of coronary stenosis decreased coronary blood flow and mean distal coronary pressure by 33% and 40 mmHg, respectively. This was associated with a 4.6% reduction in left ventricular dP/dt. In the presence of stenosis, VIP (1 and 2 μg/min) decreased coronary blood flow by 10.7% and 39.5%, reduced mean distal coronary pressure by 17.5 and 25 mmHg, and produced a 1.6-fold and 2.7-fold intensification in stenosis resistance, respectively (Figs. 2a, 3).

Effects of Substance P

In the absence of coronary stenosis, intracoronary administration of SP (0.001 to 1 pmol/kg per min) increased coronary blood flow in a dose-dependent manner by 73 ± 10.7% at the highest dose without any significant changes in systemic hemodynamic variables (Fig. 2b). The increases in coronary blood flow induced by each dose of the peptide were transient and began to decline during the infusion period.

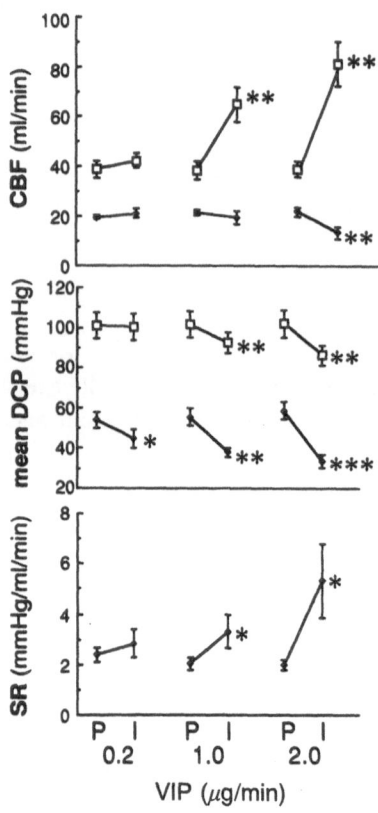

Fig. 3. Effects of vasoactive intestinal peptide (*VIP*) on coronary blood flow (*CBF*), mean distal coronary pressure (*DCP*) and stenosis resistance (*SR*) with (*closed diamonds*) and without (*open squares*) coronary stenosis. The data are expressed as the values (mean ± SEM) at preinjection (*P*) and at the end of injection (*I*). *$P < 0.05$ compared with preinjection; **$P < 0.005$ compared with preinjection; ***$P < 0.001$ compared with preinjection

Application of coronary stenosis decreased coronary blood flow and mean distal coronary pressure by 21% and 34 mmHg, respectively. This was associated with a 5.8% reduction in left ventricular dP/dt. In the presence of stenosis, SP at doses > 0.1 pmol/kg per min significantly increased coronary blood flow by 31% and 50%, reduced mean distal coronary pressure by 5 and 7 mmHg, and reduced stenosis resistance by 18% and 15%, respectively (Figs. 2b, 4).

Effects of Endothelial Denudation in vivo on the Response to VIP and Substance P

There were no significant differences in basal systemic and coronary variables before and after endothelial removal. Coronary blood flow increment in response to intracoronary VIP and SP infusion without coronary stenosis was not altered significantly before and after endothelial denudation. In the presence of coronary stenosis, intracoronary VIP-induced change in coronary blood flow, mean distal coronary pressure, and stenosis resistance was similar before and after endothelial denudation (Fig. 5a). In contrast, coronary blood flow increment induced by SP during coronary stenosis was almost completely abolished after this procedure

Fig. 4. Effects of substance P (*SP*) on coronary blood flow (*CBF*), mean distal coronary pressure (*DCP*) and stenosis resistance (*SR*) with (*closed diamonds*) and without (*open squares*) coronary stenosis. *P < 0.05 compared with preinjection; **P < 0.01 compared with preinjection. Other abbreviations and symbols as in Fig. 3

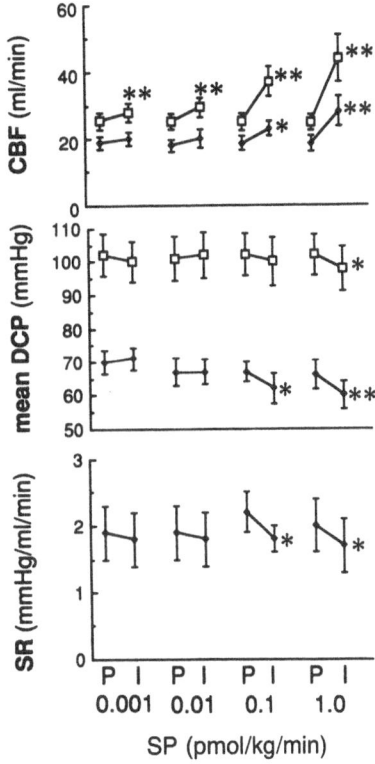

(before: 25 ± 4.4%; after: 8 ± 2.2% (P < 0.01) for 0.1 pmol/kg per min, and before: 40 ± 3.2%; after: 10 ± 3.8% (P < 0.01) for 1 pmol/kg per min) (Fig. 5b). Mean distal coronary pressure decreased further, and the decrement of stenosis resistance was completely abolished.

Intracoronary infusion of nitroglycerin (1 µg/kg per min) significantly increased coronary blood flow and mean distal coronary pressure and decreased stenosis resistance after endothelial denudation to the same extent as before the procedure (data not shown).

Discussion

The results of this study demonstrate that, in the absence of coronary stenosis, both VIP and SP increased coronary blood flow in a dose-dependent manner. In the presence of pliable flow-limiting coronary stenosis, VIP decreased coronary blood flow and intensified stenosis resistance, whereas SP increased coronary blood flow and reduced stenosis resistance. The endothelial denudation of the obstructed site had no effect on the changes in coronary variables in response to VIP. In contrast, this procedure almost completely abolished the increase in coronary blood flow and the reduction in stenosis resistance in response to SP.

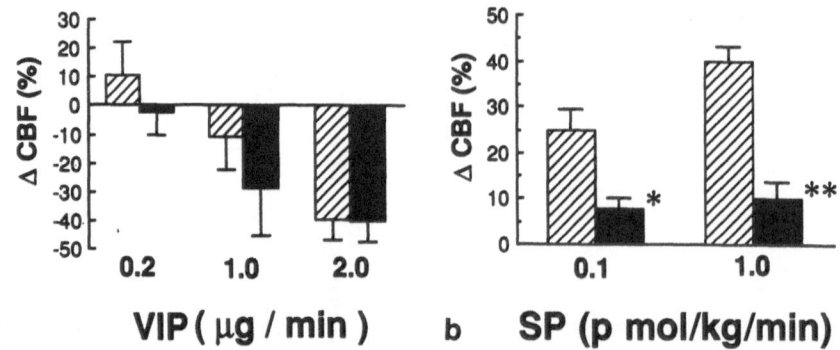

Fig. 5a,b. Percent change in coronary blood flow (ΔCBF) in response to various doses of **a** vasoactive intestinal peptide (*VIP*) and **b** substance P (*SP*) during coronary stenosis before (*cross-hatched columns*) and after (*solid columns*) endothelial denudation. *$P < 0.01$ compared with values before denudation; **$P < 0.001$ compared with values before denudation

Partial Coronary Obstruction Model

In the experimental model used in this study, coronary inflow of the left circumflex coronary artery was restricted by an inflation of the microballoon in the artery [1–3]. This coronary obstruction preserves active vasomotion in a stenosed segment and is designated as "pliable coronary stenosis". This model is capable of changing its severity as a result of alterations of large coronary arterial tone, resembling human atherosclerotic coronary lesions that contain some intact smooth muscles and maintain stenosis vasomobility [4–6]. On the other hand, in coronary stenosis produced by external application of a constrictor device around the artery and designated as "rigid coronary stenosis", active vasomotion of a stenosed segment was precluded, resembling human atherosclerotic lesions that contain little or no intact smooth muscle. Although in the normal coronary circulation the contribution of large coronary vessels to total coronary resistance appears to be trivial, less than 5% [13], in the presence of severe coronary stenosis, alterations in large epicardial arterial tone may be crucial in regulation of coronary circulation. Therefore, our pliable stenosis model is useful in testing effects of vasoactive agents on coronary vasculature (Table 1).

In this pliable stenosis model, vasoconstrictive agents like ergonovine that preferentially constrict large coronary arteries reduce coronary blood flow and mean distal coronary pressure and intensify stenosis resistance as a result of constriction of the stenosed segment [1, 2]. By contrast, vasodilators like nitroglycerin that have selective or preferential dilatory effects on large coronary arteries restore coronary perfusion and reduce stenosis severity [7]. However, in the presence of rigid stenosis produced by external application of a constrictor device around the artery, both ergonovine and nitroglycerin fail to affect coronary hemodynamics due to preclusion of active vasomotion of a stenosed segment [2, 7].

Table 1. Coronary hemodynamic effects of vasoactive agents

	Stenosis (−)		Stenosis (+)		
	CBF	mDCP	CBF	mDCP	SR
Ergonovine	→	→	↓	↓	↑
Nitroglycerin	→	→	↑	↑	↓
	H ↑				
Adenosine	↑	→	↓	↓	↑
Nifedipine	↑	→	↑	↑	↓
			H →	↓	↑
Prostacyclin	↑	→	L ↑	↑	↓
			H ↓	↓	↑
Serotonin	↑	→	↓	↓	↑

→, No change; ↑, increase; ↓, decrease; *CBF*, coronary blood flow; *mDCP*, mean distal coronary pressure; *SR*, stenosis resistance *L*, lower doses; *H*, higher doses

In addition, vasodilators like adenosine that have potent dilatory effects on small resistance coronary arteries decrease coronary blood flow and intensify stenosis resistance in the presence of pliable stenosis [8]. This deleterious response of coronary vasculature to adenosine can be explained by the mechanism of passive narrowing of the stenosed segment [14]. The potent dilation of small resistance vessels in response to adenosine causes a decrease in intraluminal distending pressure and nullifies the stenosis-dilatory effects of this agent due to an increase in large vessel diameter, resulting in a paradoxic intensification of stenosis severity and a deterioration of coronary perfusion.

In the absence of coronary stenosis, nifedipine and prostacyclin produce dose-dependent increases in coronary blood flow. During pliable stenosis, these agents cause divergent responses of coronary hemodynamic variables related to their doses [3, 8]. Lower doses of nifedipine restore coronary perfusion and reduce stenosis severity, suggesting selective or preferential dilatory effects on large coronary arteries. Lower doses of prostacyclin seem to have preferential but feeble dilatory effects on large coronary arteries. In contrast to lower doses of these agents, higher doses of nifedipine and prostacyclin had a deleterious effect on coronary vasculature due to passive narrowing caused by potent dilation of small resistance vessels.

Serotonin has divergent effects on coronary vasculature; i.e., constriction of a large coronary artery and dilation of small coronary arteries [9]. The vaso-constrictor action of serotonin is involved in the production of myocardial ischemia during pliable coronary stenosis.

Effects of Vasoactive Intestinal Peptide and Substance P

Taking these findings into consideration, we can suggest from the present results that VIP has preferential or selective dilatory action on small resistance coronary arteries and that SP has the preferential dilatory effects on the large coronary arteries as well as concomitant dilatory effects on small coronary arteries. During pliable stenosis, the deleterious change in coronary variables in response to VIP can be explained by passive narrowing of the stenosed segment. SP exerts salutary effects of restoring coronary perfusion and reducing stenosis severity through the endothelium-dependent dilatation of large coronary arteries. Although the physiological roles of VIP and SP remain to be clearly established, these neuropeptides present in VIP-containing [10, 11] and SP-containing nerves [12] in the heart and coronary arteries may participate in the physiological and pathophysiological regulation of coronary circulation.

References

1. Sakamoto S, Yokoyama M, Akita H, et al (1983) Effects of ergonovine-induced vasoconstriction on the genesis of myocardial ischemia during coronary stenosis in dogs. Jpn Heart J 24:117-125
2. Yokoyama M, Sakamoto S, Kawashima S, Okada T, Fukuzaki H (1985) Myocardial ischaemia produced by ergonovine-induced vasoconstriction during preexisting coronary stenosis: Experimental conditions for the geometric theory. Cardiovasc Res 19:237-248
3. Yokoyama M, Sakamoto S, Kusui A, Akita H, Maekawa K, Fukuzaki H (1985) Effects of nifedipine on coronary vasculature in canine models of dynamic and fixed coronary stenoses. J Pharmacol Exp Ther 233:845-852
4. Maseri A, L'Abbate A, Baroldi G, et al (1978) Coronary vasospasm as a possible cause of myocardial infarction. A conclusion derived from the study of "preinfarction" angina. N Engl J Med 299:1271-1277
5. Brown BG, Bolson EL, Dodge HT (1984) Dynamic mechanism in human coronary stenosis. Circulation 70:917-922
6. Saner HE, Gobel FL, Salomonowitz E, Erlien DA, Edwards JE (1985) The disease-free wall in coronary atherosclerosis: Its relation to degree of obstruction. J Am Coll Cardiol 6:1096-1099
7. Sakamoto S, Yokoyama M, Fukuzaki H (1985) Dilatation of coronary stenosis as the salutary effect of nitroglycerin in relief of myocardial ischemia in the dog. J Cardiovasc Pharmacol 7:562-568
8. Sakamoto S, Yokoyama M, Kashiki M, Fukuzaki H (1987) Comparative effects of intracoronary vasodilators on restoring coronary perfusion during flow-reducing coronary stenosis in the dog. J Am Coll Cardiol 9:119-126
9. Ichikawa Y, Yokoyama M, Akita H, Fukuzaki H (1989) Constriction of a large coronary artery contributes to serotonin-induced myocardial ischemia in the dog with pliable coronary stenosis. J Am Coll Cardiol 14:449-459
10. Dells NG, Papka RE, Furness JB, Costa M (1983) Vasoactive intestinal peptide-like immunoreactivity in nerves associated with the cardiovascular system of guinea-pigs. Neuroscience 9:605-619
11. Brum JM, Bove AA, Sufan Q, Reilly W, Go VLW (1986) Action and localization of vasoactive intestinal peptide in the coronary circulation: Evidence for nonadrenergic, noncholinergic coronary regulation. J Am Coll Cardiol 7:406-413
12. Brum JM, Go VLW, Sufan Q, Lane G, Reilly W, Bove AA (1986) Substance P distribution and effects in the canine epicardial coronary arteries. Regul Pept 14:41-55

13. Winbury MM, Howe BB, Hefner MA (1969) Effect of nitrates and other coronary dilators on large and small coronary vessels: An hypothesis for the mechanism of action of nitrates. J Pharmacol Exp Ther 168:70–95
14. Santamore WP, Walinsky P (1980) Altered coronary flow responses to vasoactive drugs in the presence of coronary arterial stenosis in the dog. Am J Cardiol 45: 276–285

Abstracts

Coronary Pressure-flow Relationship below the Range of Autoregulation

Shiro Iwanaga, Keiko Uno[1], Shingo Hori[2], and Yoshiro Nakamura[3]

The effects of elevated aortic pressure on the pressure-flow (P/F) relationship below the pressure range of autoregulation were investigated. The effects of myocardial contraction dysfunction on coronary blood flow were also studied. In eight dogs, the left anterior descending coronary artery (LAD) was perfused by a bypass circuit from the subclavian artery. Coronary perfusion pressure (CPP) was reduced by mechanical stenosis of the bypass tube. Aortic pressure (AoP) was elevated by aortic banding. The heart rate was kept constant by SA block and pacing. Bypass flow (CoF), myocardial shortening (%SS) of the LAD region, and instantaneous diastolic P/F relation were obtained at each CPP and AoP level. When the stenosis was mild (CPP > 60 mmHg), CoF was increased by 1.2 ml/min by 10 mmHg elevation of AoP, but when the stenosis was severe (CPP < 50 mmHg), CoF was reduced by 2.0 ml/min by AoP elevation. The slope of P/F relation was reduced by the elevation of AoP in mild stenosis. In severe stenosis, AoP elevation increased the zero flow pressure (Pzf) and did not change the slope of the P/F relation. The %SS was significantly reduced with a rise of AoP only in severe stenosis. In five other dogs, phasic LAD flow was measured, using an ultrasonic transit time flowmeter just below the first diagonal branch. The branch was cannulated proximally to infuse adenosine (500 μg/min) into the LAD to obtain maximal vasodilation. Lidocaine (0.5–4 mg) was injected into the LAD to alter the regional myocardial contraction. There was no significant change in total LAD flow, but late systolic LAD flow was increased when the contraction became akinetic.

Conclusion. When the CPP is below the range of autoregulation, the response of coronary blood flow to the elevation of aortic pressure differs significantly. AoP elevation reduces CoF and induces deterioration of myocardial contraction. This CoF reduction is not caused by the myocardial contraction abnormality itself, but by the impaired diastolic filling of the left ventricle.

Cardiopulmonary division, Department of Medicine[1], Department of Emergency Medicine[2], and Department of Geriatrics[3], Keio University, Tokyo, 160 Japan

Modulation by Aging of the Coronary Vascular Response to Endothelin-1 in the Isolated Perfused Rat Heart

Yumi Katano, Akira Ishihata, Shigeru Morinobu, and Masao Endoh[1]

Endothelin-1 (ET-1) is a potent vasoconstrictor peptide generated by vascular endothelial cells. It has been reported that ET-1 induces transient vasodilation prior to a long-lasting vasoconstrictor response in the coronary vessels of the rat. Since the dilator response to ET-1 requires intact endothelium, substances such as endothelium-delived relaxing factor (EDRF) and prostacyclin released from endothelial cells may play a role in the vasodilation. These substances may counteract the vasoconstriction induced by ET-1 in the coronary vessels. Therefore changes in endothelial function may be responsible for age-related alterations in vascular responsiveness, thereby contributing to the increased incidence of cardiovascular disorders seen in aged populations. We studied the influence of aging on the coronary vascular response to ET-1, with special attention to the involvement of EDRF, in the isolated perfused rat heart. We used 2-, 6- and 24-month-old (mo) male Fisher 344 rats in this study. Hearts were perfused at a constant pressure by the Langendorff technique, using modified Krebs-Henseleit solution (36°). ET-1 injected as a single bolus (0.3, 3, and 30 nmol) reduced the coronary flow in all groups. In half of the 2-mo group injected with 0.3 nmol of ET-1, an initial transient vasodilation prior to vasoconstriction was observed; this was abolished by hemoglobin (100 μmol/min). ET-1 (30 nmol) induced a more pronounced vasoconstriction in 6- and 24-mo than in 2-mo rats. Hemoglobin significantly enhanced the ET-1-induced vasoconstriction in 2-mo, but not in 6- and 24-mo rats. Thus the age-related change in the ET-1-induced response in the coronary resistance vessels occurred between the ages of 2 and 6 months in the rats. The response to ET-1 was not significantly different in the 6- and 24-mo rats, either in the absence or presence of hemoglobin. These results suggest that EDRF may play a role in counteracting the ET-1-induced vasoconstriction in the coronary resistance vessels of young animals, and that the effect of this mechanism may decreases with increasing age.

[1] Department of Pharmacology, Yamagata University School of Medicine, Yamagata City, 990-23 Japan

Contraction and Relaxation Responses to fMLP in Isolated Human Coronary Arteries

Mitsumasa Keitoku and Tamotsu Takishima[1]

The intimal thickening commonly found in human coronary arteries involves macrophage infiltration; this may contribute to the development of local vascular tone. We investigated the effects of the leukocyte stimulant, N-formyl-L-methionyl-L-leucyl-L-phenylalanine (fMLP), on human coronary arterial tone. For isometric tension recording, isolated human coronary arterial rings were suspended in a tissue bath filled with Krebs-Ringer solution. A single dose of fMLP (10^{-5} M) was added to rings precontracted with 10^{-5} M phenylephrine. Tension changes in response to fMLP were expressed as percentages of maximum potassium contraction or maximum papaverine relaxation. The inhibitors used were pretreated 15 min before the addition of phenylephrine. In parallel, coronary segments were histologically examined for identifying the infiltrated inflammatory cells. In most cases examined, fMLP at 10^{-5} M produced biphasic tension changes, with rapid contraction followed by relaxation ($60\pm9\%/-52\pm10\%$). Removal of the endothelium or adventitia did not alter the responses to fMLP. The contraction phase was nearly abolished by the thromboxane A_2/prostaglandin H_2 (TXA_2/PGH_2) receptor blocker, ONO3708 (10^{-6} M; $0\%/-72\pm12\%$), and the selective TXA_2 synthetase inhibitor, DP-1904 (10^{-5} M; $11\pm4\%/-55\pm11\%$). Indomethacin (2×10^{-5} M) inhibited both phases, but the contraction remained small ($11 \pm 4\%$). However, the selective PGI_2 synthetase inhibitor tranylcypromine (10^{-4} M) did not reduce the relaxation phase in the presence of ONO3708 ($-49\pm15\%$). Superoxide dismutase (150 U/ml) did not affect any phase of the responses to fMLP ($62\pm12\%/-45\pm10\%$). From these results, we conclude that fMLP, which probably activates intimal macrophages, induced contraction and relaxation responses in isolated human coronary arteries mainly through cyclooxygenase products, including TXA_2 and some vasodilator substance other than PGI_2.

[1] First Department of Internal Medicine, Tohoku University School of Medicine, Sendai, 980 Japan

Coronary Arterial Branching Structure is Highly Variable

Ed VanBavel and Jos A.E. Spaan[1]

The coronary arterial branching structure forms an important determinant for flow distribution and its heterogeneity. We analyzed this structure in two steps: (A) performing morphometric measurements on coronary casts; (B) using these measurements for the computer generation of arterial networks, in order to determine global parameters. Two porcine coronary casts were made after maximal dilation, diastolic arrest, and glutaraldehyde fixation. Unbranched lengths and diameters (L and d, μm) of successive vascular segments were measured. The length (L) was related to d: ^{10}log (L) = 1.01 + 0.72* ^{10}log (d) (n = 2366, r^2 = 0.45). For any d, L varied 100-fold. The relation between diameters of proximal (d_0) and distal segments (d_L, d_s, with $d_L > d_s$) at a node was quantified using the polar variables growth (G) and symmetry (S): G = sqrt $[(d_L/d_0)^2 + (d_s/d_0)^2]$; S = arctan $[(d_s - 5)/(d_L - 5)]$. G was slightly dependent on d_0: G = 1.131 − 0.042* ^{10}log (d_0) (r^2 = 0.02, n = 1663, P < 0.0001). Deviations of G were normally distributed with SD = 0.13. Observed values of S covered the possible range of 0 to pi/4 radians (rad). Larger vessels branched less symmetrically (P < 0.0001): d_0 < 40: S = 0.49 ± 0.20 rad (mean + SD, n = 579); 40 < d_0 < 200: S = 0.45 ± 0.19 (n = 817); d_0 > 200: S = 0.38 ± 0.19 (n 267). The above relations, including the variability, were used to construct 30 stochastic models of arterial networks, starting at 500 μm and ending with arterioles between 5 and 10 μm. Strahler ordering was applied to these generated trees. The bifurcation ratio was constant along orders 2 to 10, and equaled 3.35. Strahler ordering combines consecutive segments of equal order number to vessels. For each order, the number of segments per vessel (k) followed a wide and highly skewed geometric distribution: $N(k)/N = q*(1-q)^{(k-1)}$, with N the number of vessels, N(k) the number of vessels with k segments, and q the parameter of the distribution; q was found to decrease from 0.52 in order 2 to 0.13 in order 10. The constant bifurcation ratio and geometric nature of the number of segments per vessel suggest that the topology of the coronary arterial tree resembles that of a completely random network.

---

[1] Cardiovascular Research Institute, Department of Medical Physics and Informatics, University of Amsterdam, Meibergdreef 15, 1105 AZ Amsterdam, The Netherlands

The Inhibition of Endothelium-Dependent Vasorelaxation by Oxidized Low Density Lipoprotein and Lysophosphatidylcholine

Mitsuhiro Yokoyama, Ken-ichi Hirata, Nobutaka Inoue, Hozuka Akita, and Yuichi Ishikawa[1]

The vascular endothelium produces endothelium-derived relaxing factor(s) which regulate(s) vascular tone. In atherosclerotic arteries, endothelium-dependent relaxation is markedly reduced; this impairment is thought to play an important role in the pathogenesis of coronary spasm. The purpose of this study was to evaluate the effects of native and modified low density lipoprotein (LDL) on endothelium-dependent relaxation and to clarify the mechanism underlying the inhibitory effects of oxidized LDL. LDL(density, 1.020–1.060) was isolated by ultracentrifugation from freshly-harvested normal human plasma. Phospholipase A2-treated LDL and copper-oxidized LDL were prepared by the method of Quinn et al. For recording isometric force, rabbit aortic strips equilibrated with oxygenated Kreb's buffer were precontracted with phenylephrine. We measured phosphoinositide hydrolysis by a ^3H inositol labelling method and cytosolic Ca^{2+} level by using fura-2 fluorescence spectroscopy in cultured bovine aortic endothelial cells. Preincubation of the strips with native LDL(0.5–2 mg protein/ml) had no effect on acetylcholine(Ach)- induced endothelium-dependent relaxation. Oxidized LDL(0.1–0.5 mg protein/ml) and phospholipase A2 -treated LDL (0.01–0.1 mg protein/ml) inhibited this relaxation. Native and modified LDL did not inhibit nitroglycerin-induced relaxation. Lysophosphatidylcholine(LPC) extracted from modified LDL by thin layer chromatography inhibited Ach-induced relaxation. Exogeneous administration of synthetic 1-palmitoyl-LPC had a potent inhibitory effect on Ach-evoked relaxation. LPC and oxidized LDL inhibited both phosphoinositide hydrolyses and the elevation of cytosolic Ca^{2+} induced by bradykinin in endothelial cells. We conclude that LPC, which accumulated during the oxidation of LDL in atherosclerotic arteries, impaired endothelium-dependent relaxation through its inhibitory effects on agonist-induced phosphoinositide hydrolysis and cytosolic Ca^{2+} elevation in endothelial cells.

---

[1] First Department of Medicine, Kobe University School of Medicine, Kobe, 650 Japan

E. Pathophysiology of Coronary Circulation in Ischemic Heart Disease

Beneficial Role of Alpha-Adrenoceptor Activity in Myocardial Ischemic and Reperfusion Injury

Masatsugu Hori, Masafumi Kitakaze, Seiji Takashima, Hiroshi Sato[1], Michitoshi Inoue[2], and Takenobu Kamada[1]

Summary. The role played by alpha-adrenoceptor activity in the heart has not yet been extensively investigated. Several lines of evidence, however, support the idea that this activity is closely related to the pathogenesis of ischemic and reperfusion injury. The major effect of alpha-adrenoceptor stimulation in the coronary arteries is vasoconstriction, which may restrict myocardial oxygen supply. Alpha-adrenoceptor stimulation also increases myocardial contractility without marked increases in intracellular Ca^{2+}, which effect may increase myocardial oxygen demand. Thus, it would seem that alpha-adrenoceptor stimulation exacerbates myocardial ischemia by increasing the imbalance between myocardial oxygen demand and supply. Indeed, electrical sympathetic nerve stimulation is reported to have a deleterious effect on myocardial and coronary function. Sympathetic nerve activation during ischemia, however, may not be as potent as electrical sympathetic nerve stimulation, and mild sympathetic nerve stimulation may be beneficial for ischemia and reperfusion injury. Indeed, alpha₁-adrenoceptor stimulation maintains endocardial coronary flow at the expense of epicardial flow, and this stimulation enhances the release of adenosine, while alpha₂-adrenoceptor stimulation increases the release of endothelial-dependent relaxing factor (EDRF) and histamine, and increases the sensitivity of adenosine receptors to adenosine, all of which actions may attenuate ischemic injury. Since alpha₁-adrenoceptor activation enhances adenosine release (through the activation of 5'-nucleotidase); stimulation of alpha₁-adrenoceptors during ischemia and reperfusion produces beneficial effects on reperfusion injury via this enhanced release of adenosine. Further basic studies are necessary to gain a better understanding of the role played by alpha-adrenoceptor activity in the pathogenesis of ischemia and reperfusion injury, and further clinical studies are needed to extend these observations to human ischemic hearts.

Key words: Adenosine—A₁ receptor—A₂ receptor—Norepinephrine—Protein kinase C—Myocardial stunning

Introduction

During myocardial ischemia, norepinephrine is released from the sympathetic nerve terminals and stimulates both alpha- and beta-adrenergic receptors in the ischemic myocardium and coronary arteries [1–3]. When myocardial ischemia is

[1] The First Department of Medicine, Osaka University School of Medicine, 2-2 Yamadagaoka, Suita, 565 Japan
[2] Department of Medical Information Science, Osaka University Hospital, Osaka, 553 Japan

prolonged, norepinephrine enters the bloodstream and affects systemic vascular tone [4]. Accordingly, the release of norepinephrine during myocardial ischemia may modulate the severity of ischemia; beta-adrenergic stimulation may increase myocardial contractility and oxygen demand, and alpha-adrenergic stimulation may cause vasoconstriction. Most importantly, alpha-adrenoceptor stimulation is particularly worthy of study because of the multiplicity of its vascular effects. It is well known that alpha-adrenoceptor stimulation increases the tone of the coronary arteries [5–7]. In addition to this direct vasoactive effect, alpha-adrenoceptor stimulation influences vasoactive substances that may regulate coronary blood flow; alpha-adrenoceptor stimulation is reported to enhance the release of endothelium-dependent relaxant factor (EDRF) [8], histamine [9], and adenosine [10], and to inhibit the release of norepinephrine. Alpha-adrenoceptor stimulation increases the sensitivity of adenosine receptors in coronary arteries and increases the capacity for adenosine production [11–13]. However, there is no clear consensus on whether the net coronary effects due to alpha-adrenoceptor stimulation are beneficial or deleterious to ischemic and reperfusion injury. Here, we discuss the specific and overall effects of alpha-adrenoceptor activity on myocardial ischemia and reperfusion.

Mechanisms of Norepinephrine Release in the Heart

Myocardial ischemia is believed to be a potent trigger of norepinephrine release from sympathetic nerve terminals [3, 4, 14]. Deleterious sequelae, such as the occurrence of fatal arrhythmias [15, 16] and the acceleration of myocardial cell damage [17, 18], which are recognized as catecholamine injury, have been attributed to the excessive stimulation of both alpha- and beta-adrenergic receptors in the myocardium and coronary arteries. In the ischemic myocardium, the accumulation and release of norepinephrine are induced predominantly at the local nerve endings. During ischemia, activation of the nerve endings is reported not to be influenced by central sympathetic nervous activity [3]. Schomig et al. [3] demonstrated that massive amounts of norepinephrine were released during ischemia in the Langendorff rat heart preparation, and they showed that these amounts increased as the ischemic period was prolonged. They also reported that the release of catecholamine occurred within 10 min after the onset of ischemia. Wollenberger and Shaab [1] also observed, in isolated rabbit hearts, that 3–4 min of ischemia caused the overflow of a large amount of norepinephrine. Similarly, in in vivo dog hearts, a coronary arterial occlusion of 2.5 min also resulted in a net overflow of norepinephrine [19]. The consistency of the time course of norepinephrine release in the isolated and in vivo hearts tends to support the idea that this ischemia-induced norepinephrine release is regulated by local factors, rather being controlled by the central nervous system.

Some hormones and neurotransmitters are released from the nerve terminals; this process may depend on intracellular Ca^{2+} concentration. However, in the ischemic myocardium, the release of norepinephrine from sympathetic nerve terminals is considered to be independent of extracellular calcium [3]. Rather, the accumulation of sodium within the sympathetic neurons seems to trigger the release of norepinephrine [20]. Schomig et al. [20] identified Na^+-H^+ exchange as the predominant pathway of sodium entry into the sympathetic nerve endings during ischemia, since treatment with amiloride and ethylisopropylamiloride, inhibitors of Na^+-H^+ exchange, markedly suppressed the ischemia-induced norepinephrine release. The extrusion of protons which may be accumulated during ischemia is coupled with sodium entry, leading to intracellular sodium accumulation. During ischemia, decreases in intracellular ATP inhibit Na^+-K^+-ATPase activity, which feature may also contribute to sodium accumulation. During reperfusion following ischemia, the prompt recovery of intracellular acidosis may further increase Na^+ concentration due to the activation of Na^+-H^+ exchange, leading to the further release of norepinephrine.

Subcellular Mechanisms of Coronary Vasoconstriction due to Alpha-Adrenoceptor Stimulation

The second messengers of alpha$_1$-adrenoceptors in cells are known to be 1,2-diacyl glycerol (DG) and inositol phosphates (IP) [21], as shown in Fig. 1. DG activates protein kinase C, which phosphorylates intracellular and membrane-bound enzymes. Inositol-1,4,5-trisphosphate (IP_3), a member of the IP family, is reported to release Ca^{2+} from the sarcoplasmic reticulum. Intracellular Ca^{2+} is capable of activating the myosin light chain kinase in the myofilament in smooth muscle cells, which triggers phosphorylation of myosin light chain, and thereby causes vascular contraction [22, 23]. Although protein kinase C activation also phosphorylates myosin light chain kinase, it, unexpectedly, relaxes the coronary smooth muscles contracted by the myosin light chain kinase. This difference in action is attributed to the different sites that myosin light chain kinase and protein kinase C phosphorylate: Myosin light chain kinase phosphorylates the serine residue of the myosin light chain and protein kinase C phosphorylates the threonine residue [22–24]. Cyclic AMP, in contrast, which is increased by beta-adrenoceptor stimulation, causes vasorelaxation, since it may reduce Ca^{2+} concentration through enhancement of Ca^{2+} uptake into the sarcoplasmic reticulum. Alpha$_2$-adrenoceptor stimulation is reported to reduce intracellular cyclic AMP content, which is responsible for vasoconstriction [21]. Furthermore, alpha$_2$-adrenoceptor stimulation is reported to activate Na^+-H^+ exchange [25], which, in turn, stimulates the reverse Na^+-Ca^{2+} exchange, thereby leading to the accumulation of intracellular Ca^{2+}. These increases in Ca^{2+} may also contribute to the vasoconstriction that occurs during alpha$_2$-adrenoceptor stimulation.

Since these recent advances have clearly revealed the action of alpha$_1$- and alpha$_2$-adrenoceptors on coronary smooth muscles, we should investigate the physiological implications of the differences in location and receptor subtypes and the contribution of the subtypes to coronary flow regulation.

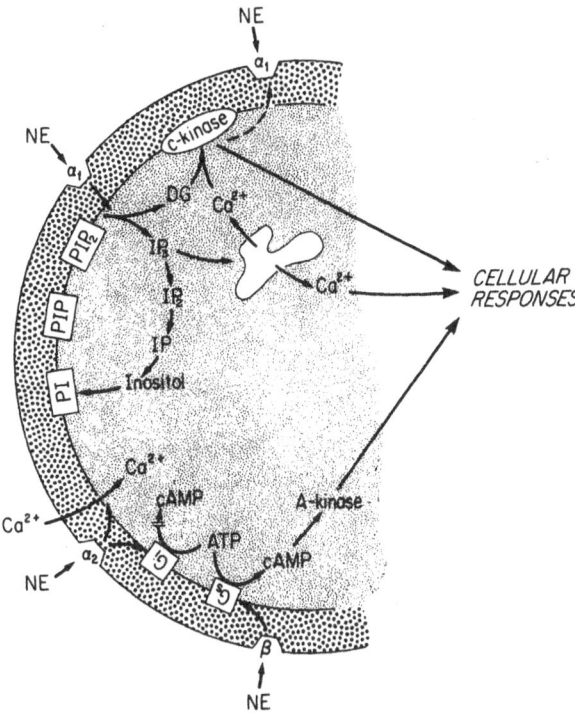

Fig. 1. Subcellular signal transductions of alpha- and beta-adrenoceptor activation. (From [21] reproduced with permission.) [Circulation Research] Copyright [1985] American Heart Association. *DG*, 1,2-Diacyl glycerol; *IP*, inositol phosphates; *NE*, norepinephrine; *PIP*, phosphatidylinositol-monophosphate

Vasoactive Responses of Coronary Arteries to Alpha-Adrenoceptor Stimulation

Alpha-adrenoceptor stimulation contracts isolated coronary arteries [5–7], while beta-adrenoceptor stimulation mediates coronary dilation. Zuberbuhler and Bohr [5], who showed that exposure to norepinephrine under beta-adrenoceptor blockade generated potent contraction of the dog isolated coronary artery, concluded that alpha-adrenoceptor stimulation mediated coronary vasoconstriction. The vasoconstrictive action induced by alpha-adrenoceptor stimulation was more potent in the large coronary arteries than in the smaller ones [5]. Murray and Vatner [6] and Vatner [7] confirmed in conscious dogs that alpha-adrenoceptor stimulation caused by exogenous and endogenous norepinephrine induced potent vasoconstriction of large epicardial coronary arteries; they carried out direct measurements of coronary arterial diameters using an ultrasonic dimension gauge. Kelley and Feigl [26], however, reported

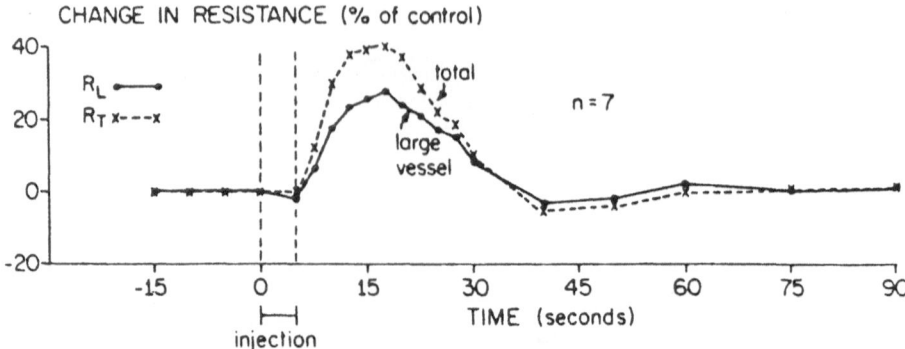

Fig. 2. Changes in coronary vascular resistance due to the contribution of large and small coronary vessels during infusion of norepinephrine. (From [26] reproduced with permission.) [Circulation Research] Copyright [1978] American Heart Association. R_L resistance of large coronary vessels; R_T resistance of total coronary vessels

that increase in coronary vascular resistance in arteries treated with propranolol during exposure to norepinephrine were due to vasoconstriction in both large and small arteries. Indeed, Chilian et al. [27] directly observed that small coronary arteries (100–300 μm) constricted due to alpha-adrenoceptor stimulation. The contribution of small coronary arteries to the increase in coronary vascular resistance during alpha-adrenoceptor stimulation is reported to be approximately 40% [26] (Fig. 2). These lines of evidence show that both small and large coronary arteries participate in the regulation of coronary blood flow when alpha-adrenergic receptors are stimulated. Heusch et al. [28] clarified the contribution of alpha$_1$- and alpha$_2$-adrenoceptor activity to vasoconstriction, using an alpha$_1$-adrenoceptor agonist and antagonist, i.e., methoxamine and prazosin, respectively, and an alpha$_2$-adrenoceptor agonist and antagonist, i.e., BHT 920 and rauwolscine, respectively, in open chest dogs. Intracoronary administration of methoxamine increased large vessel coronary arterial resistance by 19% and end-diastolic coronary resistance by 10%, whereas BHT 920 did not affect large vessel resistance but increased the end-diastolic coronary vascular resistance by 38%. End-diastolic coronary vascular resistance is determined predominantly by the tone of the small coronary arteries. Left cardiac sympathetic nerve stimulation increased large vessel resistance by 11% and end-diastolic coronary resistance by 32%. This increase in large vessel resistance was not prevented by rauwolscine, but was virtually inhibited by prazosin. The increase in end-diastolic resistance was not prevented by prazosin, but was completely inhibited by rauwolscine. These findings indicate that the alpha-adrenoceptor-mediated vasoconstriction of large coronary arteries is mediated exclusively by alpha$_1$-adrenoceptor activity, whereas the alpha-adrenergic vasoconstriction of small coronary arteries is mediated partially by alpha$_1$-adrenoceptor, but predominantly by alpha$_2$-adrenoceptor activity. Another aspect of alpha-adrenoceptor-mediated vascular effects is the redistribution of intramyo-cardial flow, since it has been reported that alpha-adrenoceptor stimulation favors an increase in endocardial flow at the expense of epicardial flow reduction during exercise and coronary hypoperfusion [29, 30]. Alpha$_1$-adrenoceptors are located predominantly in the outer region of the coronary arteries, whereas alpha$_2$-adrenoceptors are predominantly located in the inner region. Since neurons penetrate only into the outer one-third of the medial layer of the coronary arteries and arterioles, it is possible that alpha$_1$-adrenoceptors may be activated by norepinephrine secreted from sympathetic nerve endings, and that alpha$_2$-adrenoceptors may be activated predominantly by circulating catecholamines.

Interaction between Alpha-Adrenoceptor Activity and Vasoactive Substances

When we discuss the role played by alpha-adrenoceptor activity in coronary vascular tone, we need to separate vasoactive effects due to chemical vasoactive substances mediated by alpha-adrenoceptor activity, interactions of alpha-adrenoceptor activity with other receptors, and direct vasoconstriction. First of all, there are several substances whose release is mediated by alpha-adrenoceptor stimulation. Alpha$_2$-adrenoceptor stimulation enhances the release of EDRF from endothelial cells [8]. Two experiments provide evidence for this. First, when the resting tone of the vessel is low, norepinephrine concentration-vascular contraction curves in the absence of endothelium are shifted to the left and have an increased maximal response compared with endothelium-intact arteries. Second, exposure to norepinephrine relaxes arteries that are precontracted under beta-adrenoceptor blockade. This response is blocked by alpha$_2$-adrenoceptor antagonists and is mimicked by clonidine, but not by methoxamine. The release of EDRF by alpha$_2$-adrenoceptor stimulation, which release is also observed in the small coronary arteries, increases coronary blood flow [8], although its physiological significance during exercise and ischemia, and the subcellular mechanisms by which alpha$_2$-adrenoceptor stimulation increases nitric oxide (NO) production have yet to be elucidated. Recent reports suggest that the NO is the same as EDRF, which factor increases cyclic GMP and relaxes coronary smooth muscles [31]. NO also inhibits platelet aggregation. Because platelet aggregation is promoted by alpha$_2$-adrenoceptor stimulation, the linkage between NO and alpha-adrenoceptor activity may be important for the maintenance of homeostasis in coronary physiology. Although leukocytes and vascular smooth muscle cells also release NO, to our knowledge, there are no findings that alpha$_2$-adrenoceptor stimulation increases NO production in these cells. Alpha$_2$-adrenoceptor stimulation inhibits norepinephrine release from sympathetic nerve endings [26, 32], resulting in a decrease in the coronary arterial tone. Alpha$_2$-adrenoceptor stimulation also increases the release of histamine in skeletal

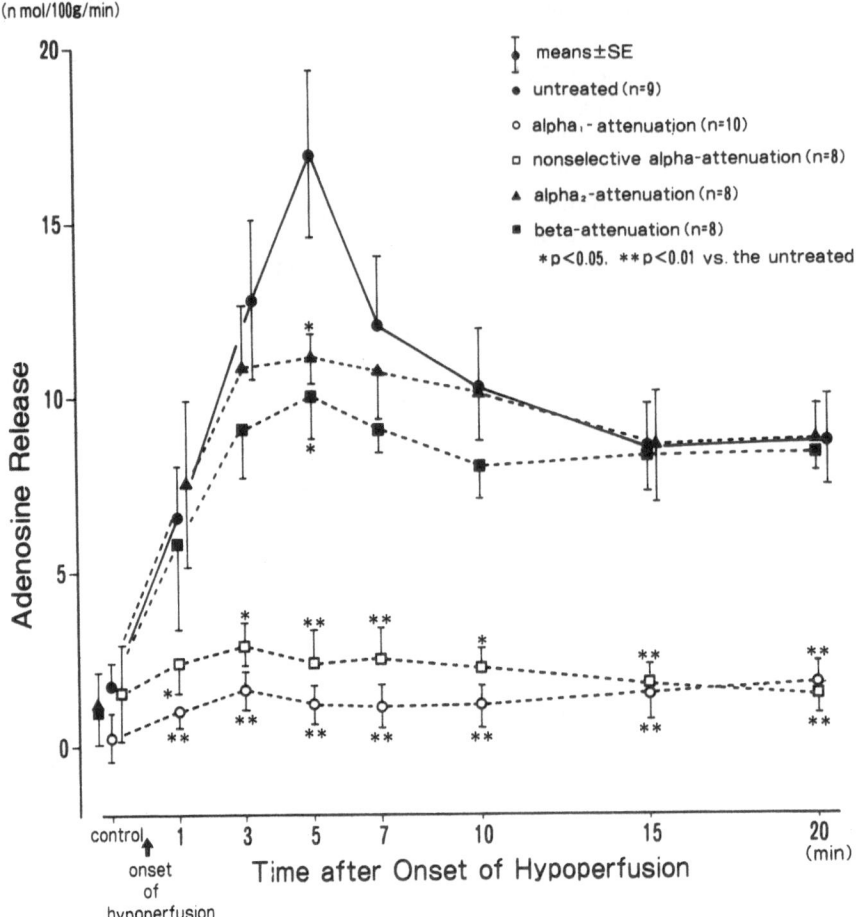

Fig. 3. Adenosine release during coronary hypoperfusion under attenuation produced by alpha₁- (prazosin), alpha₂- (yohimbine + propranolol), non-selective alpha- (phentolamine + propranolol), and beta-adrenoceptor (propranolol) blockers. Note that both prazosin and phentolamine markedly reduced adenosine release (From [11] with permission)

muscle [9]. If histamine is released due to alpha₂-adrenoceptor stimulation in the heart, this substance may act to increase coronary blood flow.

Alpha₁-adrenoceptor stimulation, on the other hand, augments the release of adenosine in the ischemic myocardium. Figure 3 depicts the effects of alpha- and beta-adrenoceptor activity on adenosine release from the ischemic myocardium in open chest dogs. Administration of a low dose of prazosin, which does not affect basal coronary blood flow, reduced adenosine release [10]. The results obtained in the ischemic myocardium [10, 33] contrast well with those in the non-ischemic myocardium [30, 34]. In non-ischemic hearts, propranolol inhibits adenosine release, and prazosin does not inhibit it during intracoronary infusion of norepinephrine. In isolated rat cardiomyocytes, intracellular adenosine production during hypoxia is increased by alpha$_1$-adrenoceptor stimulation. The enhanced production of adenosine is mimicked by phorbol 12-myristate 13-acetate (PMA), a stimulator of protein kinase C, and is inhibited by H-7, an inhibitor of protein kinase C, indicating that protein kinase C activity plays a crucial role in adenosine production in the ischemic heart [35]. These results suggest that protein kinase C may affect the enzymes responsible for adenosine production and degradation, or the production of AMP. It should be noted, however, that the enhanced release of adenosine by alpha$_1$-adrenoceptor stimulation is observed only during ischemia. Thus, the mechanism underlying this phenomenon may be different from that responsible for adenosine release when beta-adrenoceptors are stimulated.

There is an interaction between alpha-adrenoceptor activity and the vasodilatory action of adenosine. Our laboratory has recently reported that

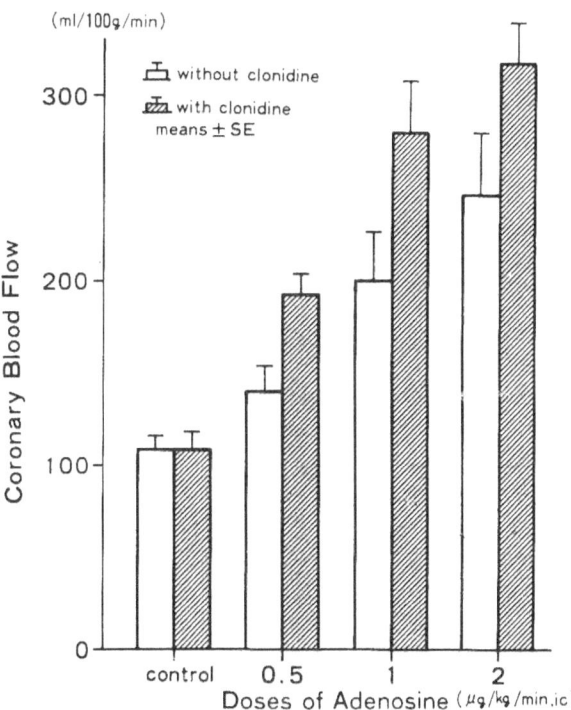

Fig. 4. Adenosine-induced coronary vasodilation with and without intracoronary clonidine administration (0.04 µg/kg per min, ic). (From [13] reproduced with permission.) [Circulation Research] Copyright [1989] American Heart Association

alpha$_2$-adrenoceptor stimulation increases the coronary vasodilatory effects of adenosine [11–13]. Adenosine is an endogenous substance released during exercise and ischemia, and it is known to increase coronary blood flow [36]. We revealed that yohimbine, an alpha$_2$-adrenoceptor antagonist, attenuated adenosine-induced coronary vasodilation in open chest dogs. Clonidine, an alpha$_2$-adrenoceptor agonist, enhances adenosine-induced coronary vasodilation (Fig. 4). These observations are consonant with the results of Nayler et al. [37].

Taken together, we should be careful to consider the effects of alpha-adrenoceptor activity on coronary blood flow in in vivo hearts.

Effects of Alpha-Adrenoceptor Activity on Myocardial Ischemia

Considering the effects of alpha-adrenoceptor activity on myocardial and coronary vascular functions, the determination of whether alpha-adrenoceptor stimulation is beneficial or deleterious to myocardial ischemia seems to be a complex matter. Since alpha$_1$- and alpha$_2$-adrenoceptor activities exert different cardiovascular effects, it would appear that dose-related action may modify the overall effects.

When levels of circulating norepinephrine are not high, alpha$_2$-adrenoceptor stimulation has a beneficial effect on ischemic injury [13]. During coronary hypoperfusion, moderate alpha$_2$-adrenoceptor stimulation, which does not cause direct coronary vasoconstriction, increases coronary blood flow in the ischemic area and restores myocardial contractile function. The increase in coronary blood flow may be attributed to the release of histamine or EDRF, to the enhanced sensitivity of adenosine receptors, and to attenuation of norepinephrine release from the sympathetic nerves. We have also observed the beneficial effects of clonidine even in denervated hearts [13], which finding indicates that attenuated norepinephrine release due to alpha$_2$-adrenoceptor stimulation is not involved. It is of interest that the beneficial effects of alpha$_2$-adrenoceptors on myocardial ischemia are abolished under treatment with 8-phenyltheophylline, an adenosine receptor antagonist. We therefore hypothesize that adenosine-induced coronary vasodilation is primarily involved in the beneficial effects of alpha$_2$-adrenoceptor stimulation. However, if the dose of clonidine is high, ischemia may be worsened, since high doses of clonidine have an adverse effect on adenosine-induced coronary vasodilation, blunting it through a direct vasoconstrictive effect.

Indeed, Heusch and Deussen [38] have demonstrated that postsynaptic alpha$_2$-adrenoceptor stimulation mediates the coronary vasoconstriction produced by sympathetic nerve stimulation during the critical reduction of coronary perfusion pressure; they concluded that alpha$_2$-adrenoceptor stimulation was deleterious for myocardial ischemia. Seitelberger et al. [39] demonstrated that alpha-adrenoceptor blockade attenuated the severity of exercise-induced myocardial ischemia. Exercise and sympathetic nerve stimulation elevates systemic norepinephrine levels to more than 1000 pg/ml, a concentration which is two- to threefold higher than that observed during regional ischemia. Norepinephrine concentration in the systemic blood was 200–300 pg/ml, even when the left anterior descending (LAD) coronary arterial flow was reduced to one-third of the control, and fractional shortening was reduced to 5% from 25% in the control condition [13]. These observations indicate that if alpha$_2$-adrenoceptor stimulation is potent, then the enhancement of adenosine-induced vasodilation by alpha$_2$-adrenoceptor stimulation is masked by direct vasoconstriction. During potent alpha$_2$-adrenoceptor stimulation, platelet aggregation may also occur, and this would contribute to the precipitation of myocardial ischemia, since platelet alpha$_2$-adrenoceptors mediate their activation and aggregation, although adenosine released from the ischemic myocardium is known to potently inhibit platelet aggregation [40, 41]. In contrast, when norepinephrine levels in the systemic arterial blood are low, alpha$_2$-adrenoceptor stimulation may enhance adenosine-induced coronary vasodilation, which action exerts a beneficial effect.

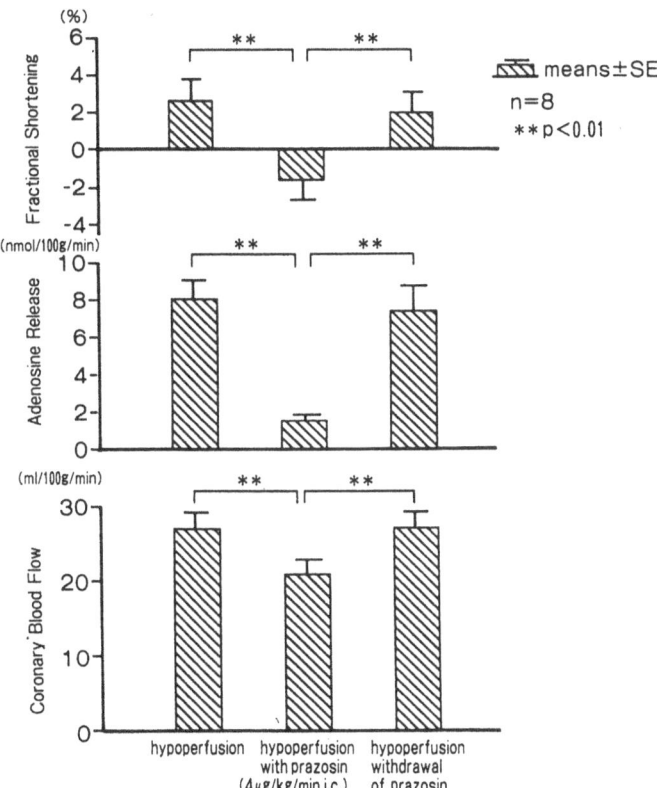

Fig. 5. Changes in fractional shortening, adenosine release and coronary blood flow after onset of alpha$_1$-adrenoceptor attenuation and withdrawal of prazosin during hypoperfusion. (From [10] reproduced with permission.) [Circulation Research] Copyright [1987] American Heart Association

Alpha$_1$-adrenoceptor stimulation is also reported to be beneficial for myocardial ischemia [10, 35]. During coronary hypoperfusion, administration of a low dose of prazosin, which does not cause direct vasodilation but inhibits the release of adenosine, further reduced coronary blood flow and exacerbated ischemia (Fig. 5) [10]. When a larger dose of prazosin was employed, this agent exerted a direct effect on coronary vasodilation, thereby attenuating the extent of ischemia [42]. Thus, it is most likely that mild to moderate alpha$_1$-adrenoceptor stimulation exerts beneficial effects on the ischemic myocardium through the augmentation of adenosine release, whereas potent alpha$_1$-adrenoceptor stimulation causes deleterious effects via direct vasoconstriction. Herrmann and Feigl [43] have also observed that adrenergic blockade blunted adenosine concentration and coronary vasodilation during hypoxia.

Effects of Alpha-Adrenoceptor Activity on Reperfusion Injury

The contractile dysfunction that occurs after a brief period of ischemia is defined as myocardial stunning [44]; although its pathophysiology has been discussed extensively, the role played by alpha-adrenoceptor activity in this abnormality is obscure. We have found that methoxamine attenuated myocardial stunning, and that this was associated with the enhanced release of adenosine; treatment with theophylline completely abolished this beneficial effect of methoxamine [45]. These results suggest that alpha$_1$-adrenoceptor stimulation can attenuate the severity of myocardial stunning through the enhanced release of adenosine (Fig. 6). This finding is compatible with those in other studies that support the protective effects of adenosine on myocardial stunning [46, 47]. However, it should be noted that a high dose of methoxamine blunted the beneficial effects of this agent. It is possible that the preservation of ATP may not be responsible for the beneficial effects of adenosine, since the replenishment of ATP does not necessarily restore contractile function [48, 49]. However this hypothesis cannot be completely discarded [50, 51].

It is possible that calcium overload may be the primary mechanism responsible for myocardial stunning [52–56]. If this is the case, then the alpha$_1$-mediated release of adenosine may attenuate myocardial stunning by inhibiting calcium influx through the stimulation of adenosine A$_1$-receptors [57–59]. The effects of higher doses of methoxamine in blunting the beneficial effects of alpha-adrenoceptor stimulation may thus be attributed to the increases in intracellular Ca^{2+} produced by higher doses of methoxamine during ischemia and reperfusion [59]. A report that prazosin attenuates Ca^{2+} increases during repersusion appears to support this hypothesis [60].

Alpha$_1$-adrenoceptor-mediated release of adenosine also attenuates neutrophil activation [61, 62]. Free radical generation is inhibited through adenosine A$_2$-receptor stimulation, and adherence to endothelial cells is attenuated through adenosine A$_1$-receptor stimulation [63]. Platelet aggregation may cause reperfusion injury, since aggregated platelets impede the microcirculation of the

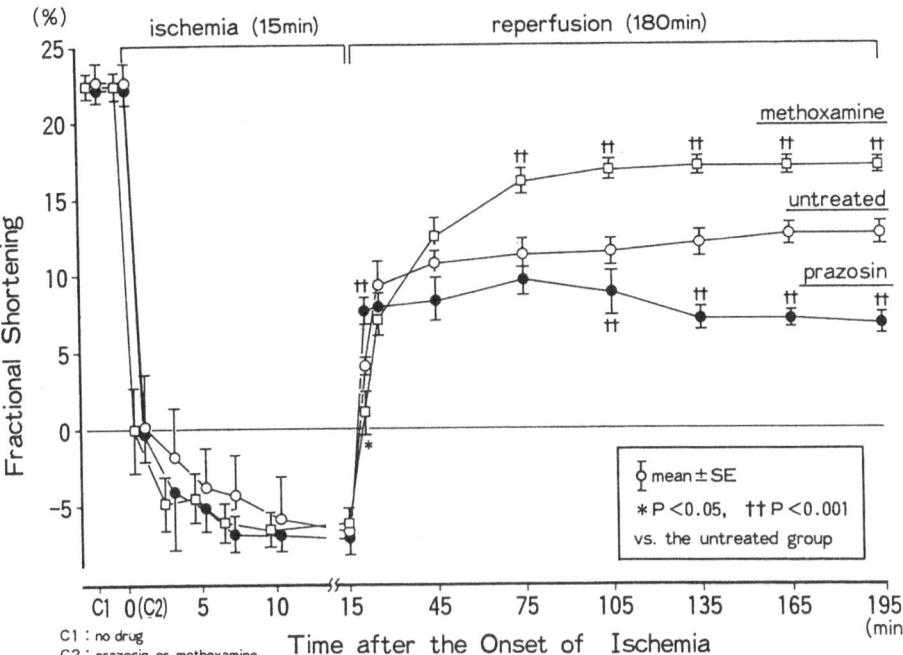

Fig. 6. Serial changes in fractional shortening of the left anterior descending coronary artery area before, during, and after 15-min myocardial ischemia. Although the pharmacological interventions do not alter the decreases in fractional shortening during myocardial ischemia, in the prazosin ($4.0\,\mu g/kg$ per min, ic) and methoxamine-treated conditions, fractional shortening at 3 h of reperfusion is attenuated and augmented, respectively, compared with the control condition.) (From [45] reproduced with permission.) [Circulation Research] Copyright [1991] American Heart Association

coronary vessels. It is known that adenosine also inhibits platelet aggregation through A_2-receptor stimulation [40, 41], and that endogenous adenosine released from the ischemic myocardium inhibits platelet aggregation in the blood-perfused canine heart [64]. When platelet aggregation occurs during periods of ischemia and reperfusion, ADP, serotonin, and several prostaglandins are released; these provoke vasoconstriction, which may further damage the myocardium [40]. Thus, inhibition of platelet aggregation by endogenous adenosine released from the ischemic myocardium may attenuate reperfusion injury. Since adenosine is a potent vasodilator, it is possible that massive releases of adenosine during ischemia and reperfusion may attenuate microcirculatory disturbances [65–67]. Thus, improvement of the coronary microcirculation by adenosine [68], enhanced by alpha$_1$-adrenoceptor stimulation, may largely contribute to the prevention of reperfusion injury.

In contrast, there is no clear consensus on whether alpha$_2$-adrenoceptor stimulation is beneficial for stunning. Although such stimulation may enhance adenosine-induced coronary vasodilation, there is no close relationship between the severity of myocardial stunning and the extent of coronary vasodilation. Alpha$_2$-adrenoceptor stimulation may cause platelet aggregation, which may worsen myocardial stunning. This question requires further investigation.

Conclusion

Briefly, we have presented and discussed here the beneficial role played by alpha-adrenoceptor activity in the pathogenesis of ischemic heart disease. Although the major effect of alpha-adrenoceptor stimulation in the coronary arteries has been shown to be vasoconstriction, an action which may restrict myocardial oxygen supply and exacerbate myocardial ischemia, this does not necessarily occur, since: (1) alpha$_1$-adrenoceptor stimulation favorably maintains endocardial coronary flow at the expense of epicardial flow reduction, (2) alpha$_1$-adrenoceptor stimulation enhances the release of adenosine and alpha$_2$-adrenoceptor stimulation increases the release of EDRF and histamine, and (3) alpha$_2$-adrenoceptor stimulation increases sensitivity to the effects of adenosine. These beneficial effects of alpha-adrenoceptor stimulation may even attenuate the severity of myocardial ischemia and reperfusion injuries. Further clinical studies are necessary to gain a better understanding of the role played by alpha-adrenoceptor activity in coronary flow regulation and myocardial function in the human ischemic heart.

References

1. Wollenberger A, Shaab L (1965) Anoxia-induced release of noradrenaline from the isolated perfused heart. Nature 207:88–89
2. Muntz KH, Hapler HK, Boulas JH, Willerson JT, Buja ML (1984) Redistribution of catecholamines in the ischemic zone of the dog heart. Am J Physiol 114:64–78
3. Schomig A, Dart AM, Dietz R, Mayer E, Kubler W (1984) Release of endogenous catecholamines in the ischemic myocardium of the rat. Part A: Locally mediated release. Circ Res 55:609–701
4. Dart AM, Schomig A, Dietz R, Mayer E, Kubler W (1984) Release of endogenous catecholamine in the ischemic myocardium of the rat. Part B: Effects of sympathetic nerve stimulation. Circ Res 55:702–706
5. Zuberbuhler RC, Bohr DF (1965) Responses of coronary smooth muscle to catecholamine. Circ Res 16:431–440
6. Murray PA, Vatner SF (1979) Alpha-adrenoceptor attenuation of the coronary vascular response to severe exercise in the conscious dog. Circ Res 45:654–660
7. Vatner SF (1983) Alpha-adrenergic regulation of the coronary circulation in the conscious dog. Am J Cardiol 52:15A–21A
8. Angus JA, Cocks TM, Satoh K (1986) The α-adrenoceptors on endothelial cells (Brief review). Fed Proc 45:2355–2359

9. Camazine B, Shannon RP, Guerrero JL, Graham RM, Powell WJ Jr (1988) Neurogenic histaminergic vasodilation in canine skeletal muscle: Mediation by alpha$_2$-adrenoceptor stimulation. Circ Res 62:871–883

10. Kitakaze M, Hori M, Tamai J, Iwakura K, Koretsune Y, Kagiya T, Iwai K, Kitabatake A, Inoue M, Kamada T (1987) Alpah$_1$-adrenoceptor activity regulates release of adenosine from the ischemic myocardium in dogs. Circ Res 60:631–639

11. Hori M, Kitakaze M, Tamai J, Koretsune Y, Iwai K, Iwakura K, Kagiya T, Kitabatake A, Inoue M, and Kamada T (1988) Alpha$_2$-adrenoceptor activity exerts dual control of coronary blood flow in canine coronary artery. Am J Physiol 255:H250–H260

12. Hori M, Kitakaze M, Tamai J, Iwakura K, Kitabatake A, Inoue M, Kamada T (1989) Alpha$_2$-adrenoceptor stimulation can augment coronary vasodilation maximally induced by adenosine in dogs. Am J Physiol 257:H132–H140

13. Kitakaze M, Hori M, Gotoh K, Sato H, Iwakura K, Kitabatake A, Inoue M, Kamada T (1989) Beneficial effects of alpha$_2$-activity on ischemic myocardium during coronary hypoperfusion in dogs. Circ Res 65:1632–1645

14. Schomig A, Fischer S, Kurz T, Richardt G, Schomig E (1987) Nonexocytotic release of endogenous noradrenaline in the ischemic and anoxic rat heart: Mechanism and metabolic requirements. Circ Res 60:194–205

15. Corr PB, Gillis RA (1978) Autonomic neural influences on the dysarrhythmia resulting from myocardial infarction. Circ Res 43:1–9

16. Penny WJ (1984) The deleterious effects of myocardial catecholamines on cellular electrophysiology and arrhythmias during ischemia and reperfusion. Eur Heart J 5:960–973

17. Rona G (1985) Catecholamine cardiotoxicity. J Mol Cell Cardiol 17:291–306

18. Gauduel Y, Karagueuzian HS, Leiris JD (1979) Deleterious effects of endogenous catecholamines on hypoxic myocardial cells following reoxygenation. J Mol Cell Cardiol 11:717–731

19. Shaab L, Wollenberger A, Haase M, Schiller U (1969) Noradrenalinaggabe aus dem Hundeherzen nach vorubergehender Okklusion einer Koronararterie. Acta Biol Med Gem 22:135–143

20. Schomig A, Kurz T, Richardt G, Schomig E (1988) Neuronal sodium homeostasis and axoplasmic amine concentration determine calcium-independent noradrenaline release in normoxic and ischemic rat heart. Circ Res 63:214–226

21. Homcy CJ, Graham RM (1985) Molecular characterization of adrenergic receptors. Circ Res 56:635–650

22. Endo T, Naka M, Hidaka H (1982) Ca^{2+}-phospholipid dependent phosphorylation of smooth muscle myosin. Biochem Biophys Res Comm 105:942–948

23. Nishikawa M, Hidaka H, Adelstein RS (1983) Phosphorylation of smooth muscle heavy meromyosin by calcium-activated, phospholipid-dependent protein kinase C. The effect on actin-activated MgATPase activity. J Biol Chem 258:14069–14072

24. Nishikawa M, Sellers JR, Adelstein RS, Hidaka H (1984) Protein kinase C modulates in vitro phosphorylation of smooth muscle heavy meromyosin by myosin light chain kinase. J Biol Chem 259:8808–8814

25. Connolly TN, Limberd LE (1983) The influence of Na^+ on the alpha$_2$ adrenergic receptor system of human platelets. A method for removal of extraplatelet Na^+: Effect of Na^+ removal on aggregation, secretion, and cAMP accumulation. J Biol Chem 258:3907–3912

220   M. Hori et al.

26. Kelley KO, Feigl EO (1978) Segmental alpha-receptor-mediated vasoconstriction in the canine coronary circulation. Circ Res 43:908–917
27. Chilian WM, Layne SM, Eastham CL, Marcus ML (1989) Heterogeneous microvascular coronary alpha-adrenergic vasoconstriction. Circ Res 64:376–388
28. Heusch G, Deussen A, Schipke J, Thamer V (1984) α_1- and α_2-Adrenoceptor-mediated vasoconstriction of large and small canine coronary arteries in vivo. J Cardiovasc Pharmacol 6:961–968
29. Nathan HJ, Feigl EO (1986) Adrenergic vasoconstriction lessens transmural steal during coronary hypoperfusion. Am J Physiol 250:H645–H653
30. Buffington CW, Feigl EO (1983) Effect of coronary artery pressure on transmural distribution of adrenergic coronary vasoconstriction in the dog. Circ Res 53:613–621
31. Palmer RMJ, Ferrige AG, Moncada S (1987) Nitric oxide release accounts for the biological activity of endothelium-derived relaxing factor. Nature 327:524–526
32. Horeyseck G, Janig W, Kirchner F, Thamer V (1976) Activation and inhibition of muscle and cutaneous postganglionic neurons to hind-limb during hypothalamically-induced vasoconstriction and atropine-sensitive vasodilation. Pflügers Arch 361:231–240
33. Hori M, Tamai J, Kitakaze M, Iwakura K, Gotoh K, Iwai K, Koretsune Y, Kagiya T, Kitabatake A, Kamada T (1989) Adenosine-induced hyperemia attenuates myocardial ischemia in coronary microembolization in dogs. Am J Physiol 257:H244–H251
34. DeWitt DF, Wangler RD, Thompson CI, Sparks HV Jr (1983) Phasic release of adenosine during steady state metabolic stimulation in the isolated guinea pig heart. Circ Res 53:636–643
35. Kitakaze M, Hori M, Iwakura K, Sato H, Gotoh K, Tada M (1989) Protein kinase C regulates production of adenosine in hypoxic myocytes of rats (abstract). Circulation 80:II-498
36. Berne RM (1980) The role of adenosine in the regulation of coronary blood flow. Circ Res 47:807–813
37. Nayler WG, Price JM, Lowe TE (1967) Inhibition of adenosine-induced coronary vasodilation. Cardiovasc Res 1:63–66
38. Heusch G, Deussen A (1983) The effects of cardiac sympathetic nerve stimulation on perfusion of stenotic coronary arteries in the dog. Circ Res 53:8–15
39. Seitelberger R, Guth BD, Heusch G, Lee JD, Katayama K, Ross J Jr (1988) Intracoronary α_2-adrenergic receptor blockade attenuates ischemia in conscious dogs during exercise. Circ Res 62:436–442
40. Agarwal KC (1987) Adenosine and platelet function. In: Stefanovich V, Okayuz-Baklouti L (eds) Adenosine in cerebral metabolism and blood flow VNU Science Press, The Netherlands pp 107–124
41. Agarwal KC, Zielinski BA, Maitra RS (1989) Significance of plasma adenosine in the antiplatelet activity of forskolin: Potentiation by dipyridamole and dilazep. Thromb Haemost 61:106–110
42. Nayler WG, Gordon M, Stephens DJ, Sturrdock JW (1985) The protective effect of prazosin on the ischemic and reperfused myocardium. J Mol Cell Cardiol 17:685–699
43. Herrmann SC, Feigl EO (1992) Adrenergic blockade blunts adenosine concentration and coronary vasodilation during hypoxia. Circ Res 70:1203–1216
44. Braunwald E, Kloner RA (1982) The stunned myocardium: Prolonged, postischemic ventricular dysfunction. Circulation 60:1146–1149

45. Kitakaze M, Hori M, Sato H, Iwakura K, Gotoh K, Inoue M, Kitabatake A, Kamada T (1991) Beneficial effects of α_1-adrenoceptor activity on myocardial stunning in dogs. Circ Res 68:1322–1339

46. Kitakaze M, Takashima S, Sato H (1990) Stimulation of adenosine A_1 and A_2 receptors prevents myocardial stunning (abstract). Circulation 82[Suppl III]:III-37

47. Brodeur RD, Storey C, Anderson PR, Cabrera BDF, Nunnally RL (1990) Effects of adenosine on functional recovery during reperfusion of the ischemic rabbit myocardium. Circulation 82[Suppl III]:III-289

48. Taegtmeyer H, Roberts AFC, Raine AEG (1985) Energy metabolism in reperfused heart muscle: Metabolic correlates to return of function. J Am Coll Cardiol 6:864–870

49. Ambrosio G, Jacobus WE, Becker LC (1986) Effect of ATP precursor administration on post-ischemic function and metabolism in isolated rabbit hearts (abstract). J Am Coll Cardiol 7:79A

50. Nunnally RL, Hollis DP (1979) Adenosine triphosphate compartmentation in living hearts: A phosphorus nuclear magnetic resonance saturation transfer study. Biochemistry 18:3642–3646

51. Bittl JA, Ingwall JS (1985) Reaction rates of creatine kinase and ATP synthesis in the isolated rat heart: A ^{31}P-NMR magnetization transfer study. J Biol Chem 260:3512–3517

52. Kusuoka H, Porterfield JK, Weisman HF, Weisfeldt ML, Marban E (1987) Pathophysiology and pathogenesis of stunned myocardium. Depressed Ca^{2+} activation of contraction as a consequence of reperfusion-induced cellular calcium overload in ferret hearts. J Clin Invest 79:950–961

53. Steenbergen C, Murphy E, Levy L, London RE (1987) Elevation in cytosolic free calcium concentration early in myocardial ischemia in perfused rat heart. Circ Res 60:700–707

54. Kitakaze M, Weisman HF, Marban E (1988) Contractile dysfunction and ATP depletion after transient calcium overload in perfused ferret hearts. Circulation 77:685–695

55. Kitakaze M, Weisfeldt ML, Marban E (1988) Acidosis during early reperfusion prevents myocardial stunning in perfused ferret hearts. J Clin Invest 82:920–927

56. Marban E, Kitakaze M, Koretsune Y, Yue DT, Chacko VP, Pike MM (1990) Quantification of $[Ca^{2+}]_i$ in perfused hearts. Critical evaluation of the 5F-BAPTA and nuclear magnetic resonance method as applied for the study of ischemia and reperfusion. Circ Res 66:1255–1267

57. Isenberg G, Cerbai E, Klockner U (1987) Ionic channels and adenosine in isolated heart cells. In: Gerlach E, Becker BF (eds) Topics and perspectives in adenosine research. Springer, Berlin Heidelberg New York London Tokyo, pp 323–335

58. Cerbai E, Klockner U, Isenberg G (1988) Ca-antagonistic effects of adenosine in guinea pig atrial cells. Am J Physiol 255:H872–H878

59. Endo M, Blinks JR (1988) Actions of sympathomimetic amines on the Ca^{2+} transients and contraction of rabbit myocardium: Reciprocal changes in myofibrillar responsiveness to Ca^{2+} mediated through α_1- and α_2-adrenoceptors. Circ Res 62:247–265

60. Sharma AD, Saffitz JE, Lee BI, Soble BE, Corr PB (1983) Alpha adrenergic-mediated accumulation of calcium in reperfused myocardium. J Clin Invest 72:802–818

61. Cronstein BN, Kramer SB, Weissmann G, Hirschhorn R (1986) Adenosine: A physiological modulator of superoxide anion generation by human neutrophils. J Exp Med 158:1160–1177

62. Cronstein BN, Levin RI, Belanoff J, Weissmann G, Hirschhorn R (1986) Adenosine: An endogenous inhibitor of neutrophil-mediated injury to endothelial cells. J Clin Invest 78:760–770
63. Cronstein BN (1990) Adenosine is an endogenous modulator of inflammation (abstract). Jpn J Pharmacol 52[Suppl II]:57
64. Kitakaze M, Hori M, Sato H, Iwakura K, Takashima S, Komamura K, Kitabatake A, Inoue M, Kamada T (1990) Endogenous adenosine inhibits formation of microthromboembolism in ischemic myocardium (abstract). Jpn Circ J 54:922–923
65. Olafsson B, Forman MB, Puett DW, Pou A, Cates CU, Friesinger GC, Virmani R (1987) Reduction of reperfusion injury in a canine preparation by intracoronary adenosine: Importance of the endothelium and the no-reflow phenomenon. Circulation 76:1135–1145
66. Engler R (1987) Consequences of activation and adenosine-mediated inhibition of granulocytes during myocardial ischemia. Fed Proc 46:2407–2412
67. Stahl LD, Weiss HR, Becker LC (1988) Myocardial oxygen consumption, oxygen supply/demand heterogeneity, and microvascular patency in regionally-stunned myocardium. Circulation 77:865–872
68. Hori M, Inoue M, Kitakaze M, Koretsune Y, Iwai K, Tamai J, Ito H, Kitabatake A, Sato T, Kamada T (1986) Role of adenosine in hyperemic response of coronary blood flow in microembolization. Am J Physiol 250:H509–H518

Ischemic Preconditioning of Myocardium: Effect of Adenosine

Donna M. Van Winkle[1], *James M. Downey*[2], *Jon D. Thornton*[2], and *Richard F. Davis*[1]

Summary. Recently, it was suggested that adenosine is involved in the cardioprotection of ischemic preconditioning (PC). If PC is mediated via adenosine, then administration of adenosine antagonists should abolish the cardioprotection, and exogenous administration of adenosine agonists should provide tolerance to ischemia. We tested these hypotheses in anesthetized rabbits ($n = 40$) which underwent 30 min coronary artery occlusion and 3 h reperfusion. PC was elicited by 5 min coronary artery occlusion and 10 min reperfusion before the 30-min coronary artery occlusion. PD115,199 (3 mg/kg iv), a nonselective adenosine antagonist (henceforth referred to as PD), was given 5 min before the first ischemic episode. $R(-)$ N^6-(2-phenylisopropyl)-adenosine (R-PIA, 200 μg/kg/min iv) an adenosine A_1 agonist, or CGS 21680 (CGS, 16 μg/kg/min iv) an adenosine A_2 agonist, were administered in place of PC. Infarct size (IS) was measured with tetrazolium and expressed as a percentage of the area-at-risk (mean ± SEM). Blockade of adenosine receptors attenuated the PC-induced infarct reduction (IS 34.5 ± 5.1% control [CON], 5.3 ± 1.8%* PC, 30.5 ± 9.0% PD, and 22.6 ± 5.3% PC + PD; *$P < 0.05$ *vs* CON), while administration of R-PIA resulted in a reduction in infarct size (R-PIA 18.7 ± 4.7% *vs* CON 34.5 ± 5.1%, $P < 0.05$) that was not different from PC hearts (R-PIA 18.7 ± 4.7% *vs* PC 5.3 ± 1.8%, $P =$ NS). CGS did not confer any cardioprotection (IS CGS 40.9 ± 14.8% *vs* CON 34.5 ± 5.1%, $P =$ NS). Since (a) administration of an adenosine antagonist attenuated the infarct-reducing effect of PC, and (b) administration of an A_1—but not of an A_2—agonist in place of transient ischemia mimicked the infarct-reducing effect of PC, we conclude that the activation of adenosine A_1 receptors is involved in ischemic preconditioning.

Key words: A_1 receptor—Myocardial infarction

Introduction

Ischemic preconditioning (PC), which can be defined as transient myocardial ischemia providing tolerance to subsequent ischemia, was first described by Murry and colleagues at Duke University [1]. It has subsequently been

[1] Departments of Anesthesiology and Physiology, Oregon Health Sciences University, Anesthesiology Service, Portland Veterans Administration Medical Center, P.O. Box 1034, Portland, OR 97207-1034, USA
[2] Department of Physiology, University of South Alabama, Mobile, Alabama, USA
This study was supported in part by grants from the American Heart Association Oregon Affiliate Grant-in-Aid, and the Medical Research Foundation of Oregon.

documented to occur in all species tested and has invariably elicited a marked reduction in infarct size (IS) [2–5]. Controversy still exists, however, over the biochemical mediators and mechanisms responsible for the cardioprotection. Downey and colleagues found that blockade of cell-surface adenosine receptors in rabbits with the polar adenosine antagonist 8-(p-sulfophenyl)-theophylline (8-SPT) resulted in loss of PC [6]. They also described cardioprotection in isolated blood-perfused rabbit hearts which had received adenosine or an A_1 adenosine receptor (A_1AR) agonist in the perfusate (Fig. 1) [6]. These data suggest that activation of adenosine receptors is the proximate event mediating PC. In a recent abstract, however, Schwarz et al. reported that while the A_1 selective adenosine receptor antagonist 8-cyclopentyl-1,3-dipropylxanthine (CPDPX) abolished the infarct-limiting effect of PC, administration of the A_1AR agonist N^6-cyclohexyladenosine (CHA) did not exhibit any infarct-reducing effect [7].

This study was undertaken to answer the questions: (a) is the infarct-limiting effect of PC attenuated by adenosine receptor blockade, and (b) can antecedent intravenous infusion of an A_1AR agonist render the myocardium more tolerant to subsequent coronary occlusion? If PC is indeed mediated by adenosine, one would expect that adenosine receptor antagonism would diminish the cardio-protection and exogenous administration of adenosine agonists would mimic the cardioprotection. In the present study, the highly potent nonselective adenosine receptor antagonist PD115,199, the A_1AR agonist R(−) N^6-(2-phenylisopropyl)-adenosine (R-PIA), and the A_2AR agonist CGS 21680 were given before oc-clusion and reperfusion of a left coronary artery in rabbits.

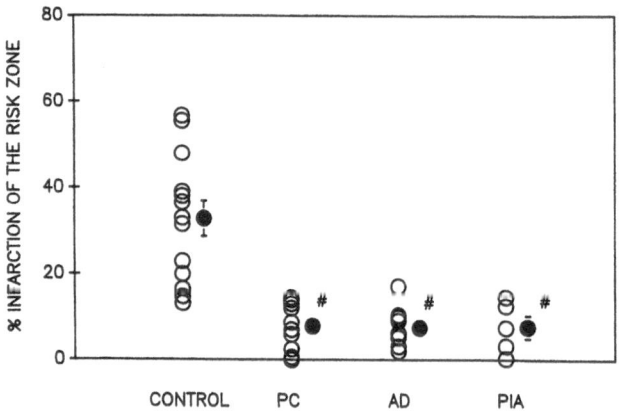

Fig. 1. Infarct size in isolated blood-perfused rabbit hearts. Adenosine (*AD*) is rapidly removed from blood by uptake and deamination; this problem was overcome with the isolated heart preparation in which adenosine was infused into the aortic root. Adenosine was infused for 5 min in place of transient ischemia (0.28 mg/min), beginning 10 min before a 45-min occlusion of a left coronary artery. R(−)N^6-(2-phenylisapropyl)-adenosine (*R-PIA*) (5.3 µg/min) was administered following the same schedule. Data are plotted as mean ± SEM. *$P < 0.01$ *vs* controls. (From [4], by permission of the American Heart Association, Inc.)

Material and Methods

Surgical Preparation

New Zealand white rabbits of either sex, 2–3 kg, were anesthetized with 30 mg/
kg sodium pentobarbital, administered via a 23-gauge butterfly in a marginal ear
vein. The neck was opened with a ventral midline incision and a tracheotomy
was performed. The animals were mechanically ventilated (MD Industries,
Mobile, Ala.) with 100% oxygen, at 30–35 breaths/min and a tidal volume of
approximately 20 ml. In some animals end-expiratory carbon dioxide tension was
monitored continuously using a mass spectrometer (Model 6000 Multi-Gas
Monitor, Ohmeda, Madison, Wis.). Ventilation was adjusted as necessary to
maintain arterial pH between 7.38 and 7.42. Core body temperature was
maintained at 37°–38°C with a heating pad and/or radiant warming. A femoral
artery was isolated, and a catheter (PE90) was advanced into the thoracic aorta
for blood pressure and blood gas sampling. A femoral vein was isolated and
catheterized for administration of drugs.

  A left thoracotomy was performed and the pericardium opened to expose the
heart. A surface electrocardiogram was recorded in all animals, and in a subset of
rabbits ($n = 28$) an esophageal electrocardiogram was also recorded. A silk
ligature was passed underneath a prominent left coronary artery and the ends of
the ligature passed through a segment of tubing to form a snare. Myocardial
ischemia was confirmed by ECG changes and regional cyanosis of the
epicardium distal to the snare, and successful reperfusion was confirmed by
blushing of the previously cyanotic myocardium.

Measurement of Area-at-Risk and Infarct Size

At the conclusion of the experiment 500 U/kg iv sodium heparin were given, the
hearts removed and transferred to a Langendorff apparatus and flushed with
room temperature saline for 1 min. The coronary artery was then reoccluded and
fluorescent particles (3 ml of a 1.2 mg/ml suspension of zinc cadmium sulfide,
1–10 μm in diameter; Duke Scientific Corp., Palo Alto, Calif.) were infused.
These particles fluoresce bright yellow under ultraviolet light and thus delineate
the area-at-risk (AAR). The heart was removed from the Langendorff apparatus,
weighed, and frozen. The frozen heart was cut into transverse slices, 2 mm thick,
and incubated in triphenyl tetrazolium chloride (1% w/v in sodium phosphate
buffer at 37°C) for 20 min. Myocardium that did not stain red was presumed to
be infarcted [8]. The area of infarct (cm^2) was determined for each slice by
computer-assisted planimetry (SigmaScan software, Jandel Scientific, Corte
Madera, Calif. with Model 2210-0.43.C digitizer, Numonics Corp.,
Montgomeryville, Pa.). The slices were then viewed under UV light, and the
areas lacking fluorescence, which represented the perfusion defect, were also
planimetered. The volume (in cm^3) of infarcted myocardium and myocardium at
risk was calculated from the planimetered areas and slice thickness (2 mm). The

volume of the areas in cm^3 was assumed to be equal to the weight of the tissue in grams.

Infusion of Drugs

Blockade of adenosine receptors was achieved with an intravenous injection of PD115,199 (Parke-Davis Pharmaceutical Research, Warner-Lambert Co., Ann Arbor, Mich.) (vehicle = 25% v/v 0.1 N HCl, 6% v/v 95% ethanol, 69% v/v 0.9% saline; pH 6.0). Activation of A$_1$ adenosine receptors was elicited by intravenous infusion of R-PIA (Research Biochemicals Inc., Natick, Mass.), and A$_2$ adenosine receptors with CGS 21680 (Research Biochemicals Inc., Natick, Mass.). Vehicle for R-PIA and CGS 21680 was 0.9% saline. Drugs were made fresh each day.

Data Analysis

Infarct size was normalized as a percent of the AAR. Data analysis was performed with a statistical software package (Crunch 4, Crunch Software Corp., Oakland, Calif.). To evaluate differences between groups, analysis of variance (ANOVA) with a Newman-Keuls post-hoc test was utilized. Differences within groups were assessed using ANOVA with the Dunnett post-hoc test for multiple comparisons. Statistical significance was assumed for $P < 0.05$. Data are expressed as mean ± SEM.

Experimental Protocol

Non-drug-treated non-preconditioned (CON, $n = 13$) and preconditioned (PC, $n = 10$) rabbits were studied concurrently with both the antagonist and agonist studies. All rabbits underwent 30 min coronary occlusion and 3 h reperfusion. Preconditioning was elicited by a single 5-min coronary occlusion beginning 15 min before the 30-min coronary occlusion.

Experiments with Adenosine Antagonists. To test if antagonism of adenosine receptors ablates the cardioprotective effect of transient ischemia, the nonselective adenosine antagonist PD115,199 was given to non-preconditioned (PD group, $n = 5$) and preconditioned rabbits (PC + PD, $n = 8$). PD115,199 (3 mg/kg) was administered as an iv bolus 5 min prior to the first episode of ischemia. This dose of PD115,199 was tested in 4 additional rabbits, using a dose-response relationship (in half-log increments) of iv adenosine versus mean arterial pressure. PD115,199 shifted this relationship by one order of magnitude (data not shown). Since PD115,199 was prepared in a nonphysiologic vehicle, 2 groups of vehicle control rabbits were also studied: non-preconditioned plus PD115,199 vehicle (PDVEH, $n = 3$) and preconditioned plus PD115,199 vehicle (PC + PDVEH, $n = 3$).

Experiments with Adenosine Agonists. To ascertain if activation of adenosine receptors provides cardioprotection, rabbits received the adenosine agonists CGS

21680 or R-PIA. CGS 21680 possesses 100-fold selectivity for the adenosine A_2 receptor, whereas R-PIA displays 100-fold selectivity for the A_1 adenosine receptor [9]. CGS 21680-treated rabbits ($n = 4$) received 16 µg/kg/min iv for 5 min beginning 15 min before the 30-min occlusion. R-PIA-treated rabbits ($n = 11$) received 200 µg/kg/min iv for 5 min, starting 15 min prior to the 30-min coronary occlusion. This schedule mimicked the schedule for preconditioning ischemia. Since R-PIA produced marked bradycardia, we compared the current results with those obtained for 8 rabbits from a previous study [10] which had received the same dose of R-PIA but had been paced to maintain heart rate at baseline values.

Results

Experiments with Adenosine Antagonists

Hemodynamics

In control rabbits, heart rate increased modestly (as compared to baseline values) following the 30-min occlusion; this difference was statistically significant at 120 min reperfusion (262 ± 8 bpm baseline versus 281 ± 8 bpm 120 min reperfusion, $P < 0.01$). In the PC, PD, PC + PD, and PDVEH groups, heart rate did not change over the course of the experiment, as compared to baseline values for each group ($P = $ NS). In PC + PDVEH rabbits, heart rate increased at the 30-min occlusion and remained elevated at all time points as compared to baseline ($P < 0.05$). Between groups CON, PC, PD, PC + PD, PDVEH, and PC + PDVEH, there were no significant differences in heart rate at any time point ($P = $ NS).

Mean aortic pressure in control rabbits increased slightly following occlusion, becoming statistically significant at 120 min reperfusion (80 ± 3 mmHg baseline *vs* 88 ± 4 mmHg 120 min reperfusion, $P < 0.05$). Mean aortic pressure did not differ from baseline within the other groups (PC, PD, PC + PD, PDVEH, and PC + PDVEH) at any time ($P = $ NS). Between groups CON, PC, PD, PC + PD, PDVEH, and PC + PDVEH there were no significant differences in mean aortic pressure ($P = $ NS). Hemodynamic data for the adenosine antagonist studies are presented in Table 1.

Table 1. Hemodynamic data for PD115,199-treated *in situ* rabbit hearts

	HR, initial	MP, initial	HR, occl.	MP, occl.	HR, rep120	HR, rep120
No PC	262 ± 8	80 ± 3	268 ± 7	79 ± 4	281 ± 8	88 ± 4
No PC + PD115,199	255 ± 7	69 ± 5	262 ± 7	82 ± 7	260 ± 11	86 ± 5
PC	259 ± 12	76 ± 4	251 ± 8	80 ± 4	260 ± 7	82 ± 3
PC + PD115,199	275 ± 10	80 ± 7	265 ± 12	77 ± 7	279 ± 16	82 ± 6

HR, heart rate; MP, mean aortic pressure
Data are presented as mean ± SEM

Dysrhythmias

Preconditioning reduced the number of premature ventricular beats (PVBs) and couplets as compared to control rabbits (PVBs, PC 5.7 ± 2.4 *vs* CON 23.4 ± 6.0; couplets, PC 2.4 ± 1.1 *vs* CON 11.1 ± 3.1; $P < 0.05$). Administration of PD115,199 did not abolish the preconditioning-associated decrease in PVBs (PC + PD 6.7 ± 2.0 *vs* PD 19.3 ± 6.4) or couplets (PC + PD 2.4 ± 0.8 *vs* PD 16.9 ± 5.5). In this study, preconditioning did not reduce the incidence of ventricular fibrillation (VF): VF occurred in 57% of CON rabbits and 43% of PC rabbits. PD115,199 did not alter the incidence of VF (57% PD and 43% PC + PD).

Infarct Data

The risk zone, expressed as a percentage of the left ventricle, did not differ significantly between groups (CON 10.3 ± 1.2%), (PC 10.1 ± 1.6%), (PD 8.3 ± 0.4%), (PC + PD 8.1 ± 0.6%), (PDVEH 5.5 ± 0.1%), and (PC + PDVEH 7.5 + 0.5%); P = NS. The vehicle for PD115,199 did not change infarct size in non-preconditioned or preconditioned animals (CON 34.5 ± 5.1% *vs* PDVEH 42.3 ± 6.4%, and PC 5.3 ± 1.8% *vs* PC + PDVEH 9.2 ± 6.2%; P = NS). However, pretreatment with PD115,199 attenuated the infarct-reducing effect of preconditioning (PC + PD 22.6 ± 5.3% *vs* PC 5.3 ± 1.8%; $P < 0.05$). PD115,199 alone did not change infarct size as compared to the control (PD 30.5 ± 9.0% *vs* CON 34.5 ± 5.1%; P = NS). Infarct size data for the PD115,199 studies are shown in Fig. 2.

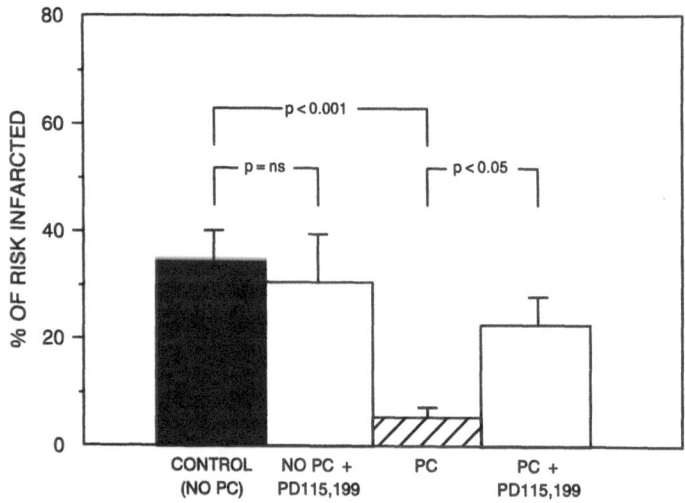

Fig. 2. Effect of adenosine receptor blockade on non-preconditioned and preconditioned hearts. Ischemic preconditioning resulted in an 85% reduction of infarct size from the control. However, preconditioning in the face of adenosine receptor blockade reduced infarct size by only 34%. Data are plotted as mean ± SEM

Experiments with Adenosine Agonists

Hemodynamics

Heart rate in R-PIA-treated rabbits decreased from 256 ± 9 bpm at baseline to 164 ± 10 bpm during the R-PIA infusion. Although it gradually returned towards baseline, heart rate remained depressed for the remainder of the protocol ($P < 0.01$). In the eight rabbits which were electrically paced during and after R-PIA infusion, heart rate was maintained at baseline values (P = NS). CGS 21680-treated rabbits did not exhibit changes in heart rate at any time as compared to baseline (P = NS). Although heart rate was comparable between CON, PC, R-PIA, and CGS groups at baseline, it was significantly depressed in the R-PIA group at all times following infusion, as compared to CON ($P < 0.05$). There were no significant differences in heart rate between CON, PC, and CGS groups at any time (P = NS).

Mean aortic pressure decreased with agonist infusion in both the R-PIA and CGS groups and remained depressed for the remainder of the protocol (R-PIA 77 ± 3 mmHg baseline to 29 ± 2 mmHg during infusion: $P < 0.01$; CGS 69 ± 4 mmHg baseline to 34 ± 1 mmHg during infusion: $P < 0.05$). Mean aortic pressure in the R-PIA and CGS groups was significantly lower than that of the CON rabbits at all times except baseline ($P < 0.05$). Hemodynamic data for the agonist studies are presented in Table 2.

Infarct Data

Risk zone sizes, expressed as a percentage of the left ventricle, did not differ significantly between groups (CON 10.3 ± 1.2%), (PC 10.1 ± 1.6%), (R-PIA 8.5 ± 0.6%), and (CGS 8.3 ± 2.7%); P = NS. Atrial pacing did not affect area-at-risk in R-PIA-treated hearts (R-PIA 8.5 ± 0.6% *vs* 10.2 ± 1.0% R-PIA-Paced; P = NS). R-PIA infusion resulted in a marked reduction in infarct size as compared to the control (CON 34.5 ± 5.1% *vs* R-PIA 18.7 ± 4.7%; $P < 0.05$), which was not markedly different from the cardioprotection of preconditioning (R-PIA 18.7 ± 4.7% *vs* PC 5.3 ± 1.8%; P = NS). Infarct size was comparable in paced rabbits receiving R-PIA as compared to unpaced rabbits (R-PIA-Paced 12.1 ± 3.7% versus R-PIA 18.7 ± 4.7%; P = NS). Infarct size data for the R-PIA and CGS studies are displayed in Fig. 3.

Table 2. Hemodynamic data for R-PIA- and CGS 21680-treated *in situ* rabbit hearts

	HR, initial	MP, initial	HR, occl.	MP, occl.	HR, rep120	MP, rep120
Control	262 ± 8	80 ± 3	268 ± 7	79 ± 4	281 ± 8	88 ± 4
PC	255 ± 7	76 ± 4	251 ± 8	80 ± 4	260 ± 7	82 ± 3
R-PIA	256 ± 9	77 ± 3	167 ± 9*	37 ± 4*	220 ± 13*	56 ± 6*
CGS 21680	268 ± 16	69 ± 4	270 ± 13	43 ± 3*	265 ± 17	55 ± 11*

HR, heart rate; MP, mean aortic pressure
Data are presented as mean ± SEM; * $P < 0.05$ as compared to control

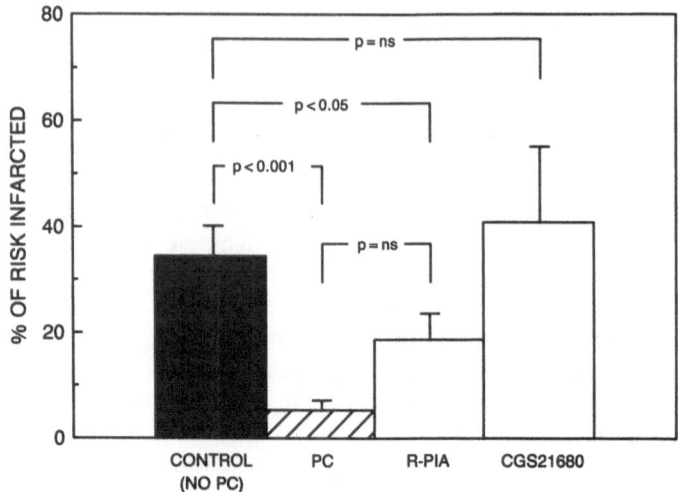

Fig. 3. Effect of antecedent adenosine agonist infusion on myocardial infarct size. Administration of *R-PIA* in place of preconditioning ischemia resulted in a 54% reduction of control infarct size (*P* < 0.05). Pretreatment with *CGS 21680* did not change infarct size as compared to control (*P* = NS). Data are plotted as mean ± SEM

Discussion

In open-chest anesthetized rabbits, pretreatment with the adenosine receptor antagonist PD115,199 attenuated the infarct-reducing effect of transient antecedent ischemia, but it had no effect on the reduction of ischemic dysrhythmias. Antecedent administration of the A_1AR agonist R-PIA, but not of the A_2AR agonist CGS 21680, resulted in marked limitation of infarct size.

As mentioned above, ischemic preconditioning can be defined as transient ischemia providing tolerance to subsequent coronary occlusion, assessed by measurement of infarct size. Ischemic preconditioning occurs in dogs [1, 2, 11–13], rats [5, 14, 15], rabbits [4, 6, 10, 16–19], and swine [3, 7]. Preconditioning has also been reported to reduce ischemia/reperfusion-induced dysrhythmias [14–15]. The phenomenon was first described in 1986, when it was reported that while repeated short coronary occlusions produced no myocardial necrosis, the same total duration of ischemia as a continuous insult had resulted in infarction [1, 11]. Murry et al. subsequently reported that preconditioned hearts had exhibited a decreased ATP utilization rate and reduced accumulation of purines [12]. They theorized that either preservation of ATP or limitation of catabolite accumulation (or both) had been responsible for the cardioprotective effect.

The adenosine hypothesis of ischemic preconditioning was initially based on the following observations: (a) a certain degree of ischemia was necessary to elicit preconditioning [4]; (b) adenosine accumulates in ischemic tissue [20, 21]; and (c) adenosine had been shown to be cardioprotective when administered at the time of reperfusion [20–24]. Moreover, the nonselective adenosine receptor antagonist 8-SPT blocked the infarct-reducing effect of ischemic preconditioning [6]. Since the polar structure of 8-SPT prevented intracellular penetration, this indicated that the receptors involved were cell surface receptors. Downey and colleagues also demonstrated that in an isolated heart preparation, both adenosine and R-PIA were as effective as transient antecedent ischemia in limiting infarct size [6]. In their research, however, cardioprotection could not be demonstrated with iv adenosine in an in situ rabbit heart, probably because of the rapid deamination and cellular uptake of adenosine in blood [10]. Miura et al. demonstrated that preconditioning could be potentiated by the addition of dipyridamole, an adenosine uptake inibitor, and that this enhancement was attenuated by 8-SPT [19]. These data also support the adenosine hypothesis of preconditioning.

Recently, it was reported that hypoxic preconditioning is equipotent with ischemic preconditioning in limiting infarct size. Since Van Wylen et al. [25] did not observe increased interstitial adenosine during hypoxia, these data would discount the idea that an accumulated catabolite such as adenosine causes ischemic preconditioning. Herrmann and Feigl [26], however, saw increased adenosine levels during systemic hypoxia except when adrenergic blockade was present. Therefore hypoxic preconditioning does not per se discount the adenosine hypothesis of preconditioning.

The present data support recent research by Schwarz et al. reporting that the A_1AR antagonist CPDPX had resulted in loss of preconditioning-induced infarct limitation in swine [7]. Although they had not observed infarct reduction subsequent to intracoronary infusion of the A_1AR agonist N^6-cyclohexyladenosine (CHA), the sample size for agonist administration was low ($n = 3$), and the schedule for agonist infusion did not mimic the schedule used for ischemic preconditioning (30-min CHA infusion immediately followed by coronary occlusion, vs two periods of 10 min preconditioning ischemia separated by 30 min reperfusion and begun 80 min prior to coronary occlusion). Thus these data are difficult to interpret.

The findings of Tsuchida et al., which revealed that ischemic preconditioning could be attenuated by adenosine receptor blockade with 8-SPT [18], are also supported by our data. However, these investigators also reported that R-PIA had been ineffective in mimicking the infarct-reducing effect of ischemic preconditioning unless the hearts were paced. Since their data demonstrated an inverse correlation between diastolic blood pressure and infarct size in R-PIA animals, Tsuchida et al. speculated that the lack of significant cardioprotection in the unpaced R-PIA group had resulted from hypotension. It should be noted, however, that there had been no significant difference in infarct size between R-PIA and R-PIA-paced groups ($P = $ NS). It is possible that with an augmented animal number in the R-PIA-paced group, a statistically significant difference would have been realized vs the control. In the present data there was no correlation between infarct size and diastolic pressure in the R-PIA group ($r = 0.15$). As we did not include a group of R-PIA treated rabbits in which blood pressure was controlled, however, we cannot rule out the possibility that normotension would have augmented R-PIA's cardioprotective effect.

It is also possible that the hypotension we observed resulting from R-PIA administration may have contributed to R-PIA's cardioprotective effect. Since afterload is a major determinant of myocardial oxygen demand (MVO_2), the cardioprotection may have been due to a decrease in MVO_2 immediately before the period of occlusion. However, administration of the A_2AR agonist CGS 21680 resulted in hypotension comparable to R-PIA-treated rabbits, but did not confer any cardioprotection. Moreover, we found no correlation between rate-pressure-product at the onset of occlusion and infarct size in R-PIA-treated rabbits.

In the present study, infarct size was assessed with the triphenyl tetrazolium chloride (TTC) method. It has been reported that some interventions may reduce the ability of TTC to discriminate between reversibly and irreversibly injured cardiac myocytes [27]. Iwamoto et al., however, utilized infarct sizing by histology following 72 h of reperfusion and reported that preconditioning of rabbit hearts had resulted in a 71% reduction in infarct size [16]. This is similar to the amount of salvage reported by other investigators who utilized TTC staining for assessing infarct size [2–4, 6]. Using histology following 72 h reperfusion to assess myocyte necrosis, this group studied the effect of pretreatment with R-PIA on infarct size; they found that paced rabbits which had received R-PIA demonstrated myocardial salvage [18]. It is therefore likely that in preconditioned and adenosine- or R-PIA-treated hearts, TTC is a reliable method for assessing infarct size. The possibility of TTC-artifactual salvage cannot be excluded, however.

In summary, the present study found that intravenous pretreatment with the adenosine antagonist PD115,199 attenuated the preconditioning-induced limitation of infarct size but did not affect the preconditioning-associated reduction in dysrhythmias. Also, intravenous administration of the A_1AR agonist R-PIA, in place of preconditioning ischemia, reduced infarct size as compared to untreated animals. However, the A_2AR agonist CGS 21680 given in place of preconditioning ischemia did not limit infarct size. These data support the concept that activation of adenosine A_1 receptors is involved in the infarct-limiting effect, but not the anti-dysrhythmic effect of ischemic preconditioning.

Acknowledgments. We would like to thank Roger A. Wolff, M.S., for his invaluable technical assistance. We would also like to thank Parke-Davis for generously providing PD115,199.

References

1. Murry CE, Jennings RB, Reimer KA (1986) Preconditioning with ischemia: A delay of lethal cell injury in ischemic myocardium. Circulation 74:1124–1136

2. Li GC, Vasquez JA, Gallagher KP, Lucchesi BR (1990) Myocardial protection with preconditioning. Circulation 82:609–619
3. Schott RJ, Rohmann S, Braun ER, Schaper W (1990) Ischemic preconditioning reduces infarct size in swine myocardium. Circ Res 66:1133–1142
4. Van Winkle DM, Thornton J, Downey DM, Downey JM (1991) The natural history of preconditioning. Cardioprotection depends on duration of transient ischemia and time to subsequent ischemia. Cor Art Dis 2:613–619
5. Li Y, Whittaker P, Kloner RA (1992) The transient nature of the effect of ischemic preconditioning on myocardial infarct size and ventricular arrhythmia. Am Heart J 123:346–353
6. Liu GS, Thornton J, Van Winkle DM, Stanley AWH, Olsson RA, Downey JM (1991) Protection against infarction afforded by preconditioning is mediated by A_1 adenosine receptors in rabbit heart. Circulation 84:350–356
7. Schwarz ER, Mohri M, Sack S, Arras M (1991) The role of adenosine and its A_1-receptor in ischemic preconditioning (Abstract). Circulation 84(4):II-191
8. Fishbein MC, Meerbaum S, Rit J, Lando U, Kanmatsuse K, Mercier JC, Corday E, Ganz W (1981) Early phase acute myocardial infarct size quantification: Validation of the triphenyl tetrazolium chloride tissue enzyme staining technique. Am Heart J 101:593–600
9. Lohse MJ, Klotz K-N, Schwabe U, Cristalli G, Vittori S, Grifantini M (1988) 2-Chloro-N6-cyclopentyladenosine: A highly selective agonist at A_1 adenosine receptors. Naunyn-Schmiedebergs Arch Pharmacol 337:687–689
10. Thornton JD, Liu GS, Olsson RA, Downey JM (1992) Intravenous pretreatment with A_1-selective adenosine analogues protects the heart against infarction. Circulation 85:659–665
11. Reimer KA, Murry CE, Yamasawa I, Hill ML, Jennings RB (1986) Four brief periods of myocardial ischemia cause no cumulative ATP loss or necrosis. Am J Physiol 251:H1306–H1315
12. Murry CE, Richard VJ, Reimer KA, Jennings RB (1990) Ischemic preconditioning slows energy metabolism and delays ultrastructural damage during a sustained ischemic episode. Circ Res 66:913–931
13. Murry CE, Richard VJ, Jennings RB, Reimer KA (1991) Myocardial protection is lost before contractile function recovers from ischemic preconditioning. Am J Physiol 260:H796–H804
14. Li Y, Kloner RA (1992) Cardioprotective effects of ischemic preconditioning are not mediated by prostanoids. Cardiovasc Res 26:226–231
15. Shiki K, Hearse DJ (1987) Preconditioning of ischemic myocardium: Reperfusion-induced arrhythmias. Am J Physical 253:H1470–H1476
16. Iwamoto T, Miura T, Adachi T, Noto T, Ogawa T, Tsuchida A, Iimura O (1991) Myocardial infarct size-limiting effect of ischemic preconditioning was not attenuated by oxygen free-radical scavengers in the rabbit. Circulation 83:1015–1022
17. Miura T, Goto M, Urabe K, Endoh A, Shimamoto K, Iimura O (1991) Does myocardial stunning contribute to infarct size limitation by ischemic preconditioning? Circulation 84:2504–2512
18. Tsuchida A, Miura T, Miki T, Shimamoto K, Iimura O (1992) Role of adenosine receptor activation in myocardial infarct size limitation by ischemic preconditioning. Cardiovasc Res 26:456–461
19. Miura T, Ogawa T, Iwamoto T, Shimamoto K, Iimura O (1992) Dipyridamole potentiates the myocardial infarct size-limiting effect of ischemic preconditioning. Circulation 86:979–985

20. Foley DH, Miller WL, Rubio R, Berne RM (1979) Transmural distribution of myocardial adenosine content during coronary constriction. Am J Physiol 236:H833–H838

21. Olsson RA (1970) Changes in content of purine nucleoside in canine myocardium during coronary occlusion. Circ Res 26:301–306

22. Norton ED, Jackson EK, Virmani R, Forman MB (1989) Intravenous adenosine limits myocardial reperfusion injury in a model with low myocardial collateral blood flow (Abstract). Circulation 80 [Suppl II]:II-238

23. Homeister JW, Hoff PT, Fletcher DD, Lucchesi BR (1990) Combined adenosine and lidocaine administration limits myocardial reperfusion injury. Circulation 82:595–608

24. Olafsson B, Forman MB, Puett DW, Pou A, Cates CU, Friesinger GC, Virmani R (1987) Reduction of reperfusion injury in the canine preparation by intracoronary adenosine: Importance of the endothelium and the no-reflow phenomenon. Circulation 76:1135–1145

25. Van Wylen DGL, Willis J, Sodhi J, Weiss RJ, Lasley RD, Mentzer RM (1990) Cardiac microdialysis to estimate interstitial adenosine and coronary blood flow. Am J Physiol 258:H1642–H1649

26. Herrmann SC, Feigl EO (1992) Adrenergic blockade blunts adenosine concentration and coronary vasodilation during hypoxia. Circ Res 70:1203–1216

27. Shirato C, Miura T, Ooiwa H, Toyofuku T, Wilborn WH, Downey JM (1989) Tetrazolium artifactually indicates superoxide dismutase-induced salvage in reperfused rabbit heart. J Mol Cell Cardiol 21:1187–1193

Cardioprotective Actions of Adenosine in the Heart: New Strategy for the Treatment of Ischemic Heart Diseases

Masafumi Kitakaze, Masatsugu Hori, Toshikazu Morioka, Seiji Takashima, Tetsuo Minamino[1], Michitoshi Inoue[2], and Takenobu Kamada[1]

Summary. Adenosine is recognized as an important chemical mediator for myocardial and vascular functions, which may assist cardioprotection during ischemia and reperfusion. Adenosine receptors can be divided into two types; A_1 and A_2 receptors, coupled with G_i and G_s proteins, respectively. Stimulations of A_1 and A_2 adenosine receptors decrease or increase adenylate cyclase activity, respectively. In ischemic hearts, the major role of endogenous adenosine is to relax coronary smooth muscles and increase blood flow through A_2 receptors. The vasodilatory effects of adenosine are enhanced by alpha$_2$-adrenoceptor stimulation due to sympathetic activation, H^+, and ATP sensitive K^+ channels, all of which occur in ischemic hearts. In the cardiovascular system, coronary vasodilation is not the only effect of adenosine. Stimulation of adenosine A_2 receptors also attenuates both free radical generation by activated leukocytes and aggregation of platelets. On the other hand, adenosine A_1 receptors activation attenuates beta-adrenoceptor-mediated increases in myocardial contractility, Ca^{2+} influx into myocytes, and norepinephrine release from presynaptic vesicles. Any of these effects may attenuate ischemic and reperfusion injury. Indeed, endogenous adenosine potentially attenuates contractile dysfunction during ischemia and reperfusion through activations of both adenosine A_1 and A_2 receptors. Furthermore, adenosine substantially limits the infarct size following sustained ischemia. Endogenous adenosine has been hypothesized as an essential mediator of ischemic preconditioning, which is generally accepted to limit infarct size. We elucidated that ischemic preconditioning increases 5'-nucleotidase activity and adenosine release, which may contribute to the reduction of infarct size. Taken together, we postulate the hypothesis that adenosine is a potential modulator for attenuation of ischemic and reperfusion injury.

Key words: 5'-Nucleotidase—5'-AMP—ATP-sensitive K^+ channel—Myocardial stunning —Myocardial necrosis—Ischemic preconditioning

Introduction

Adenosine is an important endogenous regulator of adenylate cyclase activity in a wide variety of tissues and organs [1]. In the ischemic heart, the major role of released adenosine has been thought to relax coronary smooth muscle and

---

[1] The First Department of Medicine, Osaka University School of Medicine, and [2] Department of Information Science, Osaka University Hospital 2-2 Yamada, Suita, 565 Japan

increase coronary blood flow [2, 3]. Besides this, adenosine is now recognized as a multifactorial affector of myocardial cells, sympathetic nerve cells, coronary smooth muscle, endothelial cells, leukocytes, and platelets. The malfunction of these components, which potentially leads to cellular injury in the ischemic and reperfused myocardium, can be attenuated by endogenous or exogenous adenosine. Recently, we have revealed that adenosine potentially attenuates ischemic and reperfusion injury due to its various cardiovascular effects [4–9]. We discuss here the subcellular mechanisms of adenosine production, adenosine-induced coronary vasodilation in ischemic hearts, and the potential mechanisms of adenosine-induced improvements of ischemia and reperfusion injury.

Mechanisms of Adenosine Production in Ischemic Hearts

Adenosine is produced by enzymatic reactions in cells [3] (Fig. 1) through the dephosphorylation of 5′-AMP by 5′-nucleotidase, and the hydrolysis of S-adenosylhomocystyeine (SAH) by SAH-hydrolase. Although adenosine is

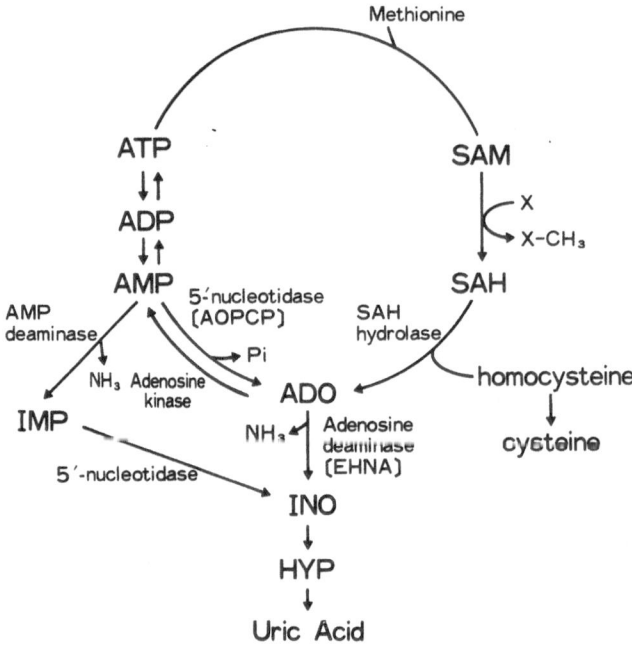

Fig. 1. Adenosine metabolism. *SAM*, S-Adenosylmethionine; *SAH*, S-adenosylhomocysteine; *ADO*, adenosine; *AMP*, adenosine 5′-monophosphate; *IMP*, inosine 5′-monophosphate; *INO*, inosine; *HYP*, hypoxanthine; *AOPCP*, alpha, beta methylene adenosine 5′-diphosphate; *EHNA*, erythro-9-(2-hydroxy-3-nonyl) adenosine. (From [3] with permission)

produced mainly through the latter pathway in normoxic hearts [10], the 5'-nucleotidase pathway of adenosine production is used during ischemia and hypoxia [11, 12]. This idea is supported by the fact that alpha-beta-methylene adenosine 5'-diphosphate (AOPCP), an inhibitor of 5'-nucleotidase, potently reduces adenosine production in the ischemic myocardium [13]. Furthermore, the extent of decrease in the reactive hyperemic flow following a brief period of coronary occlusion is comparably reduced under treatments with AOPCP and 8-phenyltheophylline (an adenosine receptor antagonist), suggesting that adenosine production during ischemia is attributable to the activity of 5'-nucleotidase [12–14]. 5'-Nucleotidase is present as ectosolic 5'-nucleotidase bound to membranes while cytosolic 5'-nucleotidase is found in the cytoplasm. Ectosolic 5'-nucleotidase activity is much higher than the cytosolic 5'-nucleotidase activity; however, 5'-AMP produced in the cell has easier access to cytosolic 5'-nucleotidase. The former result indicates that ectosolic 5'-nucleotidase is more potent in producing adenosine and the latter supports the idea that cytosolic 5'-nucleotidase is more accessible to the substrate for adenosine production. Therefore, we must conclude at present that both enzymes are thought to synergically contribute to adenosine production in the ischemic myocardium [11–13]. On the other hand, accumulation of 5'-AMP seems to be the other factor that regulates adenosine production. Cytosolic 5'-AMP concentration crucially depends on the duration and severity of ischemia and accumulates up to 1×10^{-2}–10^{-4} M. Considering that production of adenosine is 1×10^{-6}–10^{-8} M, even considerable changes in AMP concentration would not affect adenosine production during ischemia. Thus, we believe that 5'-nucleotidase activity rather than concentration of AMP is the main regulator of adenosine production in ischemic hearts.

5'-Nucleotidase activity is affected by several metabolic factors which are altered during ischemia, e.g., pH, concentrations of ATP and ADP, and Pi [1–3, 14]. Recently we have reported that adenosine itself attenuates ectosolic and cytosolic 5'-nucleotidase activities [15]. These attenuations are mimicked by N^6-cyclohexyladenosine and inhibited by pertussis toxin and by 8-phenyltheophylline, indicating that attenuations of ectosolic and cytosolic 5'-nucleotidase in cardiomyocytes are adenosine-A_1-receptor-mediated. This decreased 5'-nucleotidase activity also attenuates adenosine production in the normoxic and ischemic myocardium. Thus, adenosine causes a negative-feedback control for adenosine production through attenuation of 5'-nucleotidase activity.

Adenosine-Induced Coronary Vasodilation in the Ischemic Myocardium

Adenosine receptors can be divided into two types; A_1 and A_2 receptors, coupled with Gi and Gs proteins, respectively [16, 17]. Stimulation of A_1 and A_2 adenosine receptors decrease or increase adenylate cyclase activity, respectively.

Coronary smooth muscles only have adenosine A_2 receptors, which are responsible for coronary vasodilation in ischemic myocardium. It is widely accepted that released adenosine during ischemia benefits contractile and metabolic dysfunction by improving coronary perfusion.

Besides an increase in adenosine release, myocardial ischemia is characterized by: (1) accumulation of H^+, (2) increased norepinephrine release, and (3) decrease in ATP in myocardial and coronary smooth muscle cells. Interestingly, all of these three factors are reported to interfere with the relationship between doses of adenosine and the extent of coronary vasodilation. First of all, H^+ is reported to augment adenosine-induced coronary vasodilation [18]. Changes in pH of the solution from 7.4 to 6.8 cause about 50% augmentation of adenosine-induced coronary vasodilation in Langendorff preparations. Second, an exposure to norepinephrine can augment adenosine-induced coronary vasodilation. We have previously reported that alpha$_2$-adrenoceptor stimulation, which can be stimulated by an exposure to norepinephrine, enhances adenosine-induced coronary vasodilation [5–7]. Third, decreases in ATP in coronary smooth muscle cells during ischemia make ATP sensitive K^+ channels open, which may relax coronary vascular smooth muscles. In addition, ATP sensitive K^+ channels modulate adenosine-induced coronary vasodilation. Aversano et al. [19] reported that the adenosine-induced coronary vasodilation is substantially attenuated by blockade of ATP sensitive K^+ channels. This result suggests that adenosine A_2 receptors may also be affected by ATP sensitive K^+ channels. In our preliminary study, adenosine-induced coronary vasodilation was strongly enhanced when we additionally administered nicorandil, a K^+ channels opener [20]. This potentiation of adenosine-induced coronary vasodilation may also be beneficial for the ischemic myocardium, although further validation is necessary. Since both blockades of alpha$_2$-adrenoceptors and ATP sensitive K^+ channels modulate Na^+ outward through depolarization-induced activation of inverse Na^+/Ca^{2+} exchanges, respectively, intracellular Na^+ may be the primary regulatory factor of adenosine A_2 receptor activity. H^+ itself may affect cytosolic Na^+ concentration through Na^+/H^+ exchanges.

Is it possible to attenuate myocardial ischemia by increasing adenosine production or sensitivity of adenosine A_2 receptors? Several lines of evidence suggest that potentiation of increases in adenosine concentration reduces flow in subendocardial layer [21, 22]. However, this hypothesis may not be true when we potentiate adenosine production or sensitivity of adenosine A_2 receptor in the ischemic myocardium. There are three lines of evidence which show the beneficial effects of endogenous adenosine in ischemic hearts. First, adenosine preferentially increases blood flow in the endocardium [23]. Second, we observed that potentiation of adenosince receptor stimulation by clonidine increases coronary flow and improves myocardial function during ischemia [6], indicating that potentiation of the effects of adenosine improves myocardial perfusion. Finally, we also showed that administration of AICA-riboside, a potentiator of adenosine production from ischemic myocytes, improves coronary perfusion and contractile and metabolic dysfunction in coronary microembolization [24].

Microspheres 15 μm diameter are known to produce myocardial cellular ischemia within the size range of 0.02–0.03 mm^2 [8]. If endogenous adenosine causes redistribution of myocardial flow from the ischemic area to non-ischemic area, myocardial contractile and metabolic dysfuction may deteriorate in coronary microembolization. In our study, AICA riboside reduced the severity of myocardial ischemia in coronary microembolization [24], indicating that potentiation of adenosine production in the ischemic area favors flow from the non-ischemic myocardium. This evidence suggests that potentiation of production or increased sensitivity of adenosine receptors may improve myocardial ischemia.

Role of Adenosine in Ischemic Myocardium

Needless to say, adenosine-induced coronary vasodilation is beneficial for the contractile and metabolic dysfunction in the ischemic hearts. However, this is not the only effect of adenosine in the ischemic myocardium. Thromboembolism in the small coronary arteries can be observed in acute myocardial infarction and stimulation of adenosine A_2 receptors has been reported to inhibit platelet aggregation caused by an exposure to norepinephrine in *in vitro* experiments [25, 26]. However, there is no clear consensus that inhibition of platelet aggregation is preserved by endogenous adenosine during myocardial ischemia, leading us to test the hypothesis that endogenous adenosine plays a key role in inhibition of thromboembolism due to platelet aggregation in ischemic hearts [27]. Our observation is that blockade of adenosine receptors promotes the cascade of platelet aggregation and exerts deleterious results in small coronary arteries during severe ischemia (Fig. 2) suggesting that released adenosine during ischemia might inhibit platelet aggregation. However, when myocardial ischemia is prolonged, the capability of adenosine production may be weakened and adenosine receptors may be desensitized; both these effects attenuate the stimulatory effects of adenosine receptors in platelets. Adenosine also inhibits the production of oxygen-derived free radicals in leukocytes [28, 29] through adenosine A_2 receptors. This reduced function in leukocytes may attenuate myocardial injury due to prolonged ischemia and subsequent reperfusion. We have recently revealed that activation of human polymorphonuclear leukocytes attenuates ectosolic 5′-nucleotidase activity of the leukocytes itself, leading to the attenuation of adenosine production in the leukocyte [30]. This attenuation of adenosine release may further enhance the activation of human polymorphonuclear leukocytes and augment vascular and myocardial injury.

Adenosine also attenuates an increase in myocardial contractility induced by beta-adrenoceptor stimulation [31]. In ischemic hearts, we have revealed that this mechanism is substantially realized [32] (Fig. 3). Adenosine-induced attenuation of increases in myocardial contractility decreases the extent of metabolic dysfunction in the ischemic myocardium [32]. This observation seems to differ from the fact that adenosine inhibits norepinephrine release from the

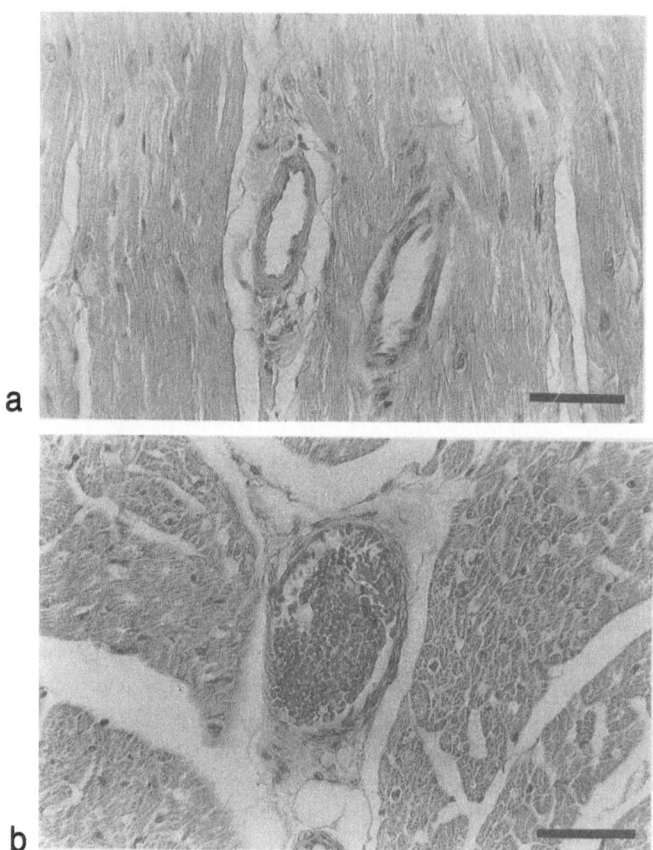

Fig. 2a,b. Hematoxylin—eosin staining **a** in untreated and **b** in 8-phenyltheophylline-treated myocardium excised after perfusion fixation after 3-min ischemia. *Bars* in the *right corner* of these figures are 50 μm. On the completion of the physiological measurements, heart tissues were prepared for light microscope analysis. (From [27] with permission)

sympathetic nerve terminals [33], because our observations were not abolished in hearts denervated with 6-hydroxydopamine. Adenosine-induced inhibition of norepinephrine release from the presynaptic vesicles may additionally contribute to the attenuation of myocardial ischemia because excess amounts of norepinephrine are known to cause catecholamine-induced injury.

Taken together, these cardiovascular effects of adenosine, including coronary vasodilation, potentially attenuate the severity of contractile and metabolic dysfunction in ischemic hearts.

Roles of Adenosine in Myocardial Stunning Following Myocardial Ischemia

Even if ischemic heart muscle is reperfused before irreversible injury occurs, contractile function remains impaired for a long period, a phenomenon known as myocardial stunning. We have reported that endogenous and exogenous adenosine attenuates myocardial stunning in dogs [34, 35]. Contractile function was assessed by the fractional shortening of regional myocardium with 3 h of reperfusion following 15 min of coronary occlusion. Administration of adenosine and 8-phenyltheophylline decreases/increases the severity of myocardial stunning, respectively. N^6-cyclohexyladenosine and 5'-N-ethylcarboxamidoadenosine synergistically improved myocardial stunning, indicating that both adenosine A_1 and A_2 receptors are responsible for attenuating myocardial stunning [34].

By what mechanisms does adenosine assist in the attenuation of myocardial stunning? Stimulation of A_1 adenosine receptors inhibits beta-adrenoceptor-mediated inotropic responses [31, 32] and intracellular Ca^{2+} influx [36]. However, the former mechanism can be eliminated because propranolol does not affect the severity of myocardial stunning, suggesting that inhibition of Ca^{2+} influx by adenosine may contribute to attenuation of myocardial stunning. Indeed, several lines of evidence support the idea that Ca^{2+} overload [37–39] is one of the potential mechanisms for myocardial stunning. On the other hand, stimulation of A_2 adenosine receptors increases hyperemic flow and inhibits activations of neutrophils [28, 29] and platelets [25, 27]. When we increased hyperemic flow with papaverine instead of administering adenosine, myocardial stunning was not improved [35], suggesting that increases in coronary blood flow due to adenosine may not be related to improvements in myocardial stunning. Microcirculatory disturbances in myocardial stunning are improved by adenosine, which attenuates reperfusion injury [40, 41]. Adenosine-induced inhibition of activation of both neutrophils and platelets may be involved in attenuating myocardial stunning. Indeed, activated neutrophils generate oxygen-derived free radicals and administration of superoxide dismutase attenuates myocardial stunning. Intriguingly, oxygen-derived free radicals are reported to attenuate ectosolic 5'-nucleotidase activity and adenosine production during ischemia [13, 41] (Fig. 4). Thus attenuation of oxygen-derived free radical generation by adenosine may preserve the activity of 5'-nucleotidase and increase adenosine production. Several lines of evidence [42–45] support the idea that oxygen-derived free radicals inactivate several enzyme activities. Ectosolic 5'-nucleotidase that is bound to the myocardial and endothelial cellular surface is thought to be a target of the injury caused by oxygen-derived free radicals. Interestingly, after AOPCP administration, treatment with SOD did not increase either the reactive hyperemic flow or adenosine release during reperfusion following ischemia, indicating that oxygen-derived free radicals may inhibit 5'-nucleotidase activity by altering the same fraction as AOPCP does. The

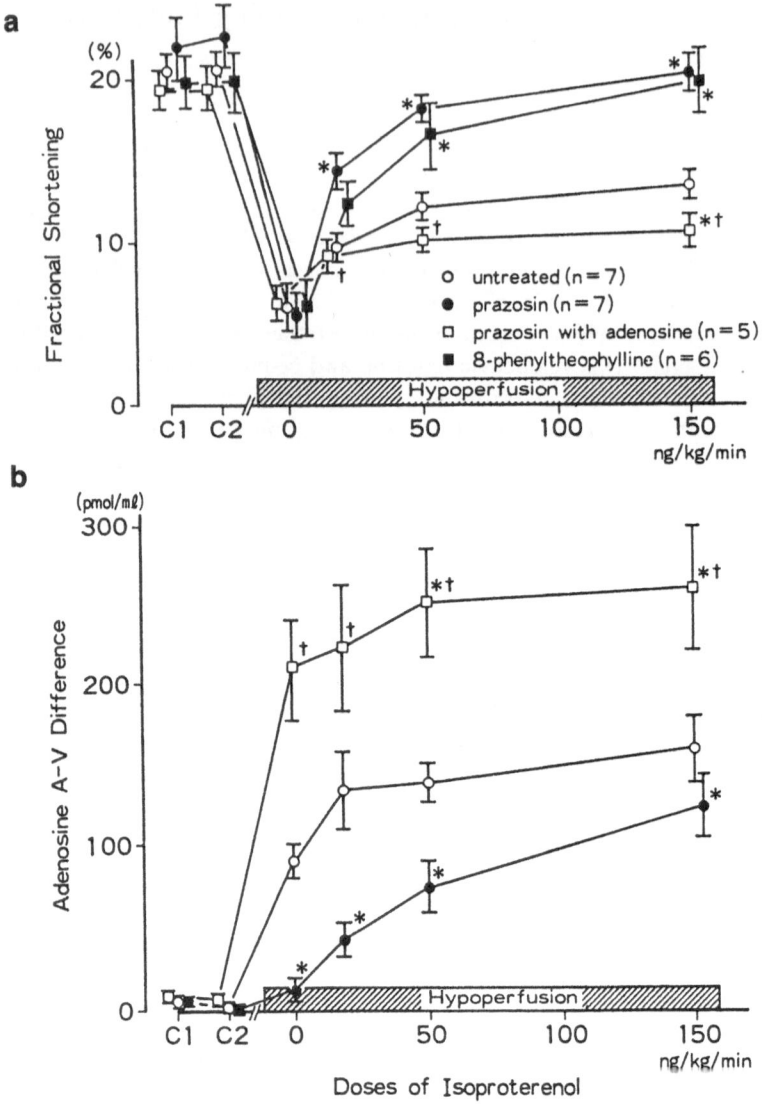

Fig. 3a,b. Dose-response relationships of isoproterenol to **a** fractional shortening and **b** coronary arteriovenous difference of adenosine in ischemic hearts. Neither an infusion of 8-phenyltheophylline nor prazosin (*C2*, pretreatment) changed fractional shortening and coronary arteriovenous differences of adenosine in the baseline condition (*C1*, no drug). In all of the groups, fractional shortening was increased by intravenous infusions of isoproterenol in a dose dependent manner, but the increments of fractional shortening were significantly ($P < 0.05$) augmented in the 8-phenyltheophylline and prazosin treated hearts. An intracoronary infusion of prazosin (4 µg/kg per min) attenuated adenosine release from the ischemic myocardium (**b**, $P < 0.05$ vs the untreated group). The effect of prazosin was completely abolished by the intracoronary infusion of adenosine (5 µg/kg per

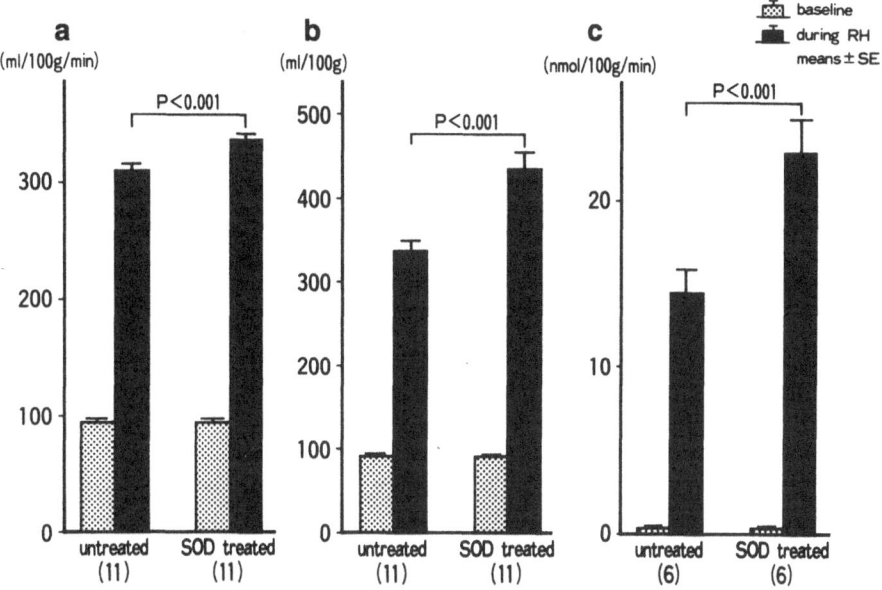

Fig. 4a–c. Effects of SOD treatment on **a,b** reactive hyperemic flow (**a** peak coronary blood flow [*CBF*], **b** debt and repayment) and **c** adenosine release following 1-min coronary occlusion. Superoxide dismutase (*SOD*) treatment significantly augmented both reactive hyperemic flow (**a,b**) and adenosine release (**c**) during reperfusion. *Numbers in parentheses* represent numbers of cases. (From [13] with permission)

multiple effects of adenosine in the cardiovascular system contribute to attenuation of the reperfusion injury, especially myocardial stunning.

Role of Adenosine in Salvage of Myocardial Mecrosis by Ischemic Preconditioning

It has been clarified that adenosine limits infarct size caused by sustained myocardial ischemia and reperfusion: intracoronary infusion of adenosine reduces myocardial infarct size in dogs by 75% [46]. The reduction of ATP during ischemia and reperfusion seems critically important because intracellular ATP concentration is one of the factors which determine whether myocardial

min), indicating that the beta-adrenoceptor-mediated inotropic response is attenuated by endogenous adenosine in the ischemic hearts. *Bars* are means ± SEM. *$P < 0.05$ vs the untreated group; +$P < 0.05$ vs the prazosin-treated group *Open circles*, Untreated ($n = 7$); *closed circles*, prazosin ($n = 7$); *open squares*, prazosin with adenosine ($n = 5$); *closed squares*, 8-phenyltheophylline ($n = 6$). (From [32] with permission)

cellular injury is reversible or irreversible. Adenosine may act by preserving ATP [47]. The myocardial ATP concentration and ATP synthesis are increased in the postischemic myocardium when exogenous adenosine is administered to hearts perfused by crystalloid solution [48, 49].

Recently ischemic preconditioning has been markedly focused, not only by basic researchers but also by clinicians in cardiovascular fields, because ischemic preconditioning is believed to limit infarct size to 10%–20% in the reperfused ischemic myocardium [50–52]. The mechanism underlying this phenomenon is being studied actively with an eye towards developing treatments to limit the size of acute myocardial infarction. Liu et al. [52] have shown that endogenous adenosine mediates ischemic preconditioning, based upon the observation that 8-phenyltheophylline abolishes the necrosis-limiting effects of ischemic precon-ditioning. Liu et al. [52] have hypothesized that ischemic preconditioning occurs via adenosine A_1 receptor activation. Increases in blood flow [1–3], inhibitions of neutrophil activations [28, 29], and platelet aggregation [25–27] are mediated through the adenosine A_2 receptor, eliminating these as mechanisms of action for ischemic preconditioning. In contrast, the inhibition of adrenergic stimulation by adenosine and norepinephrine release may account for the beneficial effects of ischemic preconditioning, since both actions are mediated via adenosine A_1-receptor activation. Indeed, myocardial oxygen consumption during reperfusion was significantly less after ischemic preconditioning [52]. Attenuation of Ca^{2+} overload by activation of adenosine A_1 receptors may also contribute to the beneficial effects of ischemic preconditioning.

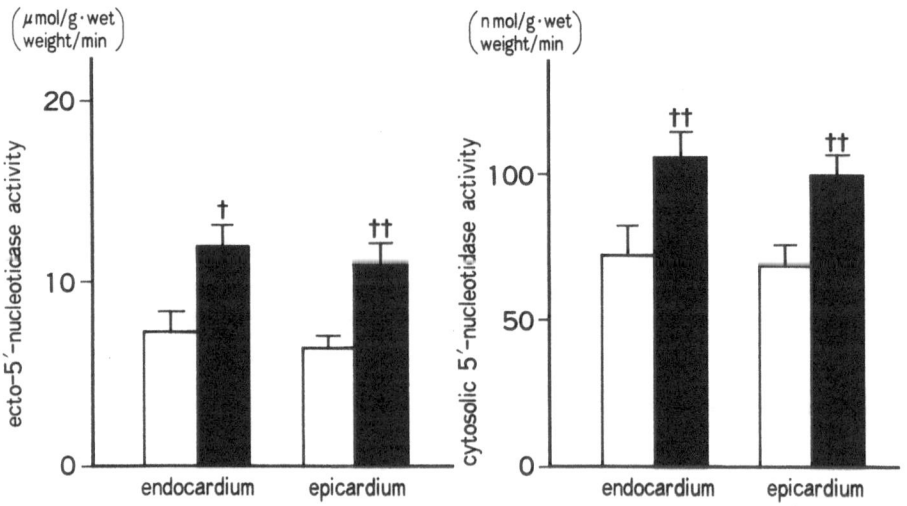

Fig. 5. Ectosolic and cytosolic 5′-nucleotidase activities in the control and ischemic preconditioned myocardium before 40-min coronary occlusion. Both activities were augmented by ischemic preconditioning. *Open bars*, Control; *closed bars*, Preconditioned. Means ± SE, $n = 5$. $+P < 0.005$; $+< 0.001$ vs control. (From [53] with permission)

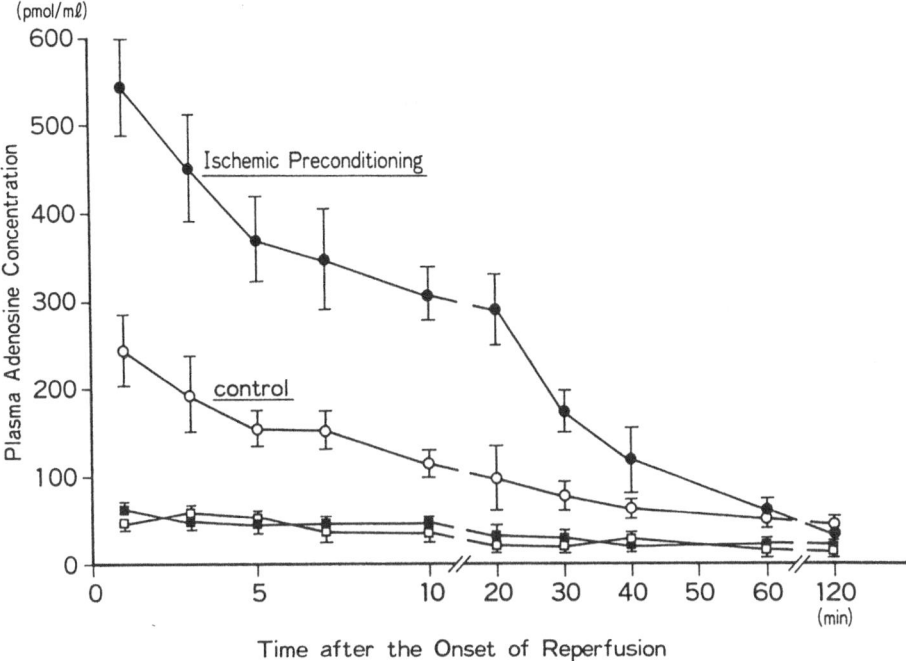

Fig. 6. Adenosine release during reperfusion after 40-min coronary occlusion with and without ischemic preconditioning. Adenosine release in the group that underwent ischemic preconditioning is enhanced significantly ($P < 0.01$) for 40 minutes compared with the untreated control condition (means ± SE). *Open circles*, Control, coronary venous adenosine ($n = 6$); *closed circles*, preconditioned, coronary venous adenosine ($n = 6$; *open squares*, control, arterial adenosine ($n = 6$); *closed squares*, preconditioned, arterial adenosine ($n = 6$)

Despite this progress in understanding of ischemic preconditioning, the mechanism relating adenosine to ischemic preconditioning is unknown. One likely mechanism is that ischemic preconditioning increases adenosine production during sustained ischemia and reperfusion, possibly by augmenting 5'-nucleotidase activity [53]. We have reported that ischemic preconditioning augments the activity of 5'-nucleotidase (Fig. 5), and thereby increases adenosine production (Fig. 6) during ischemia and reperfusion. This augmented adenosine production may contribute to the underlying mechanisms of ischemic preconditioning. Protein kinase C activity increases during ischemia and reperfusion [9, 14, 15], which also increases adenosine production. Furthermore, a transient decrease in the energy charge during ischemic preconditioning has been reported to increase cytosolic 5'-nucleotidase activity [54]. Both intracellular H^+ and inorganic phosphate concentrations, which increase during ischemia, modify 5'-nucleotidase activity [54, 55]. Recently we have observed that alpha$_1$-adrenoceptor activation mediates increases in ectosolic and cytosolic 5'-nucleotidase activity [9] and contributes to limiting the infarct size [55]. Therefore, we postulate that alpha$_1$-adrenoceptor activation during ischemic preconditioning contributes to the attenuation of the infarct size [55], although we need to continue to investigate how 5'-nucleotidase activity is enhanced by ischemic preconditioning.

Our study hints that increased adenosine release may be the mechanism by which ischemic preconditioning limits infarct size. It is possible, therefore, that intracoronary administration of adenosine or a potentiator of adenosine production, e.g., AICA riboside (5-amino-4-imidazole carboxamide-riboside), dilazep or dipyridamole, may limit infarct size in acute myocardial infarction and reduce contractile dysfunction [56, 57]. Additionally, it may be possible to enhance adenosine production during ischemia and reperfusion by activation of 5'-nucleotidase. Alpha$_1$-adrenoceptor stimulation is one approach to increasing 5'-nucleotidase activity [9]. Indeed, alpha$_1$-adrenoceptor stimulation attenuates contractile dysfunction by enhancing the release of adenosine [9, 35, 58] and ultimately limits infarct size [52, 55, 59]. Interleukin and tumor necrosis factor also have been reported to increase 5'-nucleotidase activity [60]. Further studies are required to learn how best to stimulate adenosine production and limit infarct size following acute myocardial infarction. Investigation of drugs which increase adenosine production or activate adenosine receptors is a promising approach.

References

1. Berne RM, Rubio R, Curnish RR (1974) Release of adenosine from ischemic brain: Effect on cerebral vascular resistance and incorporation into cerebral adenine nucleotides. Circ Res 35:262–271
2. Berne RM, Rubio R (1979) Coronary circulation. In: Berne RM, Sperelakis N, Geiger SR (eds) Handbook of physiology, section 2, the cardiovascular system. American Physiology Society, Washington, pp 873–952
3. Hori M, Kitakaze M (1991) Adenosine, the heart, and coronary circulation Hypertension 18:565–574
4. Kitakaze M, Hori M, Tamai J, Iwakura K, Koretsune Y, Kagiya T, Iwai K, Kitabatake A, Inoue M, Kamada T (1987) α$_1$-Adrenoceptor activity regulates release of adenosine from the ischemic myocardium in dogs. Circ Res 60:631–639
5. Hori M, Kitakaze M, Tamai J, Koretsune Y, Iwai K, Iwakura K, Kagiya T, Kitabatake A, Inoue M, Kamada T (1988) α$_2$-Adrenoceptor activity exerts dual control of coronary blood flow in canine coronary artery. Am J Physiol 255:H250–H260
6. Kitakaze M, Hori M, Gotoh K, Sato H, Iwakura K, Kitabatake A, Inoue M, Kamada T (1989) Beneficial effects of α$_2$-Adrenoceptor activity on ischemic myocardium during coronary hypoperfusion in dogs. Circ Res 65:1632–1645
7. Hori M, Kitakaze M, Tamai J, Iwakura K, Kitabatake A, Inoue M, Kamada T

(1989) α_2-Adrenoceptor stimulation can augment coronary vasodilation maximally induced by adenosine in dogs. Am J Physiol 257:H132–H140

8. Hori M, Tamai J, Kitakaze M, Iwakura K, Gotoh K, Iwai K, Koretsune Y, Kagiya T, Kitabatake A, Kamada T 1989 Adenosine-induced hyperemia attenuates myocardial ischemia in coronary microembolization in dogs. Am J Physiol 257: H244–H251

9. Kitakaze M, Kitabatake A (1991) Increased 5'-nucleotidase activity caused by protein kinase C enhances adenosine production in hypoxic cardiomyocytes of rats (abstract). Circulation 84:II-620

10. Achterberg PW, de Tombep P, Harmsen E, de Jong JW (1985) Myocardial S-adenosylhomocysteine hydrolase is important for adenosine production during normoxia. Biochem Biophys Acta 840:393–400

11. Frick GP, Lowenstein JW (1976) Studies of 5'-nucleotidases in the perfused rat heart, including measurements of the enzyme in perfused skeletal muscle and liver. J Biol Chem 251:6372–6378

12. Imai S, Nakazawa M, Imai M, Jin H (1986) 5'-Nucleotidase inhibitors and myocardial reactive hyperemia and adenosine content. In: Gerlach E, Becker BF (eds) Topics and perspectives in adenosine research. Springer, Berlin Heidelberg, pp 416–424

13. Kitakaze M, Hori M, Takashima S, Iwai K, Sato H, Inoue M, Kitabatake A, Kamada T (1992) Superoxide dismutase enhances ischemia-induced reactive hyperemic flow and adenosine release in dogs: Role of 5'-nucleotidase activity. Circ Res 71:558–566

14. Sparks HV, Bardenheuer H (1986) Regulation of adenosine formation by the heart. Circ Res 58:193–201

15. Kitakaze M, Iwai K, Kagiya T, Koretsune Y, Tada M (1992) Deactivation of 5'-nucleotidase caused by stimulation of adenosine receptors in rat cardiomyocytes (abstract). Circulation 86:I-125

16. Londos C, Cooper DMF, Schlegel W, Rodbell M (1978) Adenosine analogs inhibit adipocyte adenylate cyclase by a GTP-dependent process: Basis for actions of adenosine and methylxanthines on cyclic AMP production and lipolysis. Proc Natl Acad Sci USA 75:5362–5366

17. Bruns RF, Daly JW, Snyder SH (1980) Adenosine receptors in brain membranes: Binding of N^6-cyclohexy [^{3}H] adenosine and 1,3-diethyl-8[^{3}H] phenylxanthine. Proc Natl Acad Sci USA 77:5547–5551

18. Merrill GF, Haddy FJ, Dabney JM 1978 Adenosine, theophylline, and perfusate pH in the isolated perfused guinea pig heart. Circ Res 42:225–229

19. Aversano T, Ouyang P, Silverman H (1991) Blockade of the ATP-sensitive potassium channel modulates reactive hyperemia in the canine coronary circulation. Circ Res 69:618–622

20. Kitakaze M, Hori M, Takashima S, Morioka T, Kamada T (1993) Nicorandil enhances adenosine-induced coronary vasodilation in dogs (abstract). Ther Res 14:98–106

21. Patterson RE, Kirk ES (1983) Coronary steal mechanisms in dogs with one-vessel occlusion and other arteries normal. Circulation 67:1009–1015

22. Becker LC (1978) Conditions for vasodilator-induced coronary steal in experimental myocardial ischemia. Circulation 57:1103–1110

23. Canty J Jr (1989) Adenosine shifts the endocardial flow-function relation during ischemia in conscious dogs (abstract). Circulation 80:II-507

24. Hori M, Sato H, Koretsune Y, Kagiya T (1991) AICA-riboside (5-amino-4-imidazole carboximide riboside), a novel adenosine potentiator, attenuates myocardial ischemia in coronary microembolization (abstract). Circulation 84:II-305

25. Agarwal KC (1987) Adenosine and platelet function. In: Stefanovich V, Okayuz-Baklouti I (eds) Role of adenosine in cerebral metabolism and blood flow. VNU Science Press, Utrecht, pp 107–124

26. Newman WH, Becker BF, Heier M, Nees S, Gerlach E (1988) Endothelium-mediated coronary dilatation by adenosine does not depend on endothelial adenylate cyclase activation: Studies in isolated guinea pig hearts. Pflügers Arch 413:1–7

27. Kitakaze M, Hori M, Sato H, Takashima S, Inoue M, Kitabatake A, Kamada T (1991) Endogenous adenosine inhibits platelet aggregation during myocardial ischemia in dogs. Circ Res 69:1402–1408

28. Cronstein BN, Levin RI, Belanoff J, Weissmann G, Hirschhorn R (1986) Adenosine: An endogenous inhibitor of neutrophil-mediated injury to endothelial cells. J Clin Invest 78:760–770

29. Cronstein BN, Kramer SB, Weissmann G, Hirschhorn R (1986) Adenosine: A physiological modulator of superoxide anion generation by human neutrophils. J Exp Med 158:1160–1177

30. Kitakaze M, Takashima S, Morioka T, Kuzuya T, Tada M (1992) Attenuation of both $5'$-nucleotidase activity and adenosine production in activated human leukocytes (abstract). Circulation 86:I-24

31. Belardinelli L, Isenberg G (1983) Actions of adenosine and isoproterenol on isolated mammalian ventricular myocytes. Circ Res 53:287–297

32. Sato H, Hori M, Kitakaze M, Takashima S, Inoue M, Kitabatake A, Kamada T (1992) Endogenous adenosine blunts beta-adrenoceptor-mediated inotropic response in the hypoperfused canine myocardium. Circulation 85:1594–1603

33. Richardt G, Wassa W, Kranzhofer R, Mayer E, Schoming A (1987) Adenosine inhibits exocytotic release of endogenous noradrenaline in rat heart: A protective mechanism in early myocardial ischemia. Circ Res 61:117–123

34. Kitakaze M, Takashima S, Sato H (1990) Stimulation of adenosine A_1 and A_2 receptors prevents myocardial stunning (abstract). Circulation 82:III-37

35. Kitakaze M, Hori M, Sato H, Iwakura K, Gotoh K, Inoue M, Kitabatake A, Kamada T (1991) Beneficial effects of α_1-adrenoceptor activity on myocardial stunning in dogs. Circ Res 68:1322–1339

36. Isenberg G, Cerbai E, Klockner U (1987) Ionic channels and adenosine in isolated heart cells. In: Gerlach E, Becker BF (eds) Topics and perspectives in cardiovascular research. Springer, Berlin Heidelberg New York Tokyo, pp 323–335

37. Steenbergen C, Murphy E, Levy L, London RE (1987) Elevation in cytosolic free calcium concentration early in myocardial ischemia in perfused rat heart. Circ Res 60:700–707

38. Kitakaze M, Weisman HF, Marban E (1988) Contractile dysfunction and ATP depletion after transient calcium overload in perfused ferret hearts. Circulation 77:685–695

39. Marban E, Kitakaze M, Koretsune Y, Yue DT, Chacko VP, Pike MM (1990) Quantification of $[Ca^{2+}]_i$ in perfused hearts. Critical evaluation of the 5F-BAPTA and nuclear magnetic resonance method as applied for the study of ischemia and reperfusion. Circ Res 66:1255–1267

40. Hori M, Inoue M, Kitakaze M, Koretsune Y, Iwai K, Tamai J, Ito H, Kitabatake A,

Sato T, Kamada T (1986) Role of adenosine in hyperemic response of coronary blood flow in microembolization. Am J Physiol 250:H509–H518

41. Hori M, Gotoh K, Kitakaze M, Iwai K, Iwakura K, Sato H, Koretsune Y, Kitabatake A, Inoue M, Kamada T (1991) Role of oxygen-mediated free radicals in myocardial edema and ischemia in coronary microembolization. Circulation 84:828–840

42. Bielski BH, Chen PH (1974) Kinetic study by pulse radiolysis of the lactate dehydrogenase-catalyzed chain oxidation of nicotinamide adenine dinucleotide by HO_2 and O_2^- radicals. J Biol Chem 250:314–321

43. Kim HS, Minard P, Legoy MD, Thomas D (1986) Inactivation of 3α-hydroxydtroid dehydrogenase by superoxide radicals. Biochem J 233:493–497

44. Grover AK, Samson SE (1988) Effects of superoxide radical on Ca^{2+} pump of coronary artery. Am J Physiol 255:C297–C303

45. Kim MS, Akera T (1987) O_2 Free radicals: Cause of ischemia-reperfusion injury to cardiac Na^+-K^+-ATPase. Am J Physiol 252:H252–H257

46. Olafsson B, Forman MB, Puett DW, Pou A, Cates CU, Friesinger GC, Virmani R (1987) Reduction of reperfusion injury in a canine preparation by intracoronary adenosine: Importance of the endothelium and the no-reflow phenomenon. Circulation 76:1135–1145

47. Foker JE, Einzig E, Wang T (1980) Adenosine metabolism and myocardial preservation. J Thorac Cardiovasc Sug 80:506–516

48. Reibel DK, Rovette MJ (1979) Myocardial adenosine salvage rates and restoration of ATP content following ischemia. Am J Physiol 237:H247–H252

49. Murray CE, Jennings RB, Reimer KA (1986) Preconditioning with ischemia: A delay of lethal cell injury in ischemic myocardium. Circulation 74:1124–1136

50. Scott RJ, Rohmann S, Braun ER, Schaper W (1990) Ischemic preconditioning reduces infarct size in swine myocardium. Circ Res 66:1133–1142

51. Li GC, Vasquez JA, Gallagher KP, Lucchesi BR (1990) Myocardial protection with preconditioning. Circulation 82:609–619

52. Liu GS, Thornton J, Van Winkle DM, Stanley AWH, Olsson RA, Downey JM (1991) Protection against infarction afforded by preconditioning is mediated by A_1 adenosine receptors in rabbit heart. Circulation 84:350–356

53. Kitakaze M, Hori M, Takashima S, Sato H, Michitoshi Inoue, Kamada T (1993) Ischemic preconditioning increases adenosine release and 5'-nucleotidase activity during myocardial ischemia and reperfusion in dogs. Implications for myocardial salvage. Circulation 87:208–215

54. Itoh R, Oka J, Ozasa H (1986) Regulation of rat heart cytosol 5'-nucleotidase by adenylate energy charge. Biochem J 235:847–851

55. Kitakaze M, Takashima S, Morioka T, Sato H, Hori M (1992) Alpha₁-adrenergic activation limits infarct size by augmenting adenosine production and 5'-nucleotidase activity in ischemic preconditioning (abstract). Circulation 86:I-213

56. Gruber HE, Hoffer ME, McAllister DR, Laikand PK, Lane TA, Schmid-Schoenbein GW, Engler RL (1989) Increased adenosine concentration in blood from ischemic myocardium by AICA riboside. Effects on flow, granulocytes, and injury, Circulation 80:1400–1411

57. Hori M, Kitakaze M, Takashima S (1990) AICA-riboside (5-amino-4-imidazole carboxamide riboside 100), a novel adenosine potentiator, attenuate myocardial stunning (abstract). Circulation 82:III-466

250 M. Kitakaze et al.

58. Kitkaze M, Hori M, Kamada T (1993) Role of adenosine and its interaction with α-adrenoceptor activity in ischemic and reperfusion injury of the myocardium (review). Cardiovasc Res 27:18–27
59. Locke-Winter CR, Winter CB, Nelson DW, Bannerjee A (1991) cAMP stimulation facilitates preconditioning against ischemia-reperfusion through norepinephrine and alpha₁ mechanisms (abstract). Circulation 84:II-433
60. Savic V, Stefanovic V, Ardaillou N, Ardaillou R (1990) Induction of ecto-5'-nucleotidase of rat cultured mesengial cells by interleukin 1-beta and tumor necrotic factor-alpha. Immunology 70:321–326

Potentiation of Collateral Development in Ischemic Patients with Heparin Treatment

Masatoshi Fujita and Shigetake Sasayama[1]

Summary. A method consisting of repeated 2 min coronary occlusions every 1 h, 8 h/day, 5 days/week, developed in Dean Franklin's laboratory, allowed a quantitative evaluation of the speed of collateral growth and possible effects of some drugs on the rate of collateralization. In this canine model, the total occlusion number needed for the development of collateral circulation is an index of stimuli for collateralization.

The experimentally-identified characteristics of heparin to potentiate the mitogenic activity of ischemia-related fibroblast growth factors have been extended to the clinical setting as a possible therapeutic modality of modulating the collateral development in patients with coronary artery disease such as chronic effort angina and acute myocardial infarction.

Although these clinical studies have been conducted in small cohorts of patients, in nonrandomized and non-blinded methods, there is a possiblity that ischemic heart disease could be treated with drugs to potentiate the growth of collateral vessels. The development of such a therapeutic remedy would reduce the deleterious sequelae caused by coronary atherosclerosis.

Key words: Acute myocardial infarction,—Coronary collateral circulation,—Coronary collateral development,—Effort angina

Introduction

The primary goal of treatment for effort angina is to alleviate chest pain and to prevent the progression to acute myocardial infarction. As medical therapy, β-blockers, nitrates, and calcium-channel blockers long have been used to decrease myocardial oxygen requirements. However, some patients with intractable angina need more aggressive treatment, such as percutaneous transluminal coronary angioplasty or aorto-coronary bypass surgery. Despite the progress in these techniques, a certain number of patients are not candidates for these therapeutic approaches.

An alternative means of blood delivery to the compromised myocardium is the coronary collateral circulation. Indeed, results from some studies indicate that the collateral channels serve as significant blood-conveying conduits, at least

---

[1] The Second Department of Internal Medicine, Toyama Medical and Pharmaceutical University, 2630 Sugitani, Toyama 930-01, Japan

under resting conditions [1, 2]. However, whether the functional capacity of collaterals is sufficiently augmented to represent an adequate perfusion reserve against strenuous exercise has yet to be fully elucidated. The modulation of the development of collaterals would provide a prospect for patients with coronary artery disease.

Canine Experimental Model

An experimental model developed by Franklin et al. is shown in Fig. 1 [3]. Each dog was anesthetized with intravenous sodium pentobarbital ($30\,\mathrm{mg\,kg^{-1}}$) and ventilated with a Harvard respirator. Following the incision of the pericardium, a high fidelity micromanometer (Konigsberg P22) and a fluid filled catheter (Tygon, internal diameter 1.3 mm) were inserted into the left ventricle through a stab wound at the apex. The left circumflex coronary artery was dissected free near its origin and a miniaturized 10 MHz ultrasonic pulsed Doppler flow probe and an externally inflatable pneumatic occluder positioned around the vessel. In each dog two pairs of ultrasonic crystals (diameter 2 mm) were implanted subendocardially along the minor axis for measuring segment lengths. One pair

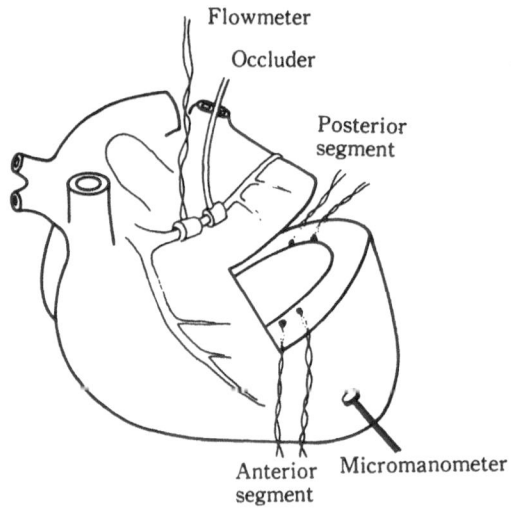

Fig. 1. Implanted instrumentation of the canine experimental model. A pulsed Doppler flow transducer was implanted near the origin of the left circumflex coronary artery. An externally controllable pneumatic occluder was implanted distal to the flow transducer. Miniature piezoelectric crystals were implanted for measurement of subendocardial, circumferential myocardial segment lengths in the area perfused by the left circumflex coronary artery (*posterior segment*) and in the area of perfusion of anterior descending coronary artery (*anterior segment*). A Konigsberg miniature pressure gauge was implanted within the left ventricle through an apical incision

was placed close to the base of the posterior papillary muscle (ischemic segment), the other pair in the anterior free wall (nonischemic segment). All wires and catheters were exteriorized at the base of the neck. Studies were initiated in the conscious state two weeks postoperatively, when the dogs appeared healthy and active. Two min occlusions of the left circumflex coronary artery were repeated at 1 h intervals 8 times daily, except at weekends, until the percentage reduction in subendocardial segment shortening in the central ischemic area at the end of the 2 min coronary occlusions was less than 5%, indicating an adequate collateral development to the area perfused by the left circumflex coronary artery [4].

Quantitative Assessment of the Rate of Collateral Development (Effect of Heparin)

The newly developed canine experimental model allowed quantitative evaluation of the speed of collateral growth and possible effects of some drugs on the rate of collateralization. In this model, the total occlusion number needed for the development of collateral circulation is an index of the rate of collateralization.

It has recently been demonstrated that heparin can facilitate angiogenesis induced by tumor extracts from human hepatoma implanted on the chorioallantoic membrane of the chick embryo [5], suggesting a new function of heparin as a positive regulator of angiogenesis.

Using this model, we documented that heparin treatment accelerates the collateral development [6]. Eight control dogs needed 129 ± 45 [mean $\pm$ standard deviation (SD)] 2 min coronary occlusions for collateralization whereas in 8 dogs with heparin treatment, the number of 2 min occlusions until the collateral development was decreased to 81 ± 33 (mean $\pm$ SD) ($p < 0.05$) (Fig. 2). Thus, heparin conclusively increases the speed of collateral development induced by the repeated 2 min coronary occlusions.

Treatment of Chronic Effort Angina with Heparin

We attempted to promote collateral development in patients with angina pectoris by repeated exercise stress combined with heparin treatment [7]. In 16 patients, with obstruction of at least one major coronary artery and angina on effort, exercise was performed according to the standard Bruce protocol twice a day. Ten patients were given an intravenous injection of heparin (5000 IU) 10–20 minutes before each exercise period, and the remaining six patients exercised without heparin treatment. Treatment with heparin increased the total exercise duration from 6.3 ± 1.9 to 9.1 ± 2.2 min ($p < 0.001$) and the maximum rate-pressure product from $18\,900 \pm 5100$ to $25\,500 \pm 6800$ mmHg $\times$ beats per min. After heparin administration, the rate-pressure product was also increased by 35% ($p < 0.01$) at the onset of angina and by 19% ($p < 0.05$) at the point at which ST segment depression ($\geqslant 0.1$ mV) first appeared. All of these variables

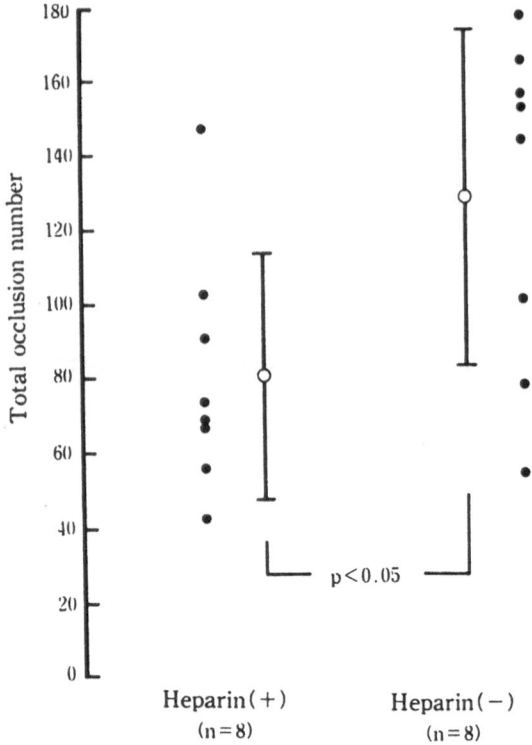

Fig. 2. Total number of 2 min left circumflex coronary artery occlusions needed for an adequate collateral development with and without heparin treatment. The number of coronary occlusions was significantly smaller in dogs with heparin treatment

remained unchanged in the 6 patients who did not receive heparin (Fig. 3). Furthermore, coronary arteriography revealed a significant increase in the extent of collateral circulation to the region perfused by the completely-obstructed coronary artery in patients treated with heparin. These findings suggest that heparin facilitates the collateral development stimulated by exercise-induced myocardial ischemia in humans. It has also been found that heparin treatment without exercise does not improve exercise capacity in patients with effort angina. Thus, it may logically be concluded that heparin does not initiate but rather accelerates the collateral development induced by myocardial ischemia.

Effect of Heparin Treatment on Collateral Development after Acute Myocardial Infarction

More recently, we examined the effects of intravenous heparin on potentiation of collateral developments in patients with acute myocardial infarction [8].

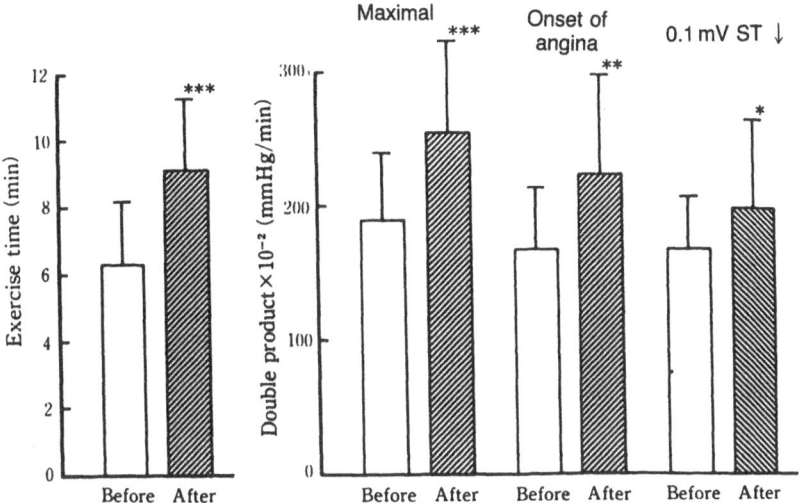

Fig. 3. Bar chart of changes in exercise time and double products (*maximal, onset of angina,* and *at 0.1 mV ST* segment depression) before and after 20 treadmill exercise periods (standard Bruce protocol) with intravenous heparin pretreatment (5000 IU) in 10 patients with chronic effort angina. *$p < 0.05$, **$p < 0.01$, ***$p < 0.001$ compared with values before heparin exercise treatment. In an additional 6 patients who had 20 exercise periods without heparin pretreatment, all of the above-mentioned variables of treadmill capacity remained unchanged

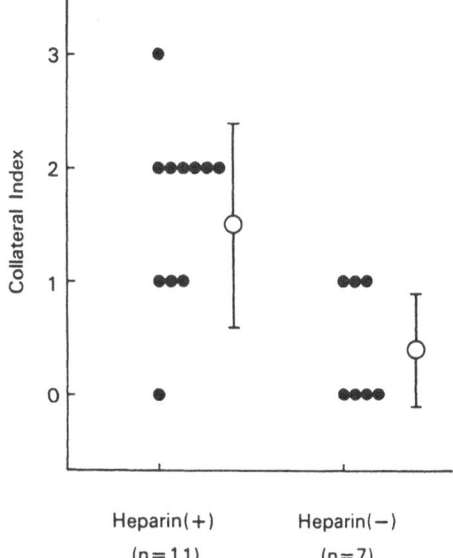

Fig. 4. Effect of heparin treatment during the acute stage of infarction on the collateral index (0–3) during the chronic stage of infarction. The collateral index was significantly higher in patients with heparin treatment ($p < 0.05$) (From [8], with permission)

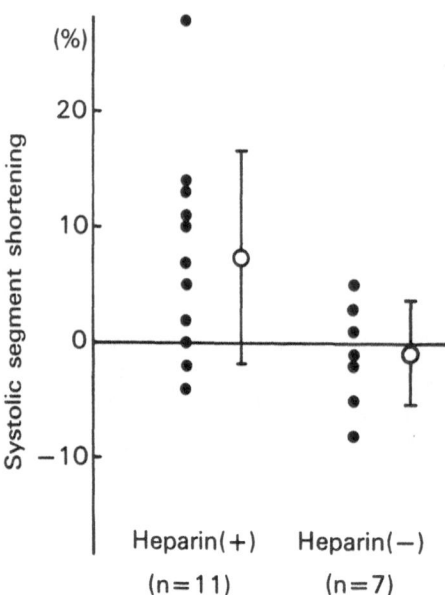

Fig. 5. Effect of heparin treatment during the acute stage of infarction on regional wall motion in the infarct areas during the chronic stage of infarction. The regional wall motion was significantly better in patients with heparin treatment ($p < 0.05$) (From [8], with permission)

Angiographically, the extent of collateral circulation was shown to be more pronounced in the patients with intravenous heparin during the acute stage of infarction (Fig. 4). Left ventricular function was also well preserved in these patients (Fig. 5).

Although these clinical investigations have been conducted in small numbers of patients, in nonrandomized and unblinded studies, there is a possibility that ischemic heart disease may be treated with drugs to increase the development of collateral blood vessels.

A Proposed Mechanism of Angiogenesis in the Heart

Over the past decade numerous angiogenic factors have been purified and their amino acid sequences were determined with subsequent gene clonings [9]. An uncharacterized small molecular weight factor (less than 1000 Da) that stimulates angiogenesis in vivo has been isolated from the heart in patients with recent myocardial infarction [10]. Since the discovery that a tumor-derived endothelial mitogen binds to heparin with high affinity [11], heparin binding growth factors have been identified in virtually all tissues examined. Recent studies have demonstrated clearly that both the acidic and basic fibroblast growth factors do exist in the heart [12].

Although heparin alone does not initiate angiogenesis, it potentiates the mitogenic activity of the acidic fibroblast growth factor [13]. The mechanisms responsible for angiogenic stimulation of heparin are avidly being sought. Until

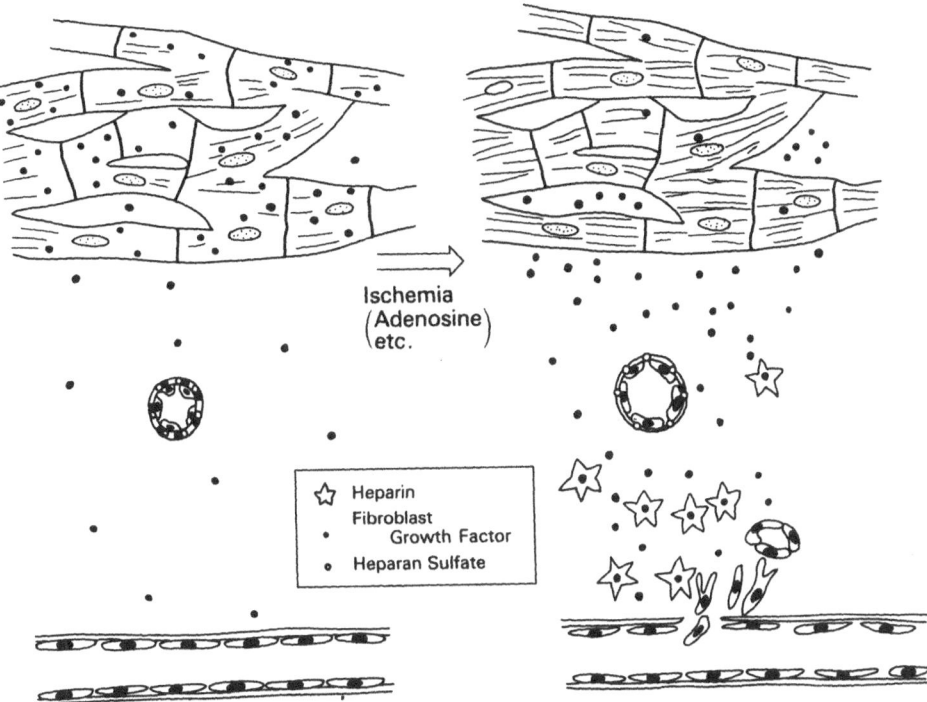

Fig. 6. Model for the process of neovascularization. Fibroblast growth factors are sequestered in myocytes, the extracellular matrix, or the endothelial cell basement membrane under normal physiologic conditions. Myocardial ischemia causes the release of these stored growth factors. When these factors are combined with heparin, the mitogenic activity is potentiated through two possible mechanisms: (1) heparin increases the binding of these growth factors to endothelial receptors, (2) heparin protects these factors from inactivation

now, two possible mechanisms have been proposed: (1) heparin increases the binding of the endothelial-cell growth factor to endothelial receptors [14], (2) heparin protects the fibroblast growth factor from inactivation [15]. We present a working hypothesis for the mechanism of angiogenesis in the heart (Fig. 6). Under normal physiologic conditions, heparin-binding factors are sequestered in myocytes, the extracellular matrix, and the endothelial cell basement membrane. Ischemia which is associated with cellular hypoxia, acidosis, and increases in lactate and adenosine, causes the release of these stored growth factors. When these factors are activated, by exposure to exogenous heparin, they occupy receptors of vascular cells and initiate angiogenesis.

Implications

Recent basic and clinical studies have elucidated the importance of heparin to myocardial ischemia for the development of coronary collateral circulation. A thorough understanding of the mechanisms of angiogenesis may lead to the development of new drugs capable of enhancing angiogenesis of collateral vessels more effectively in patients with coronary artery disease. The development of such a therapeutic remedy would reduce the deleterious sequelae due to coronary atherosclerosis.

References

1. Fujita M, Sasayama S, Ohno A, Yamanishi K, Hirai T (1987) Functional significance of coronary perfusion in preserving myocardial integrity. Clin Cardiol 10:394–398
2. Yoshida N, Fujita M, Yamanishi K, Miwa K (1993) Relation between collateral channel filling and flow grade in recipient coronary arteries in patients with stable effort angina. J Am Coll Cardiol 22:426–430
3. Franklin D, McKown D, McKown M, Hartley J, Caldwell M (1981) Development and regression of coronary collaterals induced by repeated, reversible ischemia in dogs (abstract). Fed Proc 40:339
4. Fujita M, McKown DP, McKown MD, Hartley JW, Franklin D (1987) Evaluation of coronary collateral development by regional myocardial function and reactive hyperaemia. Cardiovasc Res 21:377–384
5. Taylor S, Folkman J (1982) Protamine is an inhibitor of angiogenesis. Nature 297:307–312
6. Fujita M, Mikuniya A, Takahashi M, Gaddis R, Hartley J, McKown D, Franklin D (1987) Acceleration of coronary collateral development by heparin in conscious dogs. Jpn Circ J 51:395–402
7. Fujita M, Sasayama S, Asanoi H, Nakajima H, Sakai O, Ohno A (1988) Improvement of treadmill capacity and collateral circulation as a result of exercise with heparin pretreatment in patients with effort angina. Circulation 77:1022–1029
8. Ejiri M, Fujita M, Miwa K, Hirai T, Yamanishi K, Sakai O, Ishizaka S, Sasayama S (1990) Effects of heparin treatment on collateral development and regional myocardial function in acute myocardial infarction. Am Heart J 119:248–253
9. Folkman J, Klagsburn M (1987) Angiogenic factors. Science 235:442–447
10. Kumar S, West D, Shahabuddin S, Arnold F, Haboubi N, Reid H, Carr T (1983) Angiogenesis factor from human myocardial infarcts. Lancet 2:364–367
11. Shing Y, Folkman J, Sullivan R, Butterfield C, Murray J, Klagsburn M (1984) Heparin affinity: Purification of a tumor-derived capillary endothelial cell growth factor. Science 223:1296–1299
12. Speir E, Zhou YF, Lee M, Shrivastav S, Casscells W (1988) Fibroblast growth factors are present in adult cardiac myocytes, in vivo. Biochem Biophys Res Commun 157:1336–1340
13. Ehrlich HP, Jung WK, Costa DE, Rajaratnam JB (1988) Effects of heparin on vascularization of artificial skin grafts in rats. Exp Mol Pathol 48:244–251

14. Schreiber AB, Kenney J, Kowalski WJ, Friesel R, Mehlman T, Maciag T (1985) Interaction of endothelial cell growth factor with heparin: Characterization by receptor and antibody recognition. Proc Natl Acad Sci USA 82:6138–6142
15. Gospodarowicz D, Cheng J (1986) Heparin protects basic and acidic FGF from inactivation. J Cell Physiol 128:475–484

Abstracts

Effects of Vasodilators on Coronary Vasospasm Induced by Ergonovine Malate

Shuichi Matsumoto, Nobuo Komatsu, Takayuki Owada, Mikihiro Kijima, Kiyohiro Ikeda, Kurao Ito, and Yukio Maruyama[1]

Coronary vasospasm sometimes causes sudden cardiac death; evaluation of the effects of vasodilators would thus be of importance. Twenty patients (25 coronary arteries) with vasospastic angina were examined. Coronary vasodilators were withdrawn at least 48 h before the first coronary angiography (CAG) was performed. None of the patients had fixed stenosis of more than 50% of the coronary diameter. Total or subtotal occlusion of the coronary artery was induced in all patients by the intracoronary infusion of ergonovine malate (EM). On the 2nd day, 40 mg of long-acting isosorbide dinitrate (ISDN) was administrated at 1:00 a.m. and CAG was performed at 8:00 a.m. On the 3rd day, 20 mg of nifedipine, and on the 4th day, 60 mg of diltiazem, were added and the CAG with EM test was performed at the same time (protocol A). In protocol B, we changed the order of drug administration, i.e., on the 2nd day nifedipine, on the 3rd day diltiazem, on the 4th day nicorandil, and on the 5th day ISDN, were added. There was a significant increase in coronary diameter after the administration of vasodilators at the spasm sites in both protocols A and B. After a single administration of ISDN (protocol A) or nifedipine (protocol B), in 12 of 13 (protocol A) and in 10 of 12 coronary arteries (protocol B), spasms were again induced by EM at almost the same doses as those used before drug administration. However, after the administration of ISDN, nifedipine, and diltiazem, 9 of 11 coronary arteries (82%) in protocol A were released from EM-induced spasm, and after the combined administration of nifedipine, diltiazem, and nicorandil, 10 of 12 (84%) arteries in protocol B were released from EM spasm. These findings may be due to a synergestic effect of diltiazem and nifedipine, since the plasma concentration of nifedipine increased from 40.0 to 180.4 ng/ml after diltiazem was administered.

---

[1] Department of Cardiology, Hoshi General Hospital, First Department of Internal Medicine Fukushima Medical College, Fukushima, 151 Japan

Comparison of Intracoronary Injection of Acetylcholine and Ergonovine in All Vessel Segments in Patients with and without Vasospastic Angina Pectoris

Yoshihiro Miyazaki, Akira Suzuki, Yasuhiro Sakauchi, Akira Hirosaka, Tomiyoshi Saito, and Yukio Maruyama[1]

To compare the effects of provocative agents on coronary vascular tone, we performed intracoronary injections of acetylcholine (ACh) and ergonovine (EM) in 12 patients with vasospastic angina pectoris (VSA) and in 11 patients without coronary artery disease (control group). Incremental doses of ACh $(20-100\,\mu g)$ were injected, followed by EM $(20-50\,\mu g)$ after the disappearance of ACh action. The reduction ratio of coronary artery diameter (RR-CAD), (1-after ACh or EM CAD/after isosorbide dinitrate (ISDN CAD) $\times$ 100(%), was calculated in every AHA segment. In patients with VSA, coronary spasms (RR-CAD $>$ 99%) were induced by ACh in 10/12 patients, by EM in 10/12, and by both agents in 8/12 in the same segment, whereas in the control group, spasms were induced only by ACh, in 1 of 11 patients, and chest pain was not experienced. Responses to ACh and EM in each AHA segment in the VSA group, except in spasm-induced sites, tended to be greater than in the control group, i.e., RR-CAD values (mean $\pm$ SEM) including all American Heart Association (AHA) segments following ACh and EM administration were 47.0 $\pm$ 7.1% and 50.8 $\pm$ 6.8% respectively, in the VSA group and 35.8 $\pm$ 7.8% and 40.7 $\pm$ 5.0% respectively, in the control group. We conclude that the sensitivity to ACh and EM challenge was the same, but that both agents were needed to achieve more accurate diagnosis.

[1] First Department of Internal Medicine, Fukushima Medical College, 151 Japan

Inhibitory Effect of Free Radical Scavengers on Cyclic Flow Variations in Unsedated Dogs with Coronary Stenosis and Endothelial Injury

Hisao Ikeda, Takafumi Ueno, Hiroshi Nakayama, Kazunori Kuwano, Kohji Hiyamuta, Yoshinori Koga, Hironori Toshima, and Mitchel M. Yokoyama[1]

Severe coronary artery stenosis with endothelial injury in the canine model induces cyclic coronary flow variations (CFVs), which are partially due to spontaneous platelet aggregation and dislodgement at the stenotic site. In the present study, we used anesthetized open-chest and unsedated closed-chest dogs with CFVs to investigate whether oxidative metabolic burst (hydrogen peroxide generation) occurred in neutrophils during CFVs and whether CFVs were attenuated by superoxide dismutase (SOD) and catalase. CFVs were produced by placing a cylindrical constrictor on the left anterior descending coronary artery (LAD). LAD blood flow was monitored by means of a Doppler flow probe placed proximally to the constrictor, and the severity of CFVs was expressed by both the frequency of CFVs and mean LAD blood flow. Hydrogen peroxide generation in neutrophils was measured by flow cytometry, using single cell analysis, and was expressed as the mean fluorescence intensity of 2', 7'-dichlorofluorescein. Dogs received an intravenous infusion of saline ($n = 8$), SOD (5 mg/kg, $n = 7$), catalase (5 mg/kg, $n = 7$), or a combination of SOD and catalase (same doses, $n = 7$). Although the mean fluorescence intensity did not change in sham-operated dogs without CFVs (61.7 ± 15.8 to 60.4 ± 11.8; NS), the intensity in the dogs with CFVs was significantly increased during CFVs (62.2 ± 13.7 to 79.8 ± 9.8; $P < 0.005$). These results indicate that hydrogen peroxide is generated in neutrophils. The frequency of CFVs and mean LAD blood flow was not altered following the infusion of saline. After each treatment with SOD, catalase, or the combination of SOD and catalase, the frequency of CFVs was significantly reduced, while the mean LAD blood flow was significantly increased. Treatment with the combination of SOD and catalase or with catalase alone was more effective than treatment with SOD in attenuating the frequency of CFVs. These findings suggest that oxidative metabolic burst (hydrogen peroxide generation) contributes to the progression of CFVs. We conclude that activated neutrophils may play an important role in mediating the interaction between platelets and vascular endothelium that occurs during CFVs in this experimental canine model.

---

[1] Kurume University School of Medicine, Kurume, 830 Japan

Microcirculation and Alternating Transmural Flow Patterns in the Ischemic Left Ventricular Vascular Bed

Rafael Beyar, Reuven Kamminker, Dan Manor, Samuel Sideman[1]

A three-layer electrical analog model of the myocardium and four compartments of the coronary bed (large arteries $> 200\,\mu$, small arteries $< 200\,\mu$, small venules $< 150\,\mu$, large veins $> 150\,\mu$) was used to explore microcirculation and alternating transmural flow patterns in the ischemic left ventricular vascular bed. The myocardial compartments are represented by capacitances and resistances, accounting for the relationship between the vessels' cross-sectional area and the transmural pressure. Collaterals and autoregulation are allowed for. The intramyocardial pressure is assumed, for simplicity, to decrease linearly from the cavity pressure at the endocardium to zero pressure at the epicardium. Calculations show that the phasic diastolic flow evident in the coronary arteries converts to systolic venular flow. Calculated inlet flows to the three layers in an ischemic bed are shown in the Figure indicating that retrograde flow at the subendocardium coexists with positive subepicardial flow at the beginning of systole, while the flow in diastole is redirected from the subepicardium to the subendocardium. Tracer washouts indicate an unexplained "hidden" component of perfusion in the ischemic bed. The alternating flow during coronary ischemia may explain the suspected "hidden" perfusion that enhances metabolic washout in the ischemic bed beyond the average perfusion level.

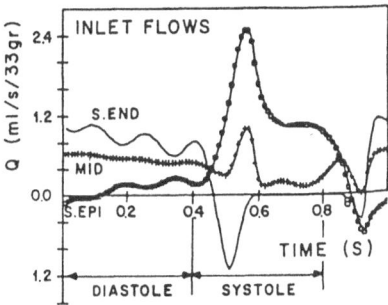

Fig. 1. Calculated inlet flows in the three layers of the ischemic myocardium. *S. END*, Subendocardium; *S. EPI*, Subepicardium

[1] Julius Silver Institute, Department of Biomedical Engineering, Cardiac System Research Center, Technion-IIT, Haifa, Israel

Can Aminophylline Improve Exercise Capacity in Patients with Coronary Artery Disease?

Tetsuo Minamino, Masatake Fukunami, Masaharu Ohmori, Kazuaki Kumagai, Takahisa Yamada, Nobuhiko Kondoh, Eiitiro Tujimura, and Noritake Hoki[1]

Aminophylline (Am) a potent adenosine receptor blocker, may increase transmural blood flow in the ischemic myocardium by preventing the maldistribution of coronary flow, although this agent is known to elevate myocardial oxygen consumption. To determine whether aminophylline does have this action, symptom-limited ramp-fashioned treadmill stress testing, combined with gas exchange analysis, was performed in 20 patients with coronary heart disease. In the first stress testing, Am (7 mg/kg) was administered over a 20-min period just before exercise. The second stress testing, the control study, was carried out without Am. There were no significant differences in basal heart rate, systolic blood pressure, rate-pressure product or oxygen consumption (VO2) before exercise, in the two studies with without Am. After the administration of Am, heart rate increased (72 ± 13 to 80 ± 16 bpm; mean $\pm$ SD, $P < 0.05$), as did rate-pressure product (96 ± 28 to 108 ± 33 mmHg $\times$ bpm $\times$ 100, $P < 0.05$), although the systolic blood pressure and VO2 did not show significant changes. During exercise, Am effected a significant increase in exercise duration (8.9 ± 2.6 versus 10.4 ± 2.8 min, $P < 0.01$). The rate-pressure product recorded at peak exercise was also significantly higher after Am (212 ± 56 versus 251 ± 74 mmHg $\times$ bpm $\times$ 100, $P < 0.05$). However, there was no significant difference in ST segment change in ECG at the end-point of exercise between the two study groups, with and without Am. Moreover, exercise maximal VO2 (20.5 ± 5.1 versus 21.6 ± 6.5 ml/min) and anaerobic threshold (14.8 ± 2.7 versus 14.8 ± 2.8 ml/min) were not significantly different in the two study groups. These findings indicate that administration of Am to patients with coronary artery disease does not induce deterioration, but rather improves exercise capacity.

---

[1] Division of Cardiology, Osaka Prefectural Hospital, Osaka, Japan

Decreased Aortic Compliance Aggravates Myocardial Ischemia with Impaired Coronary Reserve Flow

Hideki Watanabe, Sadanori Ohtsuka, Masaaki Kakihana, and Yasuro Sugishita[1]

Although decreased aortic compliance has been reported in patients with coronary artery disease, the effects of this decrease on coronary circulation and myocardial function have not been investigated sufficiently. Accordingly, to elucidate the effects of decreased aortic compliance on myocardium perfused by a stenosed coronary artery, we measured regional segment length, subendocardial ECG, and myocardial tissue oxygen tension (PO_2) in ten anesthetized open-chest dogs with coronary stenosis. A fixed coronary stenosis was made at the proximal portion of the left circumflex coronary artery (LCx). The grade of coronary stenosis was adjusted not to reduce the baseline LCx flow more than 10% but to eliminate LCx reactive hyperemia following a 10-s occlusion. To reduce aortic compliance, the thoracic aorta was sufficiently constricted with bandages without changing mean aortic pressure. Measurements were made at the baseline and in rapid pacing states to compare the two experimental conditions, with and without the bandages.

The aortic constriction produced with the bandages reduced total arterial compliance and increased pulsatile pressure without changing the mean aortic pressure or systemic vascular resistance. In the LCx area, where coronary reserve was impaired by the coronary stenosis, when myocardial oxygen demand was increased by rapid pacing, decreased aortic compliance induced significant ST elevation in the subendocardial ECG (4.2 ± 1.8 without *vs* $5.3 \pm 2.5\,mV$ with bandages; $P < 0.05$) and significant decrease of regional segment shortening (11.3 ± 5.8 without *vs* $8.3 \pm 3.4\%$ with bandages; $P < 0.01$). Moreover, subendocardial PO_2 (Endo) was decreased and subepicardial PO_2 (Epi) was increased (Endo, 43.2 ± 9.8 without *vs* $36.8 \pm 10.0\,mmHg$ with bandages; $P < 0.05$; Epi, 36.8 ± 10.0 without *vs* $44.4 \pm 7.9\,mmHg$ with bandages; $P < 0.05$), whereas, in the area perfused by the LAD artery, there was no significant change in any variables after the aortic constriction. These results indicate that decreased aortic compliance causes deterioration in myocardial perfusion in the presence of coronary stenosis and aggravates subendocardial ischemia under conditions in which myocardial oxygen demand is increased.

[1] Cardiovascular Division, Department of Internal Medicine, Institute of Clinical Medicine, University of Tsukuba, Ibaraki, Japan

Inability of Left Ventricular Decompression to Limit Myocardial Infarct Size

Donna M. Van Winkle, James M. Downey[1]

Although the restoration of blood flow is the treatment of choice for myocardial ischemia, uncontrolled reperfusion itself may be injurious. We wished to determine whether unloading the beating left ventricle at the onset of reperfusion could limit myocardial infarct size. We studied isolated rabbit hearts, perfused with blood from a support rabbit and electrically paced at 200 beats/min. Rabbits were chosen because of the lack of preformed coronary collateral vessels. Four groups of rabbits were studied; all hearts underwent 60-min coronary artery occlusion and 2 h of reperfusion. The experimental conditions were as follows: In group 1, hearts contracted isovolumetrically on a fluid-filled balloon in the left ventricle during both occlusion and reperfusion. In group 2, the balloon was present only during occlusion, and the heart was vented during reperfusion. Group 3 hearts were vented during both occlusion and developed pressure during reperfusion. Hearts in group 4 were vented during both occlusion and reperfusion. Total coronary flow was measured by timed collection of effluent blood. Infarct size was assessed by the triphenyl tetrazolium method, and expressed as a percent of the area at risk, as determined with fluorescent particles. Perfusion pressure and coronary flow were not different between groups. Left ventricular systolic and diastolic pressures during loaded conditions were similar between groups at each time period (mean of systolic and diastolic pressures for all groups during loaded periods, 87 ± 1.5 mmHg and 6.5 ± 0.6 mmHg, respectively). Venting during both ischemia and reperfusion ($n = 10$) did result in smaller infarcts than those in unvented controls ($n = 10$) ($13 \pm 5\%$ versus $41 \pm 6\%$, respectively; $P < 0.02$). However, venting during reperfusion only ($n = 10$) or during occlusion only ($n = 11$) did not limit infarct size ($57 \pm 6\%$ and $32 \pm 5\%$, respectively; P, ns) as compared to controls. Thus, in an experimental preparation with rigid control of the degree of ischemia, myocardial oxygen demand (e.g., in terms of heart rate) and certain complete ventricular decompression, i.e., left ventricular venting during reperfusion, was not associated with a decrease in myocardial infarct size.

---

[1] Department of Physiology, University of South Alabama, Mobile, Alabama, USA.

Altered Metabolic Response Accounts for Reduced ST-Elevation to Subsequent Coronary Occlusion in Ischemia-Sensitized Myocardium

Yutaka Kimura, Takashi Saito, Akira Nakagomi, Tohru Abe, Kazuhito Takahashi, Etsuko Fushimi, Yasutsugu Kudo, and Mamoru Miura[1]

Schroeder et al. reported that a preceding 5-min coronary occlusion caused earlier deterioration of myocardial function (MF) in subsequent ischemia. This study was designed to reveal the mechanism responsible for this latent abnormality and associated ECG response. In 26 dogs, we measured MF by sonomicrometry, tissue PCO_2, pH, extracellular K^+(KC), and epicardial surface electrocardiograms. The experimental protocol was as follows: the left anterior descending coronary artery (LAD) was occluded for 2 min, followed by 15-min reperfusion (Trial 1: Tr-1). In the control group (Gr-C), reperfusion was continued for another 95 min, while in the other groups LAD was occluded for 5 or 15 min, followed by 80- or 90-min reperfusion after Tr-1. The LAD was then occluded again, for 2 min (Tr-2), in all groups. In the Gr-C, there were no significant differences between the trials in any of the above variables, while in the 5-min group, MF (% systolic shortening) recovered to the control level prior to Tr-2, but showed earlier deterioration in Tr-2 than in Tr-1. Further, reduced ST-elevation was noted in Tr-2, associated with the decreases in KC, PCO_2, and pH. Similar reductions in metabolic and electrical parameters were also observed in the 15-min occlusion group. CO_2 and protons are the end-products of cardiac metabolism, so their reduced production rate reflects depressed metabolic viability in reperfused myocardium; this phenomenon indicates a sensitization-like effect produced by the preceding ischemia ("ischemia-sensitized myo-cardium"). This phenomenon may be related to limited substrate use in energy metabolism and may also play a crucial role in the preconditioning effect.

	ST-elevation (mV)		Delta KC (mmol/l)		Delta CO_2 (mmHg)	
	Tr-1	Tr-2	Tr-1	Tr-2	Tr-1	Tr-2
Control group	6.5 ± 1.0	6.0 ± 0.7	1.0 ± 0.1	1.1 ± 0.2	60.9 ± 3.6	59.3 ± 3.7
5-Min group	6.5 ± 0.6	3.6 ± 0.8**	1.1 ± 0.2	0.4 ± 0.1**	59.3 ± 4.0	48.0 ± 3.5**
15-Min group	3.7 ± 0.6	1.2 ± 0.5**	1.1 ± 0.3	0.4 ± 0.1**	71.0 ± 5.5	50.3 ± 3.3**

* $P < 0.05$; ** $P < 0.01$

---

[1] The Second Department of Internal Medicine, Akita University, Akita, Japan

F. Small Vessel Disorder in Coronary Circulation

Segmental Distribution and Control of Coronary Microvascular Resistance

David V. DeFily, Lih Kuo, Michael J. Davis, and William M. Chilian[1]

Summary. Until recently, the majority of studies examining the regulation of coronary vascular resistance have considered the coronary microcirculation as a single homogeneous vascular bed. It is becoming apparent that regulation of coronary microvascular resistance is not distributed uniformly, but varies across different segments of the vasculature. Studies utilizing a variety of techniques have begun to discover that the coronary microcirculation can be divided into several functional segments with respect to regulatory mechanisms. Generally, small arterioles, those less than 100 μm in diameter, respond differently from large arterioles or small arteries. There is a significant segmental distribution between arterioles less than 100 μm, arterioles greater than 100 μm, and small arteries. A major portion of coronary microvascular resistance is located in arterioles that are less than 100 μm. Also, there are significant differences in autoregulation, myogenic activity, α-adrenergic responses, and endothelial cell-mediated mechanisms among the various segments. From these studies we can conclude that the majority of the regulation of coronary microvascular resistance occurs in small arterioles. Furthermore, with a myriad of regulatory mechanisms at different sites in the vasculature, there must be specific mechanisms that integrate this information for coordinated responses to facilitate myocardial oxygen delivery. This segmental distribution of regulation suggests an integrative hypothesis of regulation, whereby a variety of mechanisms play a role in the overall response.

Key words: Coronary microcirculation—Autoregulation

Introduction

The regulation of blood flow to the myocardium is controlled by a variety of mechanisms, including autoregulation, comprising both metabolic and myogenic components, α-adrenoreceptor-mediated mechanisms affecting coronary vasoconstriction, and endothelium-induced vasodilation. Many studies which have directly examined the regulation of myocardial blood flow have inferred alterations in microvascular tone from changes in pressure-flow relationships. Such an approach, however, lumps coronary resistance vessels into a single compartment. Importantly, many of the mechanisms by which the coronary

---

[1] Department of Medical Physiology and Microcirculation Research Institute, Texas A and M University Health Science Center, College Station, TX 77843-1114, USA

vasculature regulate myocardial blood flow are not uniform throughout the vascular tree. The objective of this chapter is to discuss the differences in control of the coronary circulation at various segments of the microvasculature.

Sites of Microvascular Resistance

Until recently, there has been very little, if any, direct information about the specific location of vascular resistance in the coronary circulation. By using methods originally developed by Nellis and colleagues [1] to compensate for cardiac movement, Chilian et al. have been able to utilize servo-null techniques to locate the specific site of coronary microvascular resistance under normal conditions, and during intense vasodilation in the left ventricle [2]. By measuring pressures at different segments of the coronary microcirculation in the beating left ventricle, they determined the relative contribution of each segment to the overall coronary vascular resistance (Fig. 1). From these studies, it was determined that approximately 25% of coronary vascular resistance was found in vessels proximal to arterioles of 175–200 μm in diameter, while nearly 50% could be attributed to arterioles proximal to those 75–130 μm in diameter [3]. Therefore, under normal conditions with intact vascular tone, a major portion of coronary vascular resistance is controlled by the coronary arterioles. This pattern of resistance distribution is generally similar to that reported for the rabbit right

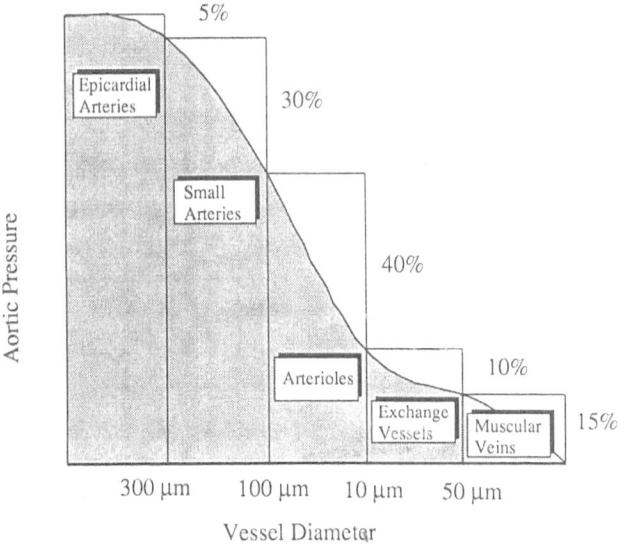

Fig. 1. Relative pressure drop across the coronary microcirculation. Although all segments of the coronary circulation contribute to overall vascular resistance, the majority of the resistance (*pressure drop*) lies in small arterioles

ventricle [1] and the rat left ventricle [4]. However, during vasodilation with papaverine [2] or dipyridamole [3], there appears to be a substantial redistribution of coronary vascular resistance to larger arterioles and arteries, as well as to larger veins. This is due to preferential arteriolar vasodilation and less dilation of the upstream and downstream segments. Whereas, under control

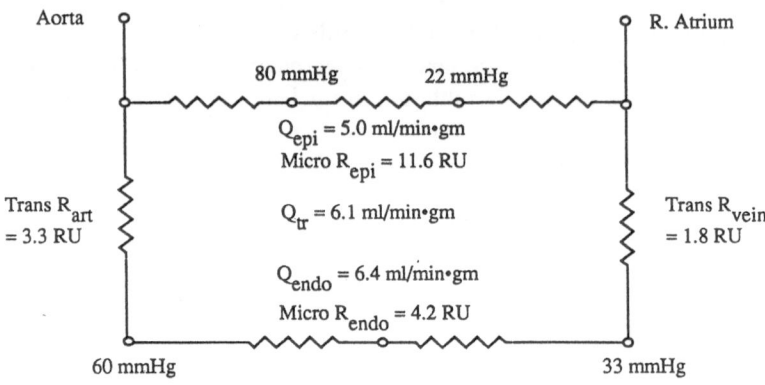

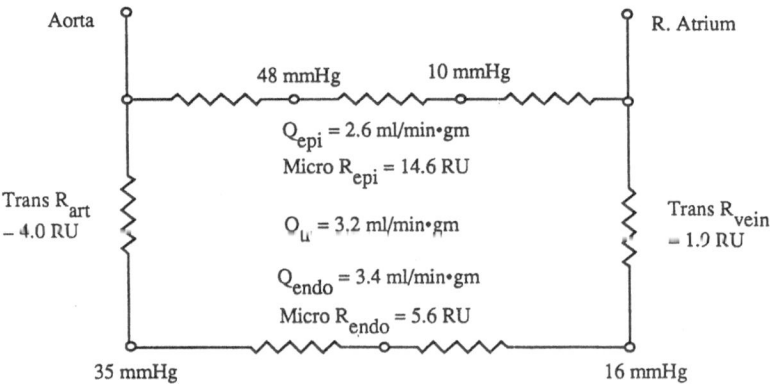

Fig. 2. Measurements of coronary blood flow and coronary microvascular pressures in the epicardium and endocardium. Q_{epi}, Q_{endo}, and Q_{tr} refer to blood flow in the subepicardium, subendocardium, and transmural blood flow, respectively. Resistance units (RU) = mmHg/(ml/min · gm). Note that, at both normal (100 mmHg) and reduced (60 mmHg) pressure, subendocardial microvascular resistance is significantly less than that in the subepicardium. *Trans R_{art}*, transmural resistance across the arterial circulation, *Trans R_{vein}*, transmural resistance across the venous circulation. (from [5] with permission)

conditions, 25% of the relative distribution of coronary vascular resistance is proximal to 170 μm arterioles, following dilation with dipyridamole this value increased to 42% [3]. So, while the larger vessels in the microvasculature do not play as significant a role as the small arterioles in the regulation of microvascular blood flow during normal conditions, their proportional contribution to total coronary vascular resistance can be greatly magnified under conditions of arteriolar vasodilation.

In addition to the segmental distribution of vascular resistance between arteries and arterioles, there is also a transmural difference in the distribution of microvascular resistance between the subepicardium and the subendocardium. In arterial and venous segments, vascular resistance is higher in the endocardium than the epicardium, whereas in the microcirculation this gradient is reversed [5]. A scheme for this is shown in Fig. 2 where microvascular resistances are calculated at the subepicardium and the subendocardium from measurements of microvascular pressure and regional myocardial blood flow [5]. It is likely that the higher arterial and venous resistance in the subendocardium is due to the contribution of transmural vessels which course from the subepicardium to the subendocardium. In the subendocardium, the relative contribution of resistance by the arteries is 30%–45%, compared to only 25% in the subepicardium. On the other hand, microvessels possess about 50% and 70% of total coronary resistance in the subendocardium and subepicardium, respectively. A consequence of the lower resistance in endocardial microvessels may be an improvement of perfusion to the region despite the higher resistance of upstream arteries, i.e., the lower subendocardial microvascular resistance may be an adaptation to improve perfusion despite lower perfusion pressure. Indeed, as shown in Fig. 2, when coronary perfusion pressure is dropped to 60 mmHg, perfusion pressure at the subendocardium is only 35 mmHg. Also, a lower resistance may partially compensate for the higher compressive forces present in the subendocardium. These transmural differences in resistance may also, in part, account for the well-documented subendocardial versus subepicardial variations in autoregulation.

Autoregulation and Myogenic Activity

Autoregulation of coronary blood flow is mediated primarily by metabolic and myogenic mechanisms [6]. It is likely that both mechanisms play a role in the regulation of blood flow; however, not necessarily to the same extent, or in the same segment of the microcirculation. The relative importance of each mechanism varies in small arteries and arterioles.

Coronary arterioles smaller than 100–150 μm in diameter dilate significantly in response to a reduced coronary perfusion pressure [7, 8]. The magnitude of this dilation is inversely proportional to the initial diameter of the vessel, i.e., smaller arterioles tend to dilate to a greater extent than larger ones. However, even at perfusion pressures as low as 30–40 mmHg, these vessels are not maximally dilated, because intracoronary infusion of adenosine is still able to produce a significant dilation of small coronary arterioles [7]. Dilation of arterioles during mild and severe coronary stenosis is also prevented by pretreatment with the ATP-sensitive potassium channel antagonist, glibenclamide [9]. ATP-sensitive potassium channels, which also mediate hypoxia-induced dilation [10], likely mediate the adenosine-induced dilation of coronary vascular smooth muscle. Furthermore, coronary arterioles less than 100–150 μm respond to coronary occlusion with significant dilation [11, 12]. Unlike the response to a critical stenosis, however, there is less pharmacologic recruitment of vasodilatory reserve [11], indicating that the vessels are nearly maximally dilated during complete occlusion. These results demonstrate that, even in situations of severe hypotension, which are below the limits of autoregulation, coronary arterioles still retain a level of vascular tone. In contrast, during severe hypotension, larger vessels (greater than 150 μm) do not dilate further to adenosine [7, 11], suggesting that vasodilator reserve during hypotension resides exclusively in arterioles. The ability to autoregulate and increase vascular diameter is more important in small arterioles than in larger vessels.

The myogenic theory of autoregulation suggests that an intrinsic property of the blood vessel regulates vascular tone in response to changes in intraluminal pressure. Specifically, vasodilation occurs in response to a decrease in intraluminal pressure, and vasoconstriction in response to an increase in pressure, potentially maintaining relatively constant levels of blood flow. Myogenic mechanisms likely play a role in autoregulation of the coronary microcirculation, but, importantly, the control of vascular tone by myogenic mechanisms is not uniform throughout the microcirculation. The response of small coronary arteries (150–300 μm) to changes in transmural pressure is passive [13], suggesting a lack of myogenic activity in these vessels. Kuo et al., however, have shown that isolated porcine coronary arterioles (40–100 μm) do exhibit significant myogenic activity [14, 15], demonstrating that myogenic mechanisms are present in the coronary microvascular segment responsible for autoregulation, and suggesting that this mechanism can contribute to the coronary autoregulation. Denudation of the endothelium from arterioles does not alter the response to changes in transmural pressure, indicating that the myogenic response is an intrinsic property of the vascular smooth muscle and is not dependent on an endothelial mechanism.

There is also a transmural difference in the myogenic activity of coronary arterioles. The myogenic activity of isolated subepicardial coronary arterioles is better than that of subendocardial arterioles [15], at both low and high pressures. This suggests that the reduced autoregulatory capacity of the subendocardium, previously attributed to differences in ventricular wall tension, myocardial metabolism, or other factors, may, at least partially, be the result of intrinsic differences in the vasculature between the inner and outer layers of the ventricular wall.

Functional and Reactive Hyperemia

Although arterioles demonstrate significant myogenic activity, local metabolic control of coronary blood flow is usually thought to be the most important regulatory mechanism [6]. As the metabolic activity of the tissue increases, so does the release of metabolites causing vasodilation, producing a reduction of vascular resistance and an increase in blood flow. Although an increase in metabolic activity or myocardial oxygen consumption causes a profound vasodilation, the increase in microvascular diameters is not homogeneous. An increase in myocardial oxygen consumption, such as that obtained during rapid pacing, produces a significant increase in microvascular diameter which is inversely proportional to the initial diameter [16], i.e., smaller arterioles dilate to a greater extent than larger vessels. In contrast, adenosine or dipyridamole infusions, which increase coronary perfusion to similar levels, do not produce an increase in diameter of coronary microvessels greater than 150 μm. These results indicate that metabolic regulation of the vasculature is significantly greater in the smaller arterioles.

Vasodilatory reserve can also be estimated as the amount of dilation present during reactive hyperemia. There is a significant dilation of all microvessels during reactive hyperemia; however, the magnitude and time course are not uniform throughout the microcirculation. In the early phase of reactive hyperemia, arterioles dilate to a greater extent than do larger arteries. This dilation is likely not mediated by adenosine, since the response is not prevented with the adenosine antagonist 8-phenyltheophylline. Dilation of larger arterioles and small arteries, however, is more prolonged than that of the arterioles, and can be inhibited with 8-phenyltheophylline [12].

α-Adrenergic Regulation

Although all coronary arteries contain some sympathetic innervation, there is both a structural and a functional heterogeneity in the distribution of nerves. There appears to be a higher density of α-adrenergic innervation in arteries than in arterioles [6]. Furthermore, a heterogeneous distribution of the effects of α-adrenergic activation is also present in the coronary microcirculation. Activation of α-receptors by exogenous norepinephrine infusion, or by sympathetic neurogenic release of norepinephrine resulting from stimulation of the stellate ganglia, produces a profound constriction of coronary arteries greater than 100 μm, and a simultaneous dilation of arterioles less than 100 μm [17]. It is speculated that the dilation of arterioles by α-adrenergic receptors is an autoregulatory response to the upstream constriction. The mechanism of this dilation may be myogenic and/or metabolic in nature. An increase in upstream resistance would result in a decrease in distal pressure and blood flow. The decrease in pressure could result in a myogenic increase in arteriolar diameter, while a decrease in blood flow may induce a decrease in arteriolar resistance via metabolic mechanisms. These responses would, in effect, be autoregulatory adjustments to the upstream constriction and imply a heterogeneous control of the coronary circulation, where neurohumoral control predominates in the arteries, while the regulation of arterioles is primarily metabolic and myogenic.

Although the primary sites for neurohumoral control appear to be in small arteries and large arterioles, there is evidence that α-adrenergic regulation of small arterioles can occur under some conditions. For instance, coronary arterioles appear to be able to escape from neurohumoral constriction [18], which gives the impression of a lack of any significant effect. Yet, by reducing coronary perfusion pressure, autoregulatory adjustments of arteriolar tone can be prevented, and the constriction induced by α_1- and α_2-adrenergic agonists can be unmasked. Similar results can also be obtained by eliminating the basal release and flow-mediated release of endothelium-dependent relaxing factor (EDRF) from endothelial cells with arginine analogues [19]. Furthermore, the constriction of arterioles to the α_2-adrenergic agonist was considerably greater than to the α_1-adrenergic agonist, and also greater than the constriction of larger vessels (greater than $100\,\mu m$), demonstrating the predominance of α_2-adrenoreceptors on coronary arterioles. These results are supported by evidence that the coronary microvascular distribution of specific α-adrenoreceptor subtypes is also not homogeneous. Whereas α_1-adrenoreceptors may be homogeneously distributed throughout the coronary microcirculation, the distribution of α_2-adrenoreceptors lies primarily in arterioles less than $100\,\mu m$ [18, 20, 21]. These findings suggest a localization of α_1- and α_2-adrenergic receptors similar to that in the skeletal muscle microcirculation [22]. However, it has not been determined to what extent specific α-adrenergic receptor subtypes such as α_{1a}- and α_{1b}- and α_{2a}- and α_{2b}-influence the regulation of coronary microvascular blood flow.

Endothelial Regulation

In addition to autoregulatory and α-adrenergic mechanisms, the endothelium also plays a major role in the modulation of coronary blood flow. There are many substances which have been shown to modulate endothelial regulation [23]. Acetylcholine, an agonist commonly used to elicit the release of EDRF or nitric oxide, induces dilation of both large coronary arteries [23] and arterioles [24]. The synthesis of nitric oxide by endothelial cells requires the presence of L-arginine, and therefore, the endothelium-dependent dilation of coronary vessels can be inhibited with L-arginine analogues [25, 26]. Vascular relaxation to acetylcholine, however, is dependent upon the presence of muscarinic receptors on the endothelium. Also, some species, such as the porcine, do not exhibit acetylcholine-dependent dilation in coronary vessels [27]. Several other agonists, however, appear to have different effects on endothelium-dependent dilation between coronary arteries and arterioles.

Serotonin produces significant constriction of large conduit coronary arteries; however, this response is modulated by an endothelium-mediated dilation. Denudation of the endothelium from large arteries significantly enhances the constriction of these vessels, indicating the presence of concurrent endothelium-mediated dilation [28]. Serotonin, however, produces a significant and selective dilation of coronary arterioles less than 90 μm in diameter [29]. Denudation of coronary arterioles both in vitro [30] and in vivo [31] eliminates the dilation to serotonin, and often reverses the response, producing a constriction. These studies demonstrate that the endothelium modulates tone in all coronary vascular segments, and may help to explain the finding that, while serotonin causes constriction of large coronary arteries, it also produces an increase in coronary flow. The increase in coronary blood flow is likely due to the selective endothelium-mediated dilation of coronary arterioles. The overall response of serotonin on an individual vessel may be the result of a balance between endothelium-mediated dilation and vascular smooth muscle-mediated constriction. The differential response between vascular segments may be due to different serotonin receptor types or sensitivities on either the endothelial cells or the smooth muscle cells. Similar differential responses have also been observed within the rat skeletal muscle microcirculation, where constriction of large arterioles is attributed to serotonin-type 2 receptors, and small arteriolar dilation to serotonin-type 1 receptors [32]. Serotonin is also a more sensitive indicator of endothelial function than acetylcholine. Studies have demonstrated a significant attenuation of serotonin-induced dilation, while the endothelium-dependent dilation to acetylcholine was either unaffected or slightly reduced [33, 34].

Vasopressin is another vasoactive compound which has differential effects on large and small coronary arteries. Vasopressin, which is a potent constrictor of coronary microvessels [24, 29], causes significant endothelium-dependent dilation of large coronary arteries [24]. Similar to the response of large vessels to serotonin, the constrictor response of vasopressin on arterioles is enhanced following inhibition of EDRF [24]. Ischemia or ischemia-reperfusion, which have been shown to selectively damage the endothelial function of microvessels [34, 35], also enhances the constrictor response of vasopressin [36].

The endothelium also modulates the response of coronary microvessels during reactive hyperemia following a short coronary occlusion. Several studies have demonstrated an attenuation of blood flow debt repayment during reactive hyperemia following treatment with L-arginine analogues which inhibit the synthesis of nitric oxide by endothelial cells [26, 37, 38]. The increase in blood flow was only partially attenuated, however, by inhibiting nitric oxide synthesis. Further attenuation of blood flow debt repayment was possible if the effects of adenosine were blocked [37]. Although these agents can reduce the repayment of flow debt and the duration of reactive hyperemia, neither has any significant effect on reducing the initial peak increase in blood flow. The mechanism of the endothelium-mediated decrease in coronary vascular resistance is not well established. It is likely that this increase is, in part, due to the increase in flow velocity and shear stress associated with increased blood flow. Kuo et al. have demonstrated that coronary arterioles exhibit flow-induced dilation. Furthermore, this flow-induced dilation of arterioles was dependent on the presence of an intact endothelium [14]. This suggests that the attenuation of the reactive hyperemic flow response with L-arginine analogues may be secondary to the metabolic vasodilation.

Conclusion

From the studies discussed here, we can conclude that the major site of coronary vascular regulation is at the level of the coronary arterioles. All coronary microvessels contribute to the autoregulatory control of coronary blood flow, but the predominant level of regulation lies in arterioles less than 100 μm. This is also the primary site of microvascular resistance. Both metabolic and myogenic mechanisms are present in coronary arterioles, but the ability of arterioles (vessels less than 150 μm in diameter) to autoregulate by either mechanism is.

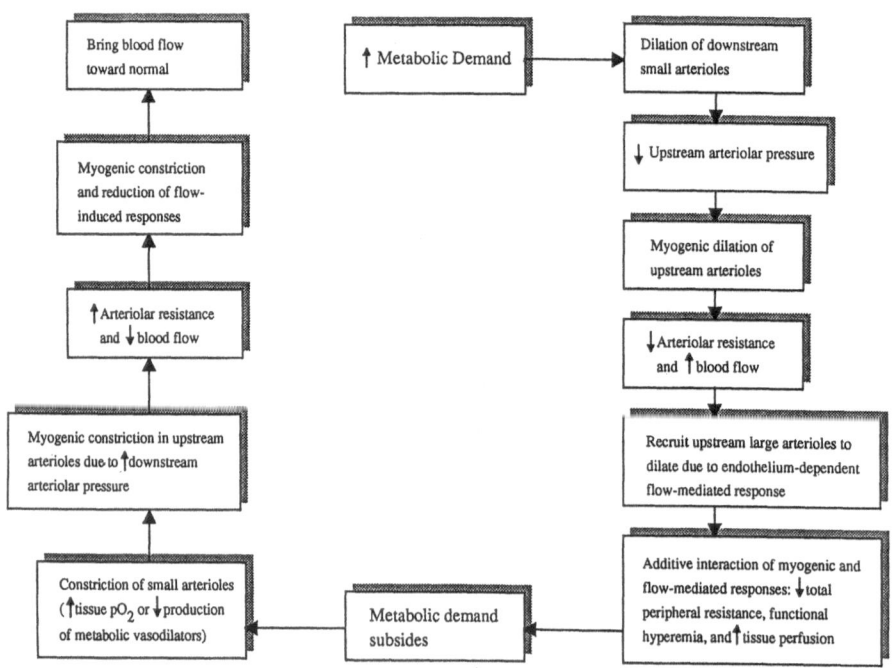

Fig. 3. Hypothesis of the integrative role of metabolic, myogenic, and flow-mediated regulation of coronary blood flow. (Modified from [14, 40] with permission)

greater than that in larger arteries. There are a variety of mechanisms which regulate blood flow at this level, including regulation by metabolities, myogenic mechanisms, adrenergic modulation of tone, and the release of nitric oxide from endothelial cells. These mechanisms are not homogeneously distributed, however, but vary between segments of the microcirculation. By measuring pressure gradients within the coronary microcirculation during the infusion of various receptor antagonists, Klassen et al. have suggested a pattern of control whereby the arterial circulation is regulated by adrenergic mechanisms, and the microcirculation and venous circulation is influenced more by a muscarinic control mechanism [39]. Kuo et al. have also proposed a heterogeneous system of control. Figure 3 provides a diagram of their integrative hypothesis for metabolic, myogenic, and flow-mediated control of coronary blood flow in the microcirculation [14, 40]. This scheme is predicated on an initial adjustment in tone by the production of vasoactive metabolites which initiates the entire sequence. First, an increase in the metabolic demand of the myocardium would preferentially dilate small arterioles through the release of vasoactive metabolites. This dilation would then result in a decreased pressure upstream from the arterioles, which would, in turn, elicit myogenic dilation of these vessels. Dilation would necessarily decrease arteriolar resistance and, hence, increase flow. The increased flow would recruit larger upstream vessels to dilate due to an endothelium-dependent flow-mediated response. The additive interaction of myogenic and flow-mediated responses would increase tissue perfusion, ultimately to the point of adequate oxygen delivery. After metabolic demands are reduced back towards the initial value, and tissue production of meta-bolites is decreased, dilation of small arterioles is reduced; this would in-crease pressure upstream as a result of arteriolar constriction, and thus elicit myogenic constriction, increasing arteriolar resistance and decreasing flow. The decreased flow will reduce the flow-mediated dilation, and return blood flow to normal.

These results imply a well-regulated system in which a heterogeneously-distributed series of mechanisms interact. In a series-coupled integrative system such as this, the importance of segmental functional differences becomes readily apparent.

References

1. Nellis SH, Liedtke AJ, Whitesell L (1981) Small coronary vessel pressure and diameter in an intact beating rabbit heart using fixed-position and free motion techniques. Circ Res 49:342–353
2. Chilian WM, Eastham CL, Marcus ML (1986) Microvascular distribution of coronary vascular resistance in beating left ventricle. Am J Physiol 251:H779–H788

3. Chilian WM, Layne SM, Klausner EC, Eastham CL, Marcus ML (1989) Redistribution of coronary microvascular resistance produced by dipyridamole. Am J Physiol (Heart Circ Physiol) 256:H383–H390
4. Tillmanns H, Steinhausen M, Leinberger H, Thederan H, Kubler W (1981) Pressure measurements in the terminal vascular bed of the epimyocardium of rats and cats. Circ Res 49:1202–1211
5. Chilian WM (1991) Microvascular pressures and resistances in the left ventricular subepicardium and subendocardium. Circ Res 69:561–570
6. Feigl EO (1983) Coronary physiology. Physiol Rev 63:1–205
7. Chilian WM, Layne SM (1990) Coronary microvascular responses to reductions in perfusion pressure: Evidence for persistent arteriolar vasomotor tone during coronary hypoperfusion. Circ Res 66:1227–1238
8. Kanatsuka H, Lamping KG, Eastham CL, Marcus ML (1990) Heterogeneous changes in epimyocardial microvascular size during graded coronary stenosis. Evidence of the microvascular site of autoregulation. Circ Res 66:389–396
9. Komaru T, Lamping KG, Eastham CL, Dellsperger KC (1991) Role of ATP-sensitive potassium channels in coronary microvascular autoregulatory responses. Circ Res 69:1146–1151
10. Daut J, Maier-Rudolph W, von Beckerath N, Mehrke G, Günther K, Goedel-Meinen L (1990) Hypoxic dilation of coronary arteries is mediated by ATP-sensitive potassium channels. Science 247:1341–1344
11. Dellsperger KC, Janzen DL, Eastham CL, Marcus ML (1990) Effects of acute coronary artery occlusion on the coronary microcirculation. Am J Physiol (Heart Circ Physiol) 259:H909–H916
12. Kanatsuka H, Sekiguchi N, Sato K, Akai K, Wang Y, Komaru T, Ashikawa K, Takishima T (1992) Microvascular sites and mechanisms responsible for reactive hyperemia in the coronary circulation of the beating canine heart. Circ Res 71:912–922
13. Nakayama K, Osol G, Halpern W (1988) Reactivity of isolated porcine coronary resistance arteries to cholinergic and adrenergic drugs and transmural pressure changes. Circ Res 62:741–748
14. Kuo L, Davis MJ, Chilian WM (1990) Endothelium-dependent, flow-induced dilation of isolated coronary arterioles. Am J Physiol (Heart Circ Physiol) 259:H1063–H1070
15. Kuo L, Davis MJ, Chilian WM (1988) Myogenic activity in isolated subepicardial and subendocardial coronary arterioles. Am J Physiol (Heart Circ Physiol) 255:H1558–H1562
16. Kanatsuka H, Lamping KG, Eastham CL, Dellsperger KC, Marcus ML (1989) Comparison of the effects of increased myocardial oxygen consumption and adenosine on the coronary microvascular resistance. Circ Res 65:1296–1305
17. Chilian WM, Layne SM, Eastham CL, Marcus ML (1989) Heterogeneous microvascular coronary α-adrenergic vasoconstriction. Circ Res 64:376–388
18. Chilian WM (1991) Functional distribution of α_1- and α_2-adrenergic receptors in the coronary microcirculation. Circulation 84:2108–2122
19. Jones CJH, DeFily DV, Patterson J, Chilian WM (1993) Endothelium-dependent relaxation competes with α_1- and α_2-adrenergic constriction in the coronary microcirculation. Circulation
20. Muntz KH, Garcia CA, Hagler HK (1985) α_1-Receptor localization in rat heart and kidney using autoradiography. Am J Physiol 249:H512–H519

21. Muntz KH, Meyer L, Gadol S, Calianos TA (1986) Alpha-2 adrenergic receptor localization in the rat heart and kidney using autoradiography and tritiated rauwolscine. J Pharmacol Exp Ther 236:542–547

22. Faber JE (1988) In situ analysis of α-adrenoceptors on arteriolar and venular smooth muscle in rat skeletal muscle microcirculation. Circ Res 62:37–50

23. Bassenge E, Busse R (1988) Endothelial modulation of coronary tone. Prog Cardiovasc Dis 30(5):349–380

24. Myers PR, Banitt PF, Guerra R Jr, Harrison DG (1989) Characteristics of canine coronary resistance arteries: Importance of endothelium. Am J Physiol 257:H603–H610

25. Amezcua JL, Palmer RMJ, de Souza BM, Moncada S (1989) Nitric oxide synthesized from L-arginine regulates vascular tone in the coronary circulation of the rabbit. Br J Pharmacol 97:1119–1124

26. Parent R, Paré R, Lavallée M (1992) Contribution of nitric oxide to dilation of resistance coronary vessels in conscious dogs. Am J Physiol (Heart Circ Physiol) 262:H10–H16

27. Tschudi M, Richard V, Bühler FR, Lüscher TF (1991) Importance of endothelial-derived nitric oxide in porcine coronary resistance arteries. Am J Physiol 260:H13–H20

28. Lamping KG, Marcus ML, Dole WP (1985) Removal of endothelium potentiates canine large coronary artery constrictor responses to 5-hydroxytryptamine in vivo. Circ Res 57:46–54

29. Lamping KG, Kanatsuka H, Eastham CL, Chilian WM, Marcus ML (1989) Nonuniform vasomotor responses of the coronary microcirculation to serotonin and vasopressin. Circ Res 65:343–351

30. Kuo L, Davis MJ, Cannon MS, Chilian WM (1992) Pathophysiological consequences of atherosclerosis extend into the coronary microcirculation: Restoration of endothelium-dependent responses by L-arginine. Circ Res 70:465–576

31. DeFily DV, Layne SM, Chilian WM (1990) Endothelial-dependent coronary microvascular vasodilation in vivo (abstract). Circulation 82:III-703

32. Alsip NL, Harris PD (1991) Receptor mediation of microvascular responses to serotonin in striated muscle. Am J Physiol (Heart Circ Physiol) 261:H1525–H1533

33. Lamping KG, Dole WP (1987) Acute hypertension selectively potentiates constrictor responses of large coronary arteries to serotonin by altering endothelial function in vivo. Circ Res 61:904–913

34. DeFily DV, Chilian WM (1993) Preconditioning protects coronary arteriolar endothelium from ischemia-reperfusion injury. Am J Physiol (Heart Circ Physiol 34) 265:H700–H706

35. Quillen JE, Sellke FW, Brooks LA, Harrison DG (1990) Ischemia-reperfusion impairs endothelium-dependent relaxation of coronary microvessels but does not affect large arteries. Circulation 82:586–594

36. Sellke FW, Quillen JE (1992) Altered effects of vasopressin on the coronary circulation after ischemia. J Thorac Cardiovasc Surg 104:357–363

37. Yamabe H, Okumura K, Ishizaka H, Tsuchiya T, Yasue H (1992) Role of endothelium-derived nitric oxide in myocardial reactive hyperemia. Am J Physiol (Heart Circ Physiol) 263:H8–H14

38. Kostic MM, Schrader J (1992) Role of nitric oxide in reactive hyperemia of the guinea pig heart. Circ Res 70:208–212

39. Klassen GA, Armour JA, Garner JB (1989) The effects of propranolol, phentolamine, and atropine on canine coronary vascular gradients. Can J Physiol Pharmacol 67:140–151
40. Kuo L, Davis MJ, Chilian WM (1992) Endothelial modulation of arteriolar tone. NIPS 7:5–9

Characteristics and Possible Cause of Instantaneous Velocity Waveforms in Small Coronary Arteries and Veins

Katsuhiko Tsujioka, Masami Goto, Yasuo Ogasawara, Osamu Hiramastu, and Fumihiko Kajiya[1]

Summary. Instantaneous blood flow velocities were measured in coronary arteries and veins of both the left and right ventricles and the left atrium by our laser Doppler velocimeter (LDV) with an optical fiber. The phase opposition of velocity waveforms between coronary arteries and veins was observed consistently in all coronary arteries and veins if the velocity waveform reflected an intramyocardial velocity waveform by minimizing the effect of capacitance of large epicardial vessels. The systolic component of the velocity waveform in the right coronary artery was slightly larger than that in the left coronary artery. A sharp decrease in atrial arterial flow was observed during atrial contraction and atrial vein flow was systolic predominant. The nature of force of myocardial contraction may be closely related to myocardial contractility, as the velocity waveforms are similar despite large differences in cavity pressure.

Key words: Coronary artery flow coronary vein flow—Right and left ventricles—Atrium— Laser Doppler velocimeter

Introduction

The instantaneous blood flow in coronary arteries and veins has unique phasic characteristics. The coronary arterial flow is diastolic-predominant, while the coronary venous flow is systolic-predominant. This phase opposition between the coronary arterial and venous flows has been proposed to originate from the myocardial contraction which squeezes blood from vessels in the deeper myocardium into the epicardial vein during systole; thus the coronary venous flow is systolic-predominant, whereas the release of compression in intramyocardial vessels during diastole causes diastolic coronary arterial flow [1]. In order to investigate this hypothesis, the instantaneous coronary arterial and venous flows have to be measured simultaneously. It is also necessary to measure the flow in small coronary arteries and veins to avoid the effect of the capacitance of the epicardial artery and vein, which distorts the phasic nature of the coronary flow. However, it is difficult to measure the phasic blood flow in small coronary

[1] Department of Medical Engineering, Kawasaki Medical School, 577 Mastushima, Kurashiki City, Okayama, 701-01 Japan

arteries and veins because of the limitations of the measuring techniques, and thus only a few studies have been reported [2–4]. For example, the electromagnetic flowmeter, a common way to measure coronary flow, is not suitable in blood flow measurments in vessels with very small diameters [5].

We have developed the laser Doppler velocimeter (LDV) with an optical fiber, and have successfully measured instantaneous blood flow in small coronary arteries and veins with diameters less than 400–500 μm [6–8]. The most important advantage of the LDV over conventional techniques is an excellent accessibility to vessels, even to easily deformable vessels like veins [9–12]. In this paper, we briefly describe first the laser Doppler velocimeter (LDV) with an optical fiber and the way to access the vessels using an optical fiber which is the velocity sensor. Then we analyze the instantaneous blood flows in the small coronary arteries and veins of the left ventricle, the right ventricle, and the left atrium. The comparison of the phasic coronary flows in these vessels, which are in different mechanical conditions, will help to elucidate the nature of the extravascular compressive forces due to myocardial contraction and relaxation.

Principle of the Fiber Optic Laser Doppler Velocimeter

Figure 1 shows the system of the fiber optic laser Doppler velocimeter (LDV). The He-Ne laser beam (632.8 nm, 5 mW) is divided into incidence and reference beams by a beam splitter (PBS). The incident beam is focused onto the entrance of a graded-index multimode fiber (clad diameter = 62.5 μm, core diameter =

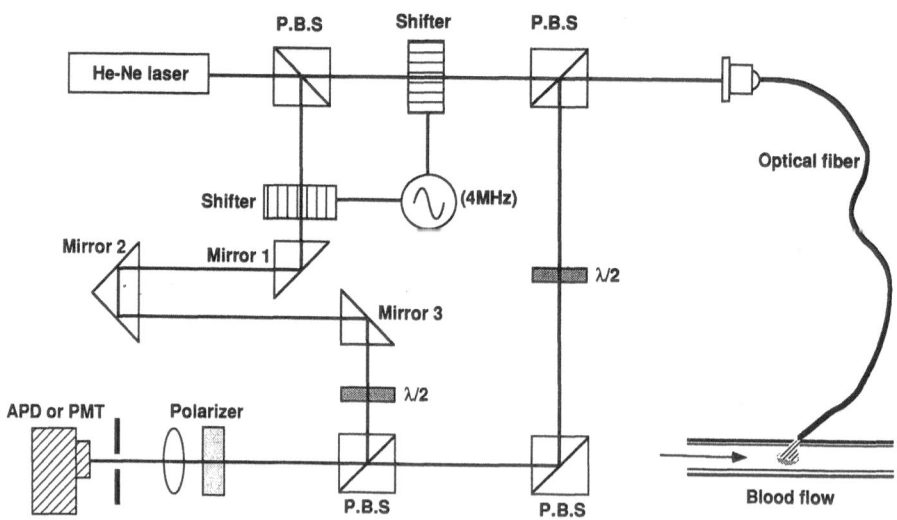

Fig. 1. Schema of the laser Doppler velocimeter with an optical fiber. *P.B.S.*, polarization beam splitter; *APD*, avalanche photodiode; *PMT*, photomultiplier

50 µm), and transmitted through the fiber into the blood stream. Part of the light back-scattered by flowing erythrocytes is re-entered into the same fiber, and transmitted back to the entrance. The optical heterodyning is obtained by mixing the Doppler-shift signal from the moving erythrocytes with the reference beam. Two frequency shifters (82 and 78 MHz) are interposed in the path of the incidence and reference beams to differentiate the forward flow from the reverse. Thus, the difference between the shifter frequencies (82 − 78 = 4 MHz) corresponds to zero flow velocity. When Doppler-shift frequency is greater than 4 MHz, the direction of blood flow is toward the fiber tip, and when it is less than 4 MHz, the direction is away from the fiber tip. The photocurrent from the photodetector (APD) is fed into a spectrum analyzer to measure the Doppler-shift frequency. The sample volume is approximately $\pi \times 0.05^2 \times 0.1\,mm^3$, and the temporal resolution is 8 ms [13].

Measurement Method of Blood Flow Velocity in Coronary Arteries and Veins Using the Laser Doppler Velocimeter (LDV)

An LDV provides excellent access to coronary vessels of the beating heart. Two different approaches are available for the LDV to evaluate the blood flow velocity in coronary arteries and veins (Fig. 2) [10]. For blood velocity measurements of small coronary arteries and veins whose vessel walls were thin enough to be transparent to laser light, we placed the fiber tip on the outer surface of the vessel and fixed it with a drop of cyanoacrylate [9, 14].

Instantaneous Blood Flow Velocity in Small Epicardial Coronary Arteries and Veins of the Left Ventricle

Coronary arterial and venous blood flow are directly affected by heart beat, since the coronary vessels are located in the myocardium to give oxygen and nutrients to cardiac muscle cells. To understand the effect of cardiac contraction and relaxation on the instantaneous myocardial perfusion throughout a cardiac cycle, it is important to analyze the instantaneous arterial inflow into the myocardium as well as the venous outflow from the myocardium. For this purpose, Chilian and Marcus [15] measured the phasic pattern of blood flow velocity in the distal coronary artery using a high-frequency ultrasound pulsed Doppler velocimeter. They characterized the velocity waveform in distal coronary arteries in two ways: (1) the rate of flow in systole decreases by more than 50%, relative to the flow in the proximal coronary artery, and (2) the midsystolic flow is retrograde. Several investigators measured blood velocities in coronary veins [16–18]; however, there are no data available for small veins at the site just after their emergence from the myocardium.

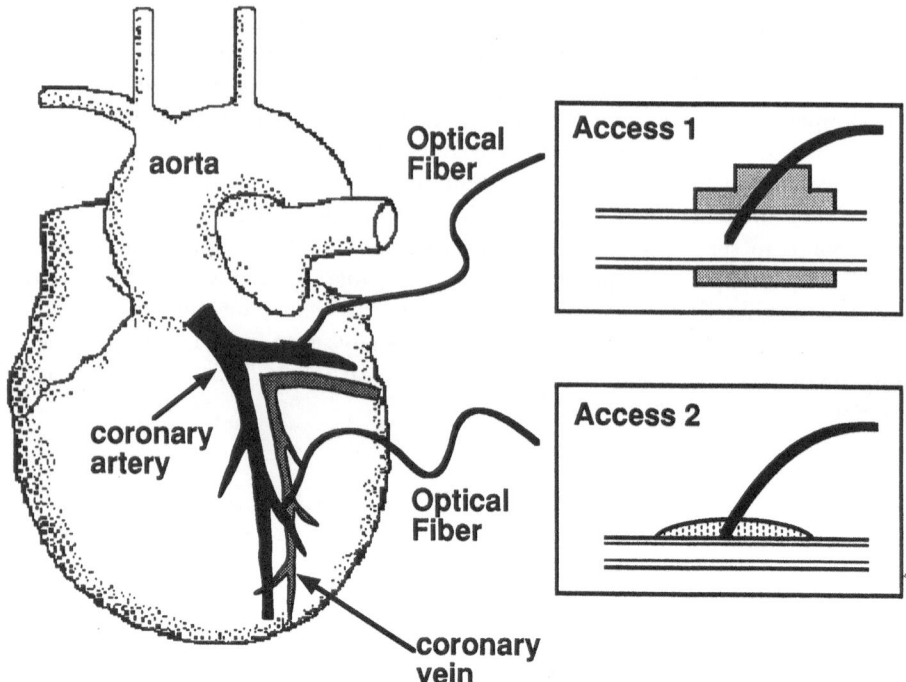

Fig. 2. Two different routes of access to blood flow. *Access 1* is used for the measurement of blood flow in relatively large epicardial coronary arteries and veins in which the wall is not transparent. *Access 2* for small coronary arteries and veins with thin, transparent walls

Using the LDV, we measured instantaneous blood flow velocity in small epicardial arteries at a position just before they penetrated into the myocardium, and in small veins just after they emerged from the myocardium. Figure 3 shows an example of the velocity waveforms in a small epicardial artery and vein of the left ventricle. As indicated by Chilian and Marcus [15], the velocity waveform in the peripheral epicardial artery was almost exclusively limited to diastole, and retrograde blood flow velocity was frequently observed during early systolic phase. On the other hand, the small epicardial veins of the left ventricle exhibited a systolic-predominant velocity waveform. Compared with the great cardiac vein (GCV) flow or coronary sinus flow, the onset of the small coronary venous flow was earlier, the flow acceleration was higher, and the diastolic flow component, which was frequently reversed, was much smaller. Therefore, the difference in the velocity waveform was more prominent between smaller peripheral arteries and veins than between proximal larger epicardial arteries and veins.

The phase shift between the arterial inflow into the myocardium and venous outflow of the myocardium should be explained based on the volume change in the myocardium. The blood volume change between diastole and systole in the myocardium has been explained by the intramyocardial pump model [19]. The

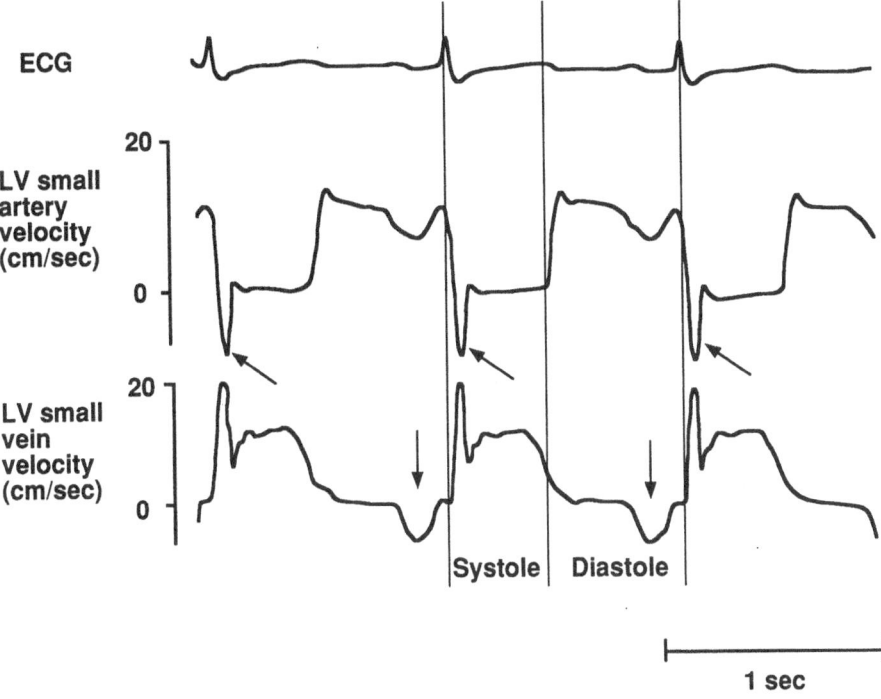

Fig. 3. A typical tracing of velocity waveforms in a small epicardial artery and vein of the left ventricle (*LV*). The arrows indicate reverse flow

intramyocardial pump model originally employed the left ventricular pressure as its pressure source. Recently, a time varying elastance model was applied to the vascular compartment in the left ventricular wall to regulate the blood volume stored in intramyocardial capacitance vessels [20]. Both models, i.e., the tissue pressure pump model and the elastance pump model, can predict systolic-dominant venous outflow and diastolic-dominant arterial inflow for the myocardial bed. Recently, Kouwenhoven et al. [21] examined the influence of left ventricular pressure on coronary arterial flow and pressure. In goats, with an cannulated and artificially-perfused left main coronary artery, the left ventricular pressure was changed by aortic occlusion. Their results showed that transmission of the left ventricular pressure to coronary arterial flow at constant pressure perfusion varies within the cardiac cycle. In the early phase of systole, this transmission was higher than in mid-systole. They suggested that stiffness of the cardiac muscle determines the influence of the left ventricular pressure on coronary flow. Thus, caution is needed in explaining the blood volume change in the myocardium with models; the determining factor of coronary blood flow varies depending on the myocardial elastance which changes during a cardiac cycle. Further studies, e.g., direct observation of the intramyocardial vessels, are needed to understand the phasic volume changes of the intramyocardial vascular compartment and its influence on coronary blood flow [22].

Instantaneous Blood Velocities in Coronary Arteries and Veins of the Right Ventricle

The instantaneous blood flow velocity pattern in the artery of the right ventricle is considered to be different from that in the left ventricle in that the systolic component is relatively larger in the right coronary artery than in the left ventricle, i.e., the phasic patterns of the blood velocities in large arteries may be distorted by large capacitance of epicardial vessels. However, it has been technically difficult so far to measure instantaneous velocities in small arteries and veins just after their penetration and emergence from the myocardium. These measurements in small vessels of the right ventricle are necessary for the same reason as are those in the left ventricle.

Recently we have succeeded in measuring coronary blood velocities in small coronary arteries and veins of the right ventricle by using our laser Doppler technique [8]. The left panel of Fig. 4 shows simultaneous recordings of blood flow velocities in the proximal and distal epicardial coronary arteries of the right ventricle under control conditions [8]. In the proximal artery the systolic

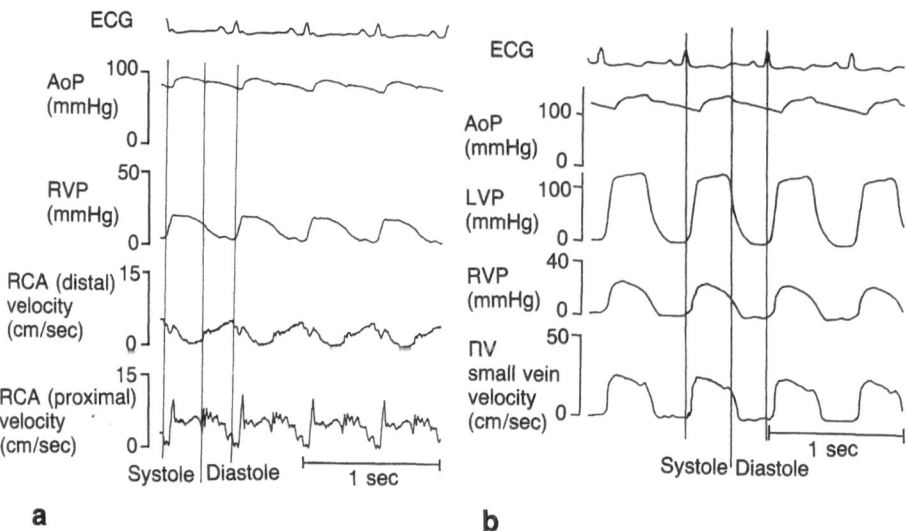

Fig. 4. An example of blood velocities in proximal and distal arteries (**a**) and veins of right ventricle (**b**). In distal right coronary artery (*RCA*), systolic component of velocity is relatively small compared to proximal RCA. Right ventricular (*RV*) vein flow is systolic-predominant. *AoP*, aortic pressure; *RVP*, right ventricular pressure; *LVP*, left ventricular pressure

component of the flow is comparable to the diastolic one, as reported by earlier studies [24, 25]. However, the instantaneous velocity pattern in the distal artery, which is minimally effected by the capacitance, is different from that in the proximal artery; the systolic component is relatively small, and the instantaneous flow is diastolic-predominant. These results indicate that the systolic compression of the right ventricular intramyocardial vessel is of a substantial amount, although the right ventricular pressure is small. The right ventricular vein instantaneous velocity pattern was systolic-predominant (Fig. 4, right panel). The blood velocity increased with a rise in the right ventricular pressure, and decreased with a fall in the right ventricular pressure. It is plausible that the systolic venous flow in the right ventricular vein is caused by the displacement of blood from intramyocardial microvessels into the epicardial large veins by contraction of the right ventricle.

In summary, the instantaneous velocity waveform in a small epicardial artery of the right ventricle is diastolic-predominant, while the waveform in small veins is systolic-predominant. These phasic characteristics of blood velocities in small epicardial arteries of the right ventricle have a similarity with those of the left venticle. It seems that the impeding effect on arterial inflow and the accelerating effect on venous outflow by myocardial contraction of the right ventiricle is more powerful than we have expected before.

Instantaneous Blood Velocities in Atrial Small Arteries and Veins

In order to examine the mechanical characteristics of the compressive force acting on the intramyocardial vessels during systole, we analyzed the phasic coronary artery and vein velocities of the left atrium. We clarified the relative importance of atrial chamber pressure, which is closely related to passive atrial intramyocardial pressure, and of atrial muscle contractility, which is closely related to the muscle elastance, in the genesis of the instantaneous velocities in the atrial artery and vein. In an open chest anesthetized dog, the heart was suspended in a pericardial cradle and the left atrial appendage was displaced gently to expose small branches of the artery or vein. The optical fiber of the LDV was glued onto the atrial vessels by a drop of cyanoacrylate.

The left panel of Fig. 5 shows a representative tracing of the blood velocity in a left atrial small artery under control conditions [7]. The blood velocity in the atrial small artery increased with aortic pressure and decreased gradually after reaching a peak during mid-systole. The blood velocity waveform showed a dicrotic notch, which resembles a notch in the aortic pressure at the end of the ejection phase. However, a sharp transient decrease in flow velocity, i.e., a dip formation, was always observed in late ventricular diastole corresponding to the atrial contraction phase. The dip appeared with rise of the left atrial pressure and disappeared at the end of the left ventricular isovolumic contraction. The right panel of Fig. 5 is a representative tracing of the instantaneous blood velocity in a

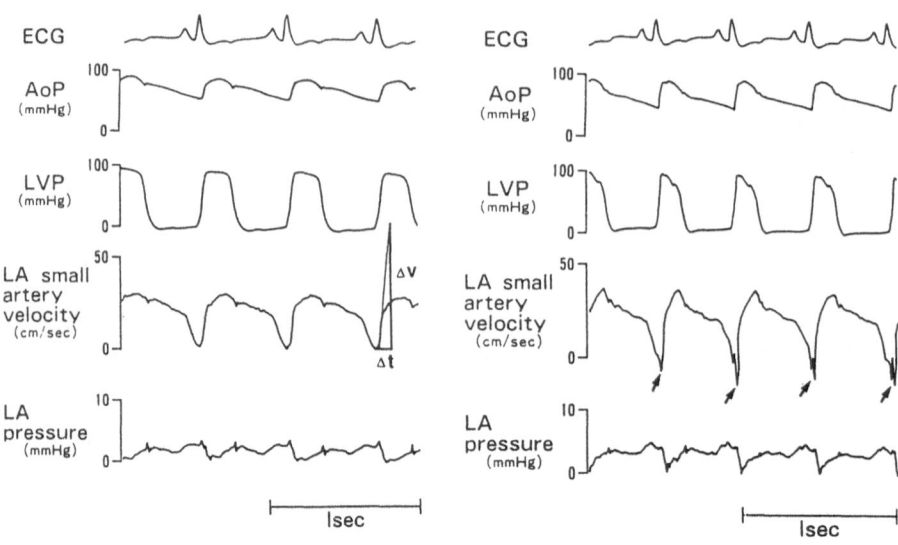

Fig. 5. Representative tracing of left atrial small arterial velocity. *left panel*, left ventricular pressure under control condition; *right panel*, after isoproterenol administration. The arrows indicate reverse flows. (from [7], with permission)

left atrial small artery during isoproterenol administration. Augmentation of atrial myocardial contraction by isoproterenol increased the rate of acceleration in systolic velocity. In addition, increase in atrial myocardial contraction induced the appearance of reverse flow during atrial contraction without a significant increase of left atrial systolic pressure. We then compared the blood velocity waveforms under control conditions with those during premature ventricular contraction. The depth of the dip after premature ventricular contraction was smaller than that of the sinus rhythm although the developed left atrial pressure was larger after premature ventricular beats.

Figure 6 shows a representative tracing of instantaneous blood velocity in a left atrial small vein [23]. The blood velocity in the left atrial small vein was characterized by a prominent forward-flow component during atrial systole. The blood velocity in the atrial vein increased rapidly with atrial contraction and decreased with atrial relaxation. Augmentation of atrial contraction by intravenous administration of isoproterenol increased the rate of rise of atrial venous flow. However the increase in left atrial systolic pressure during premature ventricular contraction decreased the rate of rise of atrial vein flow.

Thus, in terms of phase, the left atrial small artery velocity occurs mainly during atrial diastole, while the vein velocity appeared only during atrial systole. The impediment of atrial arterial flow and acceleration of atrial venous flow are closely related to atrial muscle contractility. These results indicate that the compression of the intramyocardial vessel of the left atrium is closely related to the increase of elastance of atrial muscle contractility.

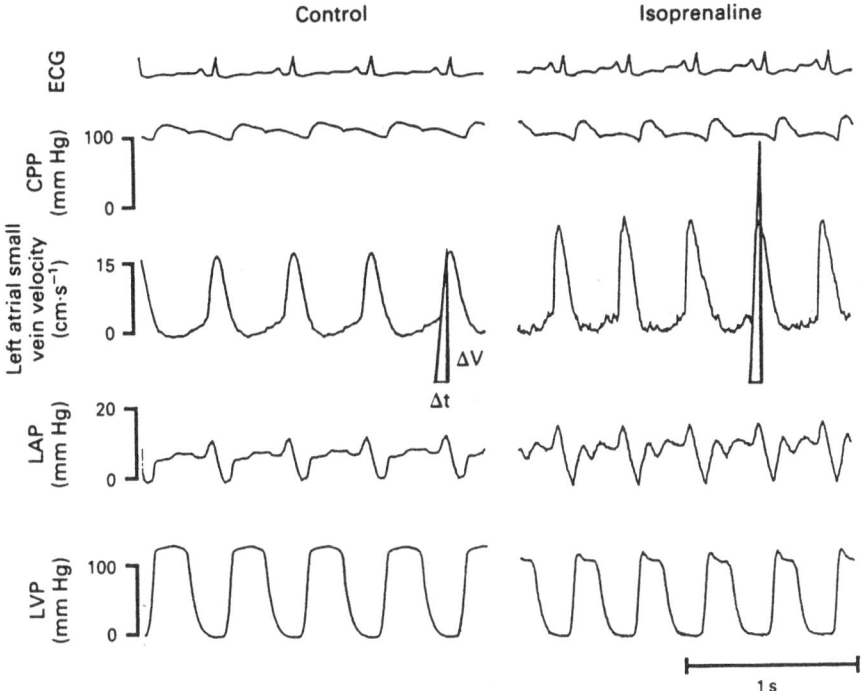

Fig. 6. Representative tracing of left atrial small vein velocity. *left panel*, under control condition; *right panel*, after isoproterenol administration. Increase of atrial myocardial contractility enhanced left atrial small vein velocity in spite of small change in left atrial pressure (*LAP*). *CPP*, coronary perfusion pressure; *LVP*, left ventricular pressure. (from [23], with permission)

Concluding Remarks

The main observations of the present study are the phase opposition of velocity waveforms between coronary arteries and veins was consistent for both the left and right ventricles, and the left atrium, when velocity was measured in a small artery just before its penetration into myocardium, and in a small vein just after its emergence from myocardium. Comparing blood velocity waveforms between the left and right ventricles, the ratio of systolic to diastolic velocity components was slightly higher in the right than in the left ventricle. In the same way as in the both ventricles, atrial contraction induced a transient sharp decrease in arterial flow (the systolic dip) and a prominent atrial systolic flow in atrial small veins.

The similar velocity waveforms in the left and right ventricles and the left atrium, despite large differences in cavity pressure, indicate that the nature of force of myocardial contraction acting on intramyocardial coronary vessels is

292   K. Tsujioka et al.

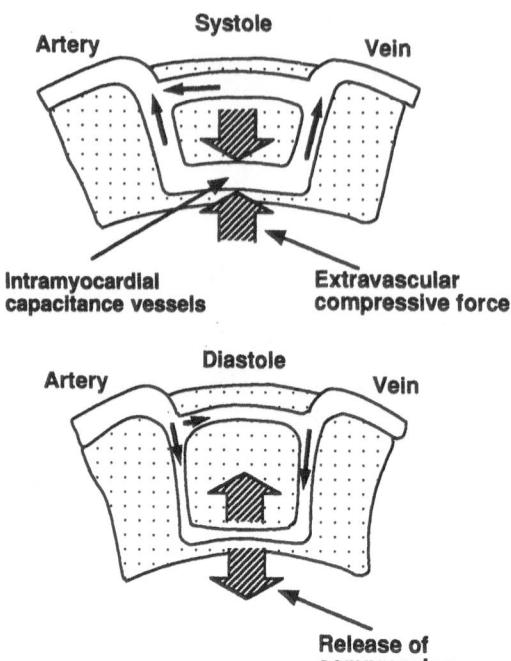

Fig. 7. Conceptual schema of the interaction between instantaneous coronary artery and vein flows, and myocardial contraction and relaxation. Extravascular compressive force squeezes out blood from intramyocardial capacitance vessels into epicardial arteries and veins during systole. The extravascular compressive force may have two components: active myocardial elastance and passive intramyocardial pressure which is related to left ventricular pressure.

closely related to myocardial contractility. Since myocardial elastance changes during the cardiac cycle, the effect of chamber pressure may change momentarily. The transmural difference of phasic flow velocity waveforms may be explained by the transmural variation of intramyocardial pressure related mainly to intracavity pressure (Fig. 7).

References

1. Porter WT (1898) The influence of the heartbeat on the flow of blood through the walls of the heart. Am J Physiol 1:145–163
2. Chilian WM, Marcus ML (1984) Coronary venous outflow persists after cessation of coronary artrial inflow. Am J Physiol 247:H984–990
3. Spaan JAE (1982) Intramyocardial compliance studies by venous outflow at arterial occlusion (abstract). Circulation 66:II-42

4. Kajiya F, Hiramatsu O, Mito K, Tadaoka S, Ogasawara Y, Tsujioka K (1990) Evaluation of coronary blood flow by fiber-optic laser Doppler velocimeter. In: Kajiya F, Klassen GA, Spaan JAE, Hoffman JIE (eds) Coronary circulation. Springer Berlin Heidelberg Tokyo, pp 43–54
5. Gregg DE, Khouri EM, Rayford CR (1965) Systematic and coronary energetics in the resting unanesthetized dog. Circ Res 16:102–113
6. Kajiya F, Hoki N, Tomonaga G, Nishihara H (1981) A laser Doppler velocimeter using an optical fiber and its application to local velocity measurement in the coronary artery. Experientia 37:1171–1173
7. Kajiya F, Tsujioka K, Ogasawara Y, Hiramatsu O, Wada Y, Goto M, Yanaka M (1989) Analysis of the characteristics of the flow velocity waveforms in left atrial small arteries and veins in the dog. Circ Res 65:1172–1181
8. Hiramatsu O, Kimura A, Yada T, Yamanoto T, Ogasawara Y, Goto M, Tsujioka K, Kajiya F (1992) Phasic characteristics of arterial inflow and venous outflow of right ventricular myocardium in dogs. Am J Physiol 262:H1422–1427
9. Tsujioka K, Goto M, Hiramatsu O, Wada Y, Ogasawara Y, Kajiya F (1990) Functional characteristics of intramyocardial capacitance vessels and their effect on coronary arterial inflow and venous outflow. In: Kajiya F, Klassen GA, Spaan JAE, Hoffman JIE (eds) Coronary circulation. Springer, Berlin Heidelberg Tokyo, pp 89–97
10. Kajiya F, Hiramatsu O, Mito K, Ogasawara Y, Tsujioka K (1987) An optical-fiber laser Doppler velocimeter and its application to measurements of coronary blood flow velocities. Med Prog Technol 12:77–85
11. Kajiya F, Tomonaga G, Tsujioka K, Ogasawara Y, Nishihara H (1985) Evaluation of local blood flow velocity in proximal and distal coronary arteries by laser Doppler method. J Biomech Eng 107:10–15
12. Kajiya F, Tsujioka K, Goto M, Wada Y, Chen X-L, Nakai M, Tadaoka S, Hiramatsu O, Ogasawara Y, Mito K, Tomonaga G (1986) Functional characteristics of intramyocardial capacitance vessels during diastole in the dog. Circ Res 58:476–485
13. Kilpatrick D, Linderer T, Sievers RE, Tyberg JV (1982) Measurement of coronary sinus blood flow by fiber-optic laser Doppler anemometry. Am J Physiol 242:H1111–H1114
14. Kajiya F, Tsujioka K, Ogasawara Y, Mito K, Hiramatsu O, Goto M, Wada Y, Matsuoka S (1989) Mechanical control of coronary artery inflow and vein outflow. Jpn Circ J 53:431–439
15. Chilian WM, Marcus ML (1985) Effects of coronary and extravascular pressure on intramyocardial and epicardial blood velocity. Am J Physiol 248:H170–H178
16. Canty JM Jr, Brooks A (1990) Phasic volumetric coronary venous outflow patterns in conscious dogs. Am J Physiol 258:H1457–H1463
17. Klassen GA, Armour JA (1984) Coronary venous pressure and flow: effect of vagal stimulation, aortic occlusion, and vasodilators. Can J Physiol Pharmacol 62:531–538
18. Hellenbrand WK, Klassen GA, Armour JA, Sezerman O, Paton B (1986) Autonomic nervous system regulation of epicardial coronary vein systolic and diastolic blood velocity as measured by a laser Doppler velocimeter. Can J Physiol Pharmacol 64:1463–1472
19. Spaan JAE, Breuls NPW, Laird JD (1981) Diastolic-systolic flow differences are caused by intramyocardial pump action in the anesthetized dog. Circ Res 49:584–593

20. Krams R, Sipkema P, Westerhof N. (1989) Varying elastance concept may explain coronary systolic flow impediment. Am J Physiol 257 (Heart Circ Physiol 26): H1471–H1479
21. Kouwenhoven E, Vergroesen I, Han Y, Spaan JAE (1992) Retrograde coronary flow is limited by time-varying elastance. Am J Physiol 263:H484–H490
22. Yada T, Hiramatsu O, Kimura A, Goto M, Ogasawara Y, Tsujioka K, Yamamori S, Ohno K, Hosaka H, Kajiya F (1993) In vivo observation of subendocardial microvessels of the beating porcine heart using a needle-probe videomicroscope with a CCO camera. Circ Res 72:939–946
23. Kimura A, Hiramatsu O, Wada Y, Yada T, Yamamoto T, Goto M, Ogasawara Y, .Tsujioka K, Kajiya F (1992) Atrial contractility affects basic blood flow velocity of atrial small vessels in the dog. Cardiovasc Res 26:1219–1225
24. Lowensohn HS, Khouri EM, Gregg DE, Pyle RL, Patterson RE (1976) Phasic right coronary artery blood flow in conscious dogs with hormal and elevated right ventricular pressures. Circ Res 39:760–766
25. Ross G (1967) Blood flow in the right coronary artery of the dog. Cardiovasc Res 1:138–144

The Coronary Circulation in Hypertrophy and in Syndrome X

Richard O. Cannon, III[1]

Summary. Although the contribution of epicardial coronary artery disease in ischemic syndromes is incontrovertible, in recent years much interest has shifted to the pathophysiologic relevance of disease or dysfunction of the coronary microcirculation (defined as intramyocardial arteries and arterioles too small to be visualized angiographically). Animal studies have shown the major contribution of intramyocardial arteries, proximal to the arteriolar bed, to coronary resistance, limitations in coronary flow reserve, and impaired coronary autoregulation in experimental left ventricular hypertrophy, and have shown evidence for microvascular endothelial dysfunction in early atherosclerosis. Despite imperfect techniques for measurement of coronary blood flow in humans, several studies have shown limitations in coronary flow reserve in familial hypertrophic cardiomyopathy and pressure-induced left ventricular hypertrophy that may be causally related to inducible myocardial ischemia in these patients. Mechanisms for impaired coronary flow reserve in hypertrophy are likely multifactorial and may include increased resting and stress-induced coronary flow requirements without neovascularization, detrimental effects of abnormal diastolic relaxation on intramyocardial coronary flow, and medial and intimal thickening of intramyocardial coronary arteries. More controversial is the presence and significance of coronary microvascular dysfunction in patients with chest pain syndromes in the absence of left ventricular hypertrophy, often with abnormal noninvasive testing, and designated as Syndrome X or microvascular angina. Although several studies have shown limited coronary flow responses to pacing or pharmacologic vasodilators, and heightened sensitivity of the coronary microcirculation to ergonovine in this patient population, proof that these altered coronary flow dynamics truly cause myocardial ischemia during stress has been elusive. Selection of patients on the basis of abnormal radionuclide studies may be more fruitful in making the connection between abnormal coronary flow dynamics and inducible myocardial ischemia than selection of patients on the basis of their exercise EKG. Coronary artery disease may also be associated with coronary microvascular dysfunction, and hypercholesterolemia and hypertension, which are commonly associated with obstructive coronary artery disease, also adversely affect the coronary microcirculation. Such involvement may explain why a subset of patients continue to have symptoms and evidence of inducible ischemia despite adequate revascularization by percutaneous transluminal coronary angioplasty or coronary bypass grafting. Further studies will undoubtedly clarify the mechanisms and the extent of coronary microvascular dysfunction in ischemic syndromes, in addition to clarifying suitable therapeutic interventions.

Key words: Angina pectoris—Cardiomyopathy—Hypertension—Hypertrophy

---

[1] Cardiology Branch, National Heart, Lung, and Blood Institute, National Institutes of Health, Bethesda, MD 301-496-9895, USA

Introduction

In ischemic cardiac syndromes, most clinical and investigative attention has focused on atherosclerotic disease of large epicardial coronary arteries. However, abnormal structure or function of the coronary microcirculation (intramyocardial arteries and arterioles too small to be visualized angiographically) has been postulated to be responsible for angina-like symptoms and noninvasive or invasive testing consistent with inducible myocardial ischemia in patients with normal coronary angiograms. This suspicion has been bolstered by animal studies showing the major contribution of pre-arteriolar intramural arteries to coronary resistance [1], limitations in coronary flow reserve in experimental left ventricular hypertrophy [2], and evidence for endothelial dysfunction of the microcirculation in animal models of early epicardial (but not microvascular) atherosclerosis [3]. This review will focus on two clinical conditions in which coronary microvascular dysfunction has been studied in humans, left ventricular hypertrophy and syndrome X (microvascular angina), in addition to addressing methodologic considerations for the study of coronary flow dynamics in humans.

Methodologic Concerns

Unlike animals, in which absolute coronary volume flow and myocardial perfusion and flow distribution can be accurately measured repeatedly in a wide variety of experimental conditions, there are ethical and technical obstacles to carrying out analogous studies in humans. Thus, investigators have had to make indirect measurement of volume flow (e.g., measuring coronary venous flow by thermodilution or intracoronary flow velocity), or they have had to use techniques that do not permit repetitive measurements (e.g., inert gas washout) or that are limited in availability (e.g., positron emission tomography [PET]). Further, in humans, no technique currently available allows the measurement of transmural distribution of flow. With the exception of the relative flow distribution determined by thallium-201 scintigraphy or technetium-99m scstamibi, indirect or direct measurement of absolute flow or perfusion during exercise is problematic with the available methodology. Assessment of flow reserve following the administration of drugs such as dipyridamole, adenosine, and papaverine has yielded a wide range of responses, even when this reserve is measured noninvasively by PET in normal controls, resulting in uncertain implications in measurement of flow reserve in a given patient.

Left Ventricular Hypertrophy—Hypertrophic Cardiomyopathy

Hypertrophic cardiomyopathy is commonly defined as unexplained myocardial hypertrophy which is familial in most cases, and is often seen with massive hypertrophy of the ventricular walls, especially the septum [4]. Although dynamic gradients within the left ventricle have been of particular clinical and research interest, most patients with hypertrophic cardiomyopathy do not have this feature in their disease. Angina-like chest pain is common in these patients, usually in the setting of angiographically normal coronary arteries. Metabolic evidence for ischemia has been demonstrated in symptomatic patients as well as asymptomatic patients, the latter observation indicating the presence of "silent ischemia" in this patient population [5–8]. O'Gara et al. [8] performed exercise thallium-201 emission computed tomographic studies in 72 patients with hypertrophic cardiomyopathy. Of these 72, 41 (57%) had abnormal scans, with complete reversibility of the perfusion abnormalities after 3 h of rest. Perfusion abnormalities were present in all regions of the left ventricle, as well as the endocardium as evidenced by the apparent dilatation of the left ventricle during exercise (radionuclide angiographic study showed no increase in left ventricular volumes). Of note, fixed or partially reversible defects were more common in patients with reduced left ventricular ejection fractions which suggest scarring. More recently, we reported that reversible thallium abnormalities are indicative of inducible myocardial ischemia during stress [7]. Of 37 patients with reversible thallium abnormalities during exercise stress, 27 (73%) had evidence of myocardial lactate production during pacing stress. In contrast, only 4 of 13 (31%) patients with normal exercise thallium studies had evidence of ischemia during pacing stress. Studies from our group have shown that thallium perfusion abnormalities can be improved or normalized by verapamil administration [9] or operative relief of left ventricular outflow obstruction [10].

Several pathophysiological features of hypertrophic cardiomyopathy could account for ischemia in this disease. Systolic compression of septal perforating (and occasionally epicardial) arteries is commonly observed during coronary angiography. However, we could not demonstrate an association between the presence and severity of septal arterial compression and either thallium perfusion abnormalities during exercise, or evidence for ischemia during pacing stress [7]. Morphologically abnormal intramural arteries are commonly found in necropsy specimens of hypertrophic cardiomyopathy [11, 12]. Of 48 patients with hypertrophic cardiomyopathy studied at necropsy, Maron et al. [11] found that 40 (83%) had abnormalities in intramural arteries involving all walls of the left ventricle, but most commonly involving the septum (33 patients). The small arteries were characterized by the thickening of the vessel wall, including both intimal and medial components, with a decrease in luminal size. Of note, abnormal intramural arteries were commonly noted in areas of fibrosis, raising the possibility of ischemic-mediated injury and scarring.

Distorted myocellular architecture, a histologic hallmark for this disease, is commonly found in substantial portions of all walls of the left ventricle in necropsy specimens, and may also contribute to ischemia by limiting appropriate perfusion and oxygen delivery during stress. Studies of myocardial metabolism during pacing stress have shown that despite myocardial ischemia during pacing stress, myocardial oxygen extraction often paradoxically decreases in this disease [6], in contrast to increased myocardial oxygen extraction during pacing stress in patients with coronary artery disease. The explanation for this phenomenon is unknown, but potential mechanisms include shunting of oxygen-rich blood directly into the coronary venous drainage because of the compressive effects of elevated filling pressures on the subendocardial microcirculation, or vasodilatation of epicardial vessels with flow in excess of epicardial demands despite subendocardial ischemia.

Other considerations undoubtedly contribute to the pathogenesis of myocardial ischemia in patients with hypertrophic cardiomyopathy, including high rest and stress-provoked coronary flow requirements which may exhaust coronary flow reserve during even relatively low levels of stress [6], and abnormal diastolic relaxation [13] which may limit the rate and extent of coronary filling of the myocardial microcirculation from epicardial arteries [14]. As a consequence of these abnormalities affecting both myocardial oxygen demand and coronary flow capacity, coronary flow reserve is often severely compromised in both hypertrophied and non-hypertrophied regions [15].

Left Ventricular Hypertrophy—Aortic Stenosis

Angina-like chest pain is common in patients with hemodynamically significant valvular aortic stenosis. Marcus et al. [16] showed a 50% reduction in the peak hyperemic left anterior descending coronary arterial flow velocity response of 14 patients with significant aortic stenosis and left ventricular hypertrophy. Measurements were made at time of surgery with an ultrasonic probe positioned over the artery, with measurements compared to eight patients undergoing cardiac surgery but without pressure-overload left ventricular hypertrophy. All measurements were made at a baseline following 20 s manual compression of the artery. Strauer [17] reported that patients with aortic stenosis commonly have reduced coronary flow response (measured by inert gas washout technique) to dipyridamole. Of note, symptomatic patients had greater impairment in pharmacologic coronary flow reserve than asymptomatic patients. Strauer also showed that basal coronary flow per unit mass is higher in symptomatic aortic stenosis patients than asymptomatic or normal patients, suggesting partial exhaustion of flow reserve because of high metabolic requirements of their pressure overload. Rakusan et al. [18] found a reduction of the capillary to myocyte density in adults with acquired aortic stenosis, indicating an absence of angiogenesis in adults to "keep up" with compensatory myocyte hypertrophy and the metabolic demands of pressure overload. Whether these changes result in a reduction in absolute peak coronary flow as opposed to a reduction in relative (to mass) peak coronary flow is unclear.

Left Ventricular Hypertrophy—Hypertension

Studies performed in patients with hypertension, left ventricular hypertrophy, and chest pain syndromes, despite angiographically normal coronary arteries, have indicated that although resting coronary flow per unit mass may be normal, coronary flow reserve may be limited. Strauer performed studies of coronary flow (inert gas washout technique) in 114 patients with hypertension, and found significant reduction in coronary flow reserve in response to dipyridamole compared to normotensive controls [19]. This investigator has emphasized the importance of wall stress in determining not only basal coronary flow requirements, but flow reserve as well. Thus, the greater the wall stress, the lower the flow reserve, presumably because of partial exhaustion of vasodilator capacity due to elevated absolute coronary flow requirements. Opherk et al. [20] also found that coronary flow responses (inert gas washout technique) to dipyridamole were significantly lower in hypertensive patients with EKG evidence of left ventricular hypertrophy following dipyridamole compared to normotensive controls.

Vogt et al. [21] measured coronary flow reserve (inert gas washout technique) following dipyridamole administration in 54 hypertensive patients with chest pain and normal coronary angiograms. Patients were reported to have resting coronary flow and myocardial oxygen consumption per unit mass similar to 12 normotensive controls. However, there was a significant reduction in the flow response to dipyridamole, with elevation of minimum coronary resistance compared to the normotensive controls. The authors found no significant correlation between minimum coronary resistance or coronary reserve, and the left ventricular muscle mass or diastolic wall stress, concluding that coronary microvascular abnormalities might contribute to impaired coronary flow reserve, independent of left ventricular hypertrophy. Houghton et al. [22] performed intracoronary Doppler flow velocity measurements in 40 hypertensive patients with chest pain and no or minor coronary artery disease, some of whom had thallium scintigraphic defects following exercise stress. Flow reserve was assessed following dipyridamole administration: patients with thallium perfusion defects were more likely to have limited coronary flow velocity responses to dipyridamole than those with normal thallium perfusion scans. Further, patients with thallium perfusion defects were more likely to have a greater degree of left ventricular hypertrophy than those with normal thallium scans. However, the decrement in flow reserve was not linearly related to the magnitude of left ventricular hypertrophy. Our group reported that 12 hypertensive patients with chest pain and normal coronary angiograms had a coronary microvascular constrictor response (great cardiac vein flow measured by thermodilution) to ergonovine administration: all had no or minimal left ventricular hypertrophy by echo [23].

In contrast, 12 normotensive subjects without pacing-induced chest pain responses had a dilator response to ergonovine. More recently, Treasure et al. [24] presented preliminary data that indicated that endothelium-mediated relaxation to acetylcholine is impaired in the microcirculation of hypertensive patients with left ventricular hypertrophy, with flow measured by an intracoronary Doppler catheter. Thus, patients with hypertension and a varying magnitude of left ventricular hypertrophy commonly have an abnormal flow reserve, with several studies suggesting an independent contribution of the coronary microcirculation, possibly related to endothelial dysfunction.

Morphologic abnormalities in coronary microvascular density and structure may also limit appropriate coronary flow in humans with pressure-induced left ventricular hypertrophy. Anatomic studies of Rakusan [25] showed a substantial reduction in capillary density in hypertrophied human hearts. Tanaka et al. [12] reported luminal narrowing of intramyocardial small arteries in hearts at necropsy of patients with left ventricular hypertrophy as a consequence of hypertension. In a preliminary report, Schwartzkopff et al. [26] found luminal narrowing of intramural arteries from right ventricular septal biopsies in patients with chest pain, hypertension, and normal coronary angiograms. They reported a correlation between the minimum coronary resistance following dipyridamole administration and the arterial medial wall area. No correlation was found between the left ventricular mass index by echocardiography and minimum coronary resistance. Abnormal structure and function of the coronary microcirculation in left ventricular hypertrophy may additionally contribute to impaired autoregulation of the coronary flow during reductions in systemic pressure, as reported by Polese et al. [27].

Although several anti-hypertensive agents have been shown to regress left ventricular hypertrophy in humans, there are few data on the impact of left ventricular hypertrophy regression on coronary flow dynamics. In a preliminary study, Vogt et al. [28] reported that 12 months of therapy with the angiotensin-converting enzyme inhibitor, enalapril, increased the coronary flow response to dipyridamole and reduced the minimum coronary resistance compared to the pretreatment study. There was only a mild reduction in echocardiographic left ventricular mass, suggesting that the response to enalapril was more a consequence of the vascular effects of the drug rather than left ventricular hypertrophy regression. Further studies in humans will be necessary in order to assess the impact of anti-hypertensive therapy on left ventricular hypertrophy regression and coronary vascular morphology, endothelial function, autoregulation, and coronary flow reserve.

Syndrome X (Microvascular Angina)

Several groups have reported that dysfunction of the coronary microcirculation may account for ischemia and symptoms in a subset of normotensive patients with chest pain, angiographically normal coronary arteries, and no structural heart disease or hypertrophy by echocardiogram. In 1981, Opherk et al. [29] reported limited increase in coronary blood flow (inert gas washout technique) after dipyridamole administration in 21 normotensive, nondiabetic patients in which 12 had ischemic-appearing EKG changes during exercise and 4 had left bundle branch block responses to exercise (the remaining 5 patients terminated exercise because of chest pain without EKG changes). Compared to 15 patients with "atypical chest pain" and normal EKG responses to exercise, the "Syndrome X" population had smaller increases in flow in response to dipyridamole administration. Left ventricular endocardial biopsies performed in 18 patients showed no vascular abnormalities. In 1983, our group reported limitations in the great cardiac vein flow response to rapid atrial pacing in patients who described their typical chest pain as being provoked by this stress [30]. Patients were not selected for participation in this study on the basis of the characteristics of their pain or results of noninvasive testing. To assess whether a dynamic component to this flow limitation was present (because patients commonly reported variation in their chest pain threshold), ergonovine (0.15 mg I.V.) was administered after the initial pacing stress, and repeat flow measurements were taken during pacing stress at the same heart rate (150 per min). Those patients who experienced their typical chest pain during pacing following ergonovine administration increased great cardiac vein flow less from the baseline than those remaining symptom free who had an increase in coronary vascular resistance compared with pacing prior to the administration of ergonovine. Because angiography demonstrated no significant change in epicardial coronary artery dimensions following ergonovine, we concluded that the increase in coronary vascular resistance was caused by coronary microvascular constriction in response to ergonovine. In a subset of patients, dipyridamole administration resulted in less vasodilatation in the same patients who demonstrated a microvascular constrictor response to ergonovine [31, 32]. In 1985 we proposed the term "microvascular angina" for this patient population [32], in view of what appeared to be a heightened sensitivity of the coronary microcirculation to vasoconstrictor stimuli associated with a limited microvascular vasodilator capacity, and proposed that dysfunction of small intramural prearteriolar coronary arteries might be the pathogenetic cause of this syndrome [33].

Other groups have also measured coronary flow by various methodologies, both invasively and noninvasively, and have concluded that a subset of their population has limited flow responses to stress or vasodilators. Noninvasive studies of flow reserve by positron emission tomography are of particular interest because of the ability to use normal healthy controls for reference. Geltman [34], Camaci [35], and Galassi [36] and coworkers have all reported subsets of patients (either selected because of an ischemic-appearing EKG response to exercise or unselected) having limited coronary flow responses to dipyridamole administration. However, the wide range of "normal" flow responses to dipyridamole makes interpretation of individual flow responses problematic in deciding whether the flow response is normal or abnormal for that patient.

The mechanism of microvascular dysfunction remains unknown. Demonstration of co-existing abnormal forearm hyperemic responses to ischemia [37], esophageal motility abnormalities [38], and bronchoconstrictor responses to methacholine inhalation [39] have suggested that some patients with coronary microvascular dysfunction may have a generalized disorder of vascular and nonvascular smooth muscle function. However, other studies have suggested the possibility of endothelial dysfunction of the microcirculation [40] or excessive sympathetic stimulation of the coronary microcirculation [41, 42]. Possibly, different pathogenetic abnormalities produce a dysfunctional coronary microcirculation in different patients with a final common pathway of abnormal coronary flow dynamics during stress.

Thus, several groups using different methodologies have come to similar conclusions with regard to abnormal flow responses to pharmacologic vasodilators or stress in subsets of patients with chest pain and normal coronary angiograms. However, demonstration that this abnormality causes ischemia has been more elusive [43]. Camici et al. [44] performed studies of myocardial metabolism during pacing stress in 11 women selected because of ischemic-appearing EKG responses during exercise stress, and found no metabolic evidence for ischemia despite their limited flow response (thermodilution technique) to pacing compared to controls who had no EKG changes during exercise stress. Other groups have noted that patients, even if selected for their ischemic-appearing EKG response to exercise, commonly have no confirmatory hemodynamic evidence for inducible ischemia during exercise [43]. This has led some investigators to conclude that there is no ischemia in this patient population as a group. We have found, however, that radionuclide angiographic left ventricular ejection fraction responses to exercise are commonly abnormal in those patients who demonstrated coronary microvascular constrictor responses to ergonovine administration [45, 46]. Approximately one-third of patients who took part in our studies and were found to have microvascular constrictor responses to ergonovine administration had either no increase or a fall in the ejection fraction in response to exercise and/or developed wall motion abnormalities during exercise, a higher prevalence than those patients who did not have a coronary microvascular constrictor response to ergonovine. Patients with ischemic-appearing EKG responses to exercise had a slightly higher prevalence of abnormal responses (53%). Whether selection of patients on the basis of abnormal radionuclide angiographic or thallium scintigraphic responses to stress will be more fruitful in defining a link between coronary microvascular dysfunction and inducible myocardial ischemia than selection of patients on the basis of their EKG response to exercise alone is at present unknown.

The observation that only a minority of patients studied have evidence for inducible myocardial ischemia during stress despite demonstration of "dysfunctional" coronary microcirculation has led us to consider an alternative mechanism for chest pain in this patient population; abnormal visceral sensitivity that might exaggerate the pain of mild ischemia or produce pain on a nonischemic basis [47]. Whether drugs successful in treating chronic neuropathic pain syndromes will also be of benefit in these patients is currently under investigation.

Conclusion

Mounting evidence for disease or dysfunction of the coronary microcirculation has shifted much of our interest to the microcirculation in ischemic heart disease. Indeed, patients with coronary artery disease may have co-existing dysfunction of the microcirculation that may directly contribute to ischemia in this patient population. Drexler et al. [48] have shown that hypercholesterolemia can lead to microvascular dysfunction in patients with "minimal" atherosclerosis, and that this dysfunction that can be reversed by infusion of the precursor to nitric oxide, l-arginine. Dysfunctional endothelium has also been demonstrated in patients with hypertension, particularly with left ventricular hypertrophy, in the absence of obstructive atherosclerotic disease [24]. As hypercholesterolemia and hypertension are commonly associated with obstructive coronary artery disease, it is likely that the microcirculation shares involvement with epicardial vessels, probably as a consequence of endothelial dysfunction and abnormal smooth muscle regulation. Microvascular involvement in coronary artery disease may explain why a subset of patients continue to have symptoms and evidence of inducible ischemia despite adequate revascularization by percutaneous transluminal coronary angioplasty or coronary bypass grafting. Further, dysfunctional microvascular endothelial and smooth muscle regulation might explain limitation of pharmacologic coronary flow reserve immediately following apparently successful coronary angioplasty [49] and in vascular territories subserved by "non-diseased" arteries in patients with single vessel coronary artery disease [50]. Further studies will undoubtedly clarify the mechanisms and extent of coronary microvascular dysfunction in ischemic syndromes in addition to clarifying suitable therapeutic interventions.

References

1. Chilian WM, Eastham CL, Marcus ML (1986) Microvascular distribution of coronary vascular resistance in beating left ventricle. Am J Physiol 251:H779–788
2. Dellsperger KC, Marcus ML (1990) Effects of left ventricular hypertrophy on the coronary circulation. Am J Cardiol 65:1504–1510
3. Selke FW, Armstrong ML, Harrison DG (1990) Endothelium-dependent vascular relaxation is abnormal in the coronary microcirculation of atherosclerotic primates. Circulation 81:1586–1593
4. Maron BJ, Bonow RO, Cannon RO, Leon MB, Epstein SE (1987) Hypertrophic cardiomyopathy. Interrelation of clinical manifestations, pathophysiology, and therapy. N Engl J Med 316:780–789, 844–852
5. Cannon RO, Rosing DR, Maron BJ, Leon MB, Bonow RO, Watson RM, Epstein SE (1985) Myocardial ischemia in patients with hypertrophic cardiomyopathy:

Contribution of inadequate vasodilator reserve and elevated left ventricular filling pressures. Circulation 71:234–243

6. Cannon RO, Schenke WH, Maron BJ, Tracy CM, Leon MB, Brush JE, et al. (1987) Differences in coronary flow and myocardial metabolism at rest and during pacing between patients with obstructive and patients with nonobstructive cardiomyopathy. J Am Coll Cardiol 10:53–62

7. Cannon RO, Dilsizian V, O'Gara PT, Udelson JE, Schenke WH, Quyyumi A, et al. (1991) Myocardial metabolic, hemodynamic, and electrocardiographic significance of reversible thallium-201 abnormalities in hypertrophic cardiomyopathy. Circulation 83:1660–1667

8. O'Gara PT, Bonow RO, Maron BJ, Damske BA, Van Lingen A, Bacharach SL, et al. (1987) Myocardial perfusion abnormalities in patients with hypertrophic cardiomyopathy: Assessment with thallium-201 emission computed tomography. Circulation 76:1214–1223

9. Udelson JE, Bonow RO, O'Gara PT, Maron BJ, Van Lingen A, Bacharach SL, et al. (1989) Verapamil prevents silent myocardial perfusion abnormalities during exercise in asymptomatic patients with hypertrophic cardiomyopathy. Circulation 79:1052–1060

10. Cannon RO, Dilsizian V, O'Gara PT, Udelson JE, Tucker E, Panza JA, et al. (1992) Impact of surgical relief of outflow obstruction on thallium perfusion abnormalities in hypertrophic cardiomyopathy. Circulation 85:1039–1045

11. Maron BJ, Wolfson JK, Epstein SE, Roberts WC (1986) Intramural ("small vessel") coronary artery disease in hypertrophic cardiomyopathy J Am Coll Cardiol 8:545–557

12. Tanaka M, Fujiwara H, Onodera T, Wu D-J, Matsuda M, Hamashima Y, et al. (1987) Quantitative analysis of narrowings of intramyocardial small arteries in normal hearts, hypertensive hearts, and hearts with hypertrophic cardiomyopathy. Circulation 75:1130–1139

13. Bonow RO, Rosing DR, Bacharach SL, Green MV, Kent KM, Lipson LC, et al. (1981) Effects of verapamil on left ventricular systolic function and diastolic filling in patients with hypertrophic cardiomyopathy. Circulation 64:787–796

14. Brutsaert DL, Housmans PR, Goethals MA: Dual control of relaxation: Its role in the ventricular function in the mammalian heart (1980) Circ Res 47:637–52

15. Camici P, Chiriatti G, Lorenzoni R, Bellina RC, Gistri R, Italiani G, et al. (1991) Coronary vasodilation is impaired in both hypertrophied and nonhypertrophied myocardium of patients with hypertrophic cardiomyopathy: A study with nitrogen-13 ammonia and positron emission tomography. J Am Coll Cardiol 17:879–886

16. Marcus ML, Doty DB, Hiratzka LF, Wright CB, Eastham CL (1982) Decreased coronary reserve. A mechanism for angina pectoris in patients with aortic stenosis and normal coronary arteries. N Engl J Med 307:1362–1366

17. Strauer BE (1990) The significance of coronary reserve in clinical heart disease. J Am Coll Cardiol 15:775–783

18. Rakusan K, Flanagan MF, Geva T, Southern J, Van Praagh R (1992) Morphometry of human coronary capillaries during normal growth and the effect of age in left ventricular pressure-overload hypertrophy. Circulation 86:38–46

19. Strauer BE (1979) Ventricular function and coronary hemodynamics in hypertensive heart disease. Am J Cardiol 44:999–1006

20. Opherk D, Mall G, Zebe H, Schwarz F, Weihe E, Manthey J, et al. (1984) Reduction of coronary reserve: A mechanism for angina pectoris in patients with arterial hypertension and normal coronary arteries. Circulation 69:1–7

21. Vogt M, Motz W, Schwartzkopff B, Strauer BE (1990) Coronary microangiography and cardiac hypertrophy. Eur Heart J 11 [Suppl B]:133–138

22. Houghton JL, Frank MJ, Carr AA, Von Dohlen TW, Prisant LM (1990) Relations among impaired coronary flow reserve, left ventricular hypertrophy, and thallium perfusion defects in hypertensive patients without obstructive coronary artery disease. J Am Coll Cardiol 15:43–51

23. Brush JE, Cannon RO, Schenke WH, Bonow RO, Leon MB, Maron BJ, et al. (1988) Angina due to coronary microvascular disease in hypertensive patients without left ventricular hypertrophy. N Engl J Med 319:1302–1307

24. Treasure CB, Klein JL, Vita JA, Selwyn AP, Alexander RW, Ganz P (1991) Left ventricular hypertrophy secondary to hypertension is associated with impaired endothelium-mediated relaxation in the coronary microvessels (abstract). J Am Coll Cardiol 17:237A

25. Rakusan K (1971) Quantitiative morphology of capillaries of the heart. Number of capillaries in animal and human hearts under normal and pathological conditions. Methods Achiev Exp Pathol 5:272–286

26. Schwartzkopff B, Vogt M, Knauer S, Motz W, Strauer BE (1991) Medial hypertrophy of intramural coronary arteries in patients with reduced coronary reserve in hypertensive heart disease (abstract). Circulation 84:II-479

27. Polese A, De Cesare N, Montorsi P, Fabbiocchi F, Guazzi M, Loaldi A, et al. (1991). Upward shift of the lower range of coronary flow autoregulation in hypertensive patients with hypertrophy of the left ventricle. Circulation 83:845–853

28. Vogt M, Motz W, Politz B, Scheler S, Strauer BE (1991) Improvement of coronary reserve by chronic treatment with ACE-inhibitors (abstract). Circulation 84:II-136

29. Opherk D, Zebe H, Weihe E, Mall G, Durr C, Gravert B, et al. (1981) Reduced coronary dilatory capacity and ultrastructural changes of the myocardium in patients with angina pectoris but normal coronary arteriograms. Circulation 63:817–825

30. Cannon RO, Watson RM, Rosing DR, Epstein SE (1987) Angina caused by reduced vasodilator reserve of the small coronary arteries. J Am Coll Cardiol 1359–1373

31. Cannon RO, Schenke WH, Leon MB, Rosing DR, Urquhart J, Epstein SE (1987) Limited coronary flow reserve after dipyridamole in patients with ergonovine-induced coronary vasoconstriction. Circulation 75:163–174

32. Cannon RO, Epstein SE (1988) "Microvascular angina" as a cause of chest pain with angiographically normal coronary arteries. Am J Cardiol 61:1338–1343

33. Epstein SE, Cannon RO (1985) Site of increased resistance to coronary flow in patients with angina pectoris and normal epicardial coronary arteries. J Am Coll Cardiol 8:459–461

34. Geltman EM, Henes CG, Senneff MJ, Sobel BE, Bergmann SR (1990) Increased myocardial perfusion at rest and diminished perfusion reserve in patients with angina and angiographically normal coronary arteries. J Am Coll Cardiol 16:586–595

35. Camici PG, Lorenzoni R, Gistri R (1990) Regional coronary vasodilator reserve in syndrome X (abstract). Circulation 82:III-478

36. Galassi AR, Araujo LI, Crea F, Kaski JC, Lammertsma AA, Yamamoto Y, et al. (1991) Myocardial blood flow is altered at rest and after dipyridamole in patients with syndrome X (abstract). J Am Coll Cardiol 17:227A

37. Sax FL, Cannon RO, Hanson C, Epstein SE (1987) Impaired forearm vasodilator reserve in patients with microvascular angina: Evidence of a generalized disorder of vascular function? N Engl J Med 317:1366–1370

38. Cannon RO, Cattau EL, Yakshe PN, Maher K, Schenke WH, Benjamin SB, Epstein SE (1990) Coronary flow reserve, esophageal motility, and chest pain in patients with angiographically normal coronary arteries. Am J Med 88:217–222
39. Cannon RO, Peden DB, Berkebile C, Schenke WH, Kaliner MA, Epstein SE (1990) Airway hyperresponsiveness in patients with microvascular angina: Evidence for a diffuse disorder of smooth muscle responsiveness. Circulation 82:2011–2017
40. Motz W, Vogt M, Rabenay O, Scheler S, Luckhoff A, Strauer BE (1991) Evidence of endothelial dysfunction in coronary resistance vessels in patients with angina pectoris and normal coronary angiograms. Am J Cardiol 68:996–1003
41. Bortone AS, Hess OM, Eberli FR, Nonogi H, Marolf AP, Grimm J, Krayenbuehl HP (1989) Abnormal coronary vasomotion during exercise in patients with normal coronary arteries and reduced coronary flow reserve. Circulation 79:516–527
42. Montorsi P, Manfredi M, Loaldi A, Fabbiocchi F, Polese A, de Cesare N, Bartorelli A, Guazzi MD (1989) Comparison of coronary vasomotor responses to nifedipine in syndrome X and in Prinzmetal's angina pectoris. Am J Cardiol 63:1198–1202
43. Cannon RO, Camici PG, Epstein SE (1991) Pathophysiologic dilemma of syndrome X. Circulation 85:883–892
44. Camici PG, Marraccini P, Lorenzoni R, Buzzigoli G, Pecori N, Perissinotto A, et al. (1991) Coronary hemodynamics and myocardial metabolism in patients with syndrome X: Response to pacing stress. J Am Coll Cardiol 17:1461–1470
45. Cannon RO, Bonow RO, Bacharach SL, Green MV, Rosing DR, Leon MB, et al. (1985) Left ventricular dysfunction in patients with angina pectoris, normal epicardial coronary arteries, and abnormal vasodilator reserve. Circulation 71:218–226
46. Cannon RO (1991) Microvascular angina. Cardiovascular investigations regarding pathophysiology and management. In: Richter JE, Cannon RO, Beitman B (eds) Unexplained chest pain. Medical Clinics of North America, vol 75. WB Saunders Philadelphia, pp 1097–1118
47. Cannon RO, Quyyumi AA, Schenke WH, Fananapazir L, Tucker EE, et al. (1990) Abnormal cardiac sensitivity in patients with chest pain and normal coronary arteries. J Am Coll Cardiol 16:1359–1366
48. Drexler H, Zeiher AM, Meinzer K, Just H (1991) Correction of endothelial dysfunction in coronary microcirculation of hypercholesterolaemic patients by L-arginine. Lancet 338:1546–1550
49. Wilson RF, Johnson MR, Marcus ML, Alyward PE, Talmen CT, White CW (1988) The effect of coronary angioplasty on coronary flow reserve. Circulation 77:873–885
50. Sambuceti G, Parodi O, L'Abbate A (1991). Evidence for global impairment of coronary vasodilating capacity even in single vessel coronary artery disease (abstract). Circulation 84:II–652

Coronary Hemodynamic Characteristics in Small Vessel Disorder: Impaired Endothelium-Dependent Vasodilation

Atsushi Mikuniya and Kogo Onodera[1]

Summary. The endothelium plays a crucial role in the control of vascular tone through the production of endothelium-derived relaxing and contracting factors. To define coronary hemodynamic characteristics in small vessel disorder that is not detected by coronary angiography, we examined endothelium-independent vasodilation with papaverine (PAP), and endothelium-dependent vasodilation with acetylcholine (ACh) in patients with syndrome X. Responses of the large coronary artery in syndrome X to papaverine and acetylcholine were similar to those in control patients. However, changes in coronary venous oxygen saturation (CSO_2-Sat) in response to acetylcholine were significantly lower in syndrome X compared with control patients, while response of CSO_2-Sat to papaverine was preserved. These results imply that small vessel disorder involves impaired endothelium-dependent relaxation at the prearteriolar levels, resulting in the development of myocardial ischemia during exercise through the coronary steal phenomenon.

Key words: Small vessel disease—Syndrome X—Papaverine—Acetylcholine—Coronary venous oxygen saturation

Introduction

The term "syndrome X" [1] or "microvascular angina" [2] characterizes the clinical picture in patients with typical angina pectoris and ECG ST depression during exercise, but normal coronary angiograms at rest. In the absence of both atherosclerotic obstruction and demonstrable epicardial artery spasm in the coronary arteries, the occasional development of myocardial ischemia in these patients is likely caused by small vessel disease. However, the mechanism by which small coronary disorder causes myocardial ischemia is not fully understood.

To define the coronary hemodynamic characteristics in small vessel disorder, the present study was carried out to examine the coronary responses to the endothelium-independent vasodilator papaverine and the endothelium-

---

[1] The Second Department of Internal Medicine, Hirosaki University School of Medicine, Hirosaki, 036 Japan

dependent vasodilator acetylcholine in patients with "syndrome X" [1] or "microvascular angina" [2].

Methods

Patient Selection

The control group of thirteen patients, 8 males and 5 females, was evaluated. Their average age was 49.9 ± 10.7 years, ranging from 39 to 65 years. All patients suffered from atypical chest pain, which was long lasting and not sensitive to nitroglycerin. They had a normal ECG at rest and a normal exercise test. These patients were referred for cardiac catheterization and coronary angiography to definitely exclude cardiac disease as the cause of their symptoms.

The syndrome X group consisted of 11 patients, 3 males and 8 females, with an average age of 52.7 ± 8.8 years, ranging from 41 to 60 years. These patients had typical angina and ECG ST depression ($> 1.0\,mV$) during the treadmill exercise test with the standard Bruce protocol despite normal coronary angiograms. No patient had valvular heart disease, left ventricular hypertrophy, hypertension, diabetes mellitus, or collagen disease.

Cardiac Catheterization

Except for sublingual nitroglycerin, all medications were discontinued at least 48 hours before cardiac catheterization. During cardiac catheterization, a fiberoptic oxygen saturation catheter (7.5 Opticath, Oximetrix, Mountain View, CA) was inserted into the right internal jugular vein and placed into the great cardiac vein through the coronary sinus under fluoroscopic control. The position was checked by injection of a small quantity of radiographic contrast medium. Coronary venous oxygen saturation was measured optically by a spectrophotometric catheter oxygen saturation system (Oximetrix, USA) and recorded continuously [3, 4]. The upper panel of Fig. 1 shows the study protocol of this investigation. After the diagnostic cardiac catheterization was completed, a dose of 10 mg papaverine was infused as a bolus into the left coronary artery with continuous measurement of coronary venous oxygen saturation. The same dosage of papaverine was again infused into the left coronary artery and left coronary angiography was performed at the peak increase in coronary venous oxygen saturation between 60 and 90 sec after the bolus injection. Following the papaverine studies, control coronary angiography was performed again. Next, a dose of 50 µg acetylcholine was infused into the left coronary artery with continuous measurement of coronary venous oxygen saturation and acetylcholine infusion was repeated for coronary angiography studies.

For estimation of coronary responses to papaverine and acetylcholine, changes in the coronary diameter of the left anterior descending artery (LAD) and in coronary venous oxygen saturation were examined in all patients. Changes in coronary venous oxygen saturation were quantatively calculated and expressed

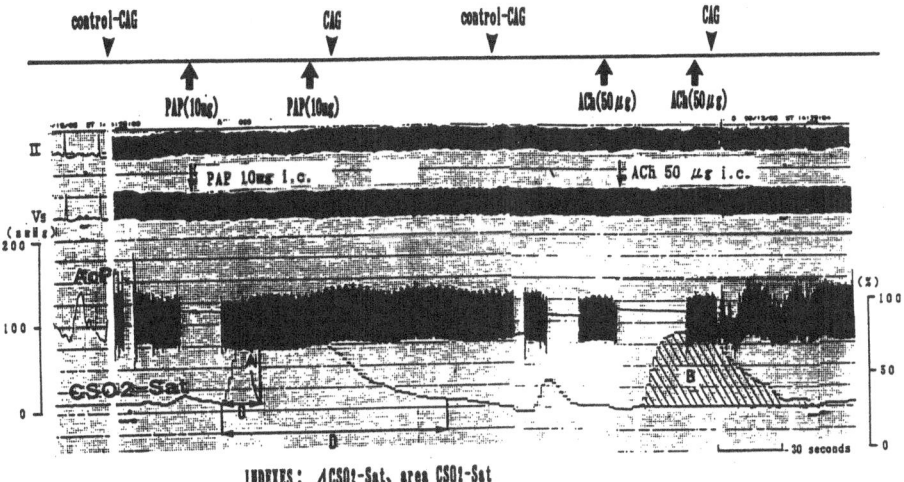

Fig. 1. Study protocol (upper panel) and examples of original tracings of ECG, aortic pressure, and continuous measurement of coronary venous oxygen saturation (lower panel). Note that both intracoronary infusions of papaverine (*PAP*) and acetylcholine (*ACh*) increased coronary venous oxygen saturation. Changes in coronary venous oxygen saturation were quantatively estimated as the time integral area between the base line and the increasing curve (B). *CAG*, coronary angiography, *i.c*, intracoronary, *AoP*, aortic pressure, *CSO₂-Sat*, coronary venous oxygen saturation

as the time integral area of coronary venous oxygen saturation (area CSO_2-Sat) between the basal line and the increasing curve [5], as shown in the lower panel of Fig. 1.

Data were analyzed with the Student's t-test for unpaired date or Wilcoxon's rank sum test. A p value < 0.05 was considered statistically significant. All values are presented as mean $\pm$ SD.

Results

Epicardial Coronary Artery Responses

In all patients, the diameter of the LAD segment 6 was measured. Intracoronary papaverine increased the coronary diameter by 10.2% from 2.9 $\pm$ 0.7 mm to 3.3 $\pm$ 0.9 mm ($p < 0.05$) in the control group and by 9.8% from 3.0 $\pm$ 0.5 mm to 3.4 $\pm$ 0.4 mm ($p < 0.05$) in the syndrome X group. No significant difference in change of the coronary diameter was found between the two groups. Intracoronary acetylcholine decreased the diameter by 1.7% from 2.9 $\pm$ 0.6 mm to 2.8 $\pm$ 0.8 mm in the control group and by 2.3% from 3.0 $\pm$ 0.6 mm to 2.9 $\pm$ 0.5 mm in the syndrome X group, showing similar behavior of coronary conduit vessels in the two groups.

Coronary Venous Oxygen Saturation Responses

As show in the lower panel of Fig. 1, both intracoronary infusions of papaverine and acetylcholine markedly increased coronary venous oxygen saturation, indicating an increase in the coronary blood flow.

Figure 2 summarizes the data of the coronary venous oxygen saturation responses. Area CSO_2-Sat, in response to intracoronary papaverine, was 3547 ± 1173% · sec in the control group, and 3314 ± 1186% · sec in the syndrome X group. There was no significant difference in papaverine-induced area CSO_2-Sat between the two groups. However, area CSO_2-Sat, in response to intracoronary acetylcholine, was 1929 ± 875% · sec in the syndrome X group and 2722 ± 890% · sec in the control group. Acetylcholine-induced area CSO_2-Sat in the syndrome X group was significantly smaller than that in the control group ($p <$ 0.05), indicating less acetylcholine-induced vasodilation in the syndrome X group.

When the coronary response was estimated as the ratio of the acetylcholine-induced area to the papaverine-induced area CSO_2-Sat (ACh/PAP), the ratio in the syndrome X group (0.62 ± 0.17), was significantly smaller than that in the control group (0.82 ± 0.17) ($p <$ 0.05).

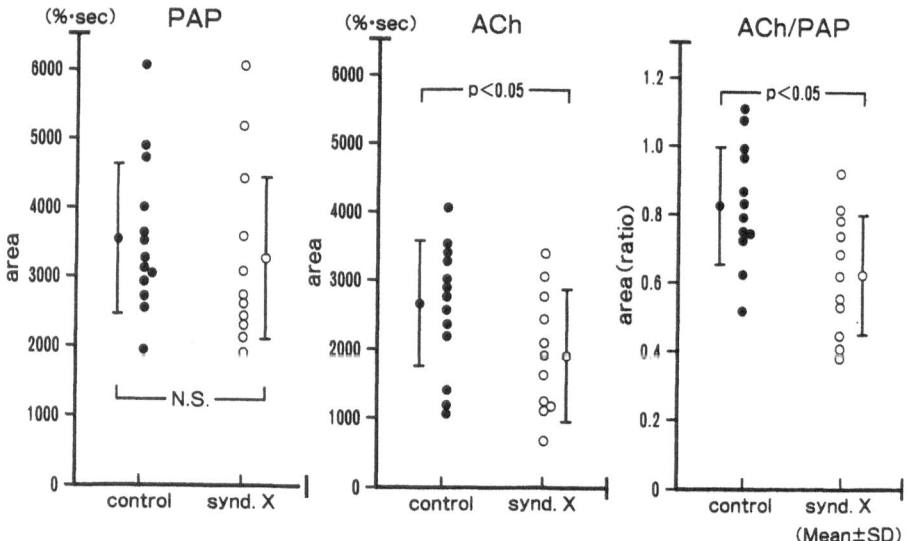

Fig. 2. Responses of coronary venous oxygen saturation to papaverine (*PAP*) and acetylcholine (*ACh*) in the control and syndrome X groups. Note that the response to ACh in syndrome X was significantly lower than that in control group. Area *CSO₂-Sat*, time integral area of coronary venous oxygen saturation, ACh/PAP, ratio of ACh-induced to PAP-induced Area CSO_2-Sat, *synd. X*, syndrome X

Discussion

The results of this study demonstrate that endothelium-dependent coronary vasodilation is impaired in patients with small vessel disease, whereas endothelium-independent vasodilation is preserved. Responses of the large coronary artery in the syndrome X group to papaverine, an endothelium-independent vasodilator, and acetylcholine, an endothelium-dependent vasodilator, were not significantly different from those in the control group. The coronary venous oxygen saturation response to papaverine was also similar in the two groups. However, response of coronary venous oxygen saturation to acetylcholine was significantly lower in the syndrome X group when compared with the control group. The ratio of acetylcholine-induced to papaverine-induced response was also lower in the syndrome X group. These findings suggest that there is an impairment in the endothelium-mediated modulation of coronary vasomotor tone in patients with small vessel disease, whereas the endothelium-mediated modulation is preserved.

"Syndrome X" [1] or "microvascular angina" [2] defines a group of patients who present with typical exertional angina pectoris and a normal coronary angiogram. Epstein and Cannon [6] proposed that in that group of patients, the impediment to blood flow may be located in the small intramural prearteriolar vessels before they give off subepicardial and subendocardial branches. They considered that the localization of the flow impediment in the intramyocardial vessels before the branching point of the subepicardial vessels was enough to account for myocardial ischemia induced during exercise. Maseri et al. [7] proposed that patients with syndrome X have a patchily-distributed abnormal constriction of the coronary prearteriolar vessels which are not involved in metabolic autoregulation of blood flow. In addition, the subjective report of a wide variability of effort tolerance with occasional occurrence of apparently spontaneous episodes and the objective findings of variable levels of heart rate at the onset of ST segment depression during Holter ECG monitoring [8, 9] suggest that the flow-limiting abnormality of small vessels is due to functional rather than organic, obstructive changes in these vessels.

It has been recognized that the endothelium plays a crucial role in the control of vascular tone through the production of endothelium-derived relaxing and contracting factors [10, 11, 12]. Therefore, based on previous reports and the results in this investigation, it would seem reasonable to postulate that small vessel disorder involves impaired endothelium-dependent relaxation at the prearteriolar levels, resulting in the development of myocardial ischemia during exercise through the coronary steal phenomenon.

References

1. Kemp HG (1973) Left ventricular function in patients with the anginal syndrome and normal coronary arteriograms. Am J Cardiol 32:375–376

2. Cannon RO, Epstein SE (1988) "Microvascular angina" as a cause of chest pain with angiographically normal coronary arteries. Am J Cardiol 61:1338–1343

3. Crake T, Canepa-Anson R, Shapiro L, Poole-Wilson PA (1988) Continuous recording of coronary sinus oxygen saturation during atrial pacing in patients with coronary artery disease or with syndrome X. Br Heart J 59:31–38

4. Mikuniya A, Fujino F, Higashiyama A, Narita T, Fukushi T, Sasaki M, Onodera K, Sawai M, Akitsu Y, Oike Y (1992) Evaluation of myocardial oxygen extraction dynamics in syndrome X by continuous measurement of coronary sinus oxygen saturation and its relation to clinical features (in Japanese). Kokyu to Junkan 40:473–480

5. Fujino, Y, Mikuniya A, Narita T, Higashiyama A, Onodera K, Oike Y (1991) Endothelium-dependent dilatation of coronary resistance vasculature is impaired in patients with exercise induced chest pain, ST-depression and normal coronary angiograms. Circulation [Suppl II] 84:II-729

6. Epstein, SE, Cannon RO (1986) Site of increased resistance to coronary flow in patients with angina pectoris and normal epicardial coronary arteries. J Am Coll Cardiol 8:459–461

7. Maseri A, Crea F, Kaski JC, Crake T (1991) Mechanism of angina pectoris in syndrome X. J Am Coll Cardiol 17:499–506

8. Kaski JC, Crea F, Nihoyannopoulos P, Hackett D, Maseri A (1986) Transient myocardial ischemia during daily life in patients with syndrome X. Am J Cardiol 58:1242–2347

9. Pupita G, Kaski JC, Galassi A, Vejar M, Crea F, Maseri A (1989) Long-term variability of angina pectoris and electrocardiographic signs of ischemia in syndrome X. Am J Cardiol 64:139–143

10. Furchgott RF (1983) Role of the endothelium in responses of vascular smooth muscle. Circ Res 53:557–573

11. Vanhoutte PM (1988) The endothelium: Modulator of vascular smooth muscle tone. N Engl J Med 319:512–513

12. Lusher TF (1989) Endothelium-derived relaxing and contracting factors: Potential role in coronary artery disease. Eur Heart J 10:847–857

Abstracts

Effects of α- and β-Adrenergic Blockade on Coronary Microcirculation in the Beating Canine Left Ventricle

Hiroshi Kanatsuka[1], Nobuyo Sekiguchi[1], Tatsuya Komaru[1], Kouichi Ashikawa[2], and Tamotsu Takishima[1]

This sudy was carried out to clarify the effects of α- and β-adrenergic blockade on coronary arterial microvessels in beating hearts. For this purpose, we used anesthetized open chest dogs ($n = 47$). Coronary microcirculation of the epimyocardium was observed with a microscope equipped with a floating objective and a high-speed video system. Diameters of arterial microvessels were measured by fluorescence microangiography. Heart rate and aortic pressure were kept nearly constant throughout the experimental period. In the following experiments, drugs were administered into the left anterior descending coronary artery (LAD). To assess the effect of α-adrenergic blockade, phentolamine (100 μg/kg) was administered in the absence or presence of β-adrenergic blockade (propranolol 50 μg/kg). To assess the effect of β-adrenergic blockade, propranolol (50 μg/kg) or three doses of ICI-118 551 a selective $β_2$-antagonist (0.1, 0.5, and 1.0 μg/kg per min) were administered. Coronary arterial microvessels were divided into three groups according to control internal diameters (ID) (small, ID < 100 μm; medium, 100 ≤ ID < 200 μm; and large, ID ≥ 200 μm). In the absence of β-adrenergic blockade, phentolamine significantly dilated vessels in all groups (small, 19.6 ± 5.6%; medium, 5.8 ± 2.3%; and large, 5.3 ± 0.9%; $P < 0.05$ vs control). In the presence of β-adrenergic blockade, the vasodilator effect of phentolamine was completely abolished. Propranolol caused constriction in all vessel groups (small, −3.6 ± 1.1%; medium, −4.8 ± 1.0%; and large, −3.5 ± 1.0%; $P < 0.05$ vs control). On the other hand, ICI-118 551 significantly constricted only vessels in the medium and large groups (large, −2.5 ± 0.6%; $P < 0.05$ vs control at the middle dose, medium, −3.1 ± 0.8% and large, −3.5 ± 1.3%; $P < 0.05$ vs control at the highest dose). These results led us to conclude that: (1) the vasodilation caused by phentolamine is mediated by the stimulation of β-adrenergic receptors, probably via the blockade of presynaptic $α_2$-adrenoceptors, (2) basal α-

---

[1] The First Department of Internal Medicine, Tohoku University School of Medicine, Sendai, Japan
[2] Department of Cardiology, National Sendai Hospital, Sendai, Japan

adrenergic tone is minimal in all size classes of coronary arterial microvessels, and (3) β_2-adrenergic tone plays a physiological role in the medium and large vessel groups, at least. The responses of the small vessel group may be modified by the autoregulatory response produced by the constriction of the medium and large vessels. The different effects of propranolol and ICI-118551 on coronary arterial microvessels, especially on the small vessel group, are thought to be related to reductions in metabolic demand.

Evidence of Myocardial Raynaud's Phenomenon in Patients with Systemic Sclerosis

Yoshio Ishida, Akihiro Tani, Takakazu Morozuki[1], Chincho Takijiri[2],
Hidetoshi Yamagami[3], Kazufumi Kimura, Masatsugu Hori, Akira Kitabatake[1],
Takahiro Kozuka[3], and Takenobu Kamada[1]

Raynaud's phenomenon and myocardial abnormalities are frequently observed in patients with systemic sclerosis (SS). Recent studies, utilizing thallium-201 (Tl-201) myocardial imaging during cold pressor stimulus, have shown reversible impairment of myocardial perfusion in these patients. In this study, we examined whether the impairment of myocardial perfusion during cold stress was linked to Raynaud's phenomenon in six female patients with SS and diffuse scleroderma (mean age, 44 ± 5 years. Serial thallium-201 myocardial single photon emission computed tomography (SPECT) imaging was performed during dipyridamole infusion (DI) (0.56 mg/kg i.v. for 4 min) and during cold stress (CS) (immersion of the hands into water at 15 °C for 2 min) over a 2-week period. Left ventricular Tl-201 activity, normalized for millicurie-injected dose, was determined in both studies, and the ratio of activity during CS to that during DI served as an index of the CS-induced change in myocardial perfusion. Within 1 week after the Tl-201 studies, contact plate thermography during and after CS was performed

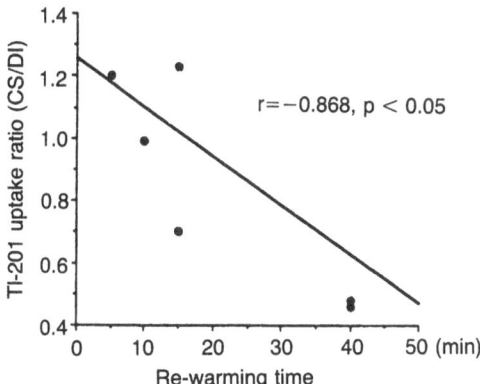

Tl-201 uptake ratio of CS to DI *vs* re-warming time

The First Department of Medicine[1], the Department of Dermatology[2], and the Department of Radiology[3], Osaka University School of Medicine, Osaka, Japan

and the re-warming time (80% recovery) was determined. All patients showed Raynaud's phenomenon after exposure to cold, although the severity differed among patients: the re-warming time was in the range of 5–40 min. The Tl-201 uptake ratio (CS/DI) was inversely correlated with the re-warming time ($r = -0.868$, $P < 0.05$), as shown in the Figure. Thus, the CS-induced decrease in myocardial perfusion was closely related to the severity of Raynaud's phenomenon. From these results, we concluded that, in patients with SS, the impairment of myocardial perfusion during CS is due to the same mechanism as that operating in Raynaud's phenomenon, this probably being microangiospasm, "myocardial Raynaud's phenomenon".

Abnormal Throughput Coronary Flow in Syndrome X

Takafumi Ueno, Akihiko Shiraishi, Kohji Hiyamuta, Hideki Yoshiyama, Akihiko Yamaga, Hidekatsu Maruyama, Masayoshi Yoh, Hisao Ikeda, Yoshinori Koga, and Hironori Toshima[1]

To elucidate the mechanism of ST segment depression in patients with syndrome X, myocardial blood flow and myocardial O_2 and lactate uptake were evaluated in 25 patients who had exhibited chest pain but had normal coronary arteriograms. Fourteen of the patients had chest pain and ST segment depression in an exercise stress test (SX), while the remaining 11 did not exhibit either of these symptoms (control). Atrial pacing induced significant ST segment depression in all patients with SX and did not do so in controls. There was no evidence of coronary artery spasm produced by intravenous administration of ergometrine maleate in any patients. Myocardial blood flow was measured with a Webster's thermodilution catheter. Coronary flow reserve was defined as the increase occurring in coronary flow after the administration of intravenous dipyridamole (0.56 mg/kg), divided by coronary flow at rest. After the administration of dipyridamole, the coronary flow reserve was 2.70 in the SX group and 2.90 in the controls; this difference was not significant. During atrial pacing, there was a reduction of myocardial lactate uptake and a rise in left ventricular end-diastolic pressure in the SX group, but not in the controls. Oxygen saturation in coronary sinus blood did not change significantly in the controls, while it rose significantly in SX patients. Consequently, the coronary A-VO$_2$ difference was reduced in the SX patients (from 7.0 to 6.3 ml (O_2)/dl; $P < 0.05$), while it remained at the same level in the controls (from 7.0 to 7.0 ml (O_2)/dl; NS). The balance of oxygen supply and demand was defined as the myocardial oxygen consumption divided by the rate-pressure product (MVO$_2$/RPP). During atrial pacing, MVO$_2$/RPP declined significantly in the SX group and did not change significantly in the controls. These findings suggested that the mechanism responsible for ST segment depression in patients with syndrome X might be myocardial ischemia, and a possible cause of myocardial ischemia in these patients might be increased throughput coronary flow, rather than impaired coronary flow reserve.

---

[1] Kurume University School of Medicine, Kurume, 830 Japan

Evaluation of Left Coronary Vascular Bed Reserve in Patients with Hypertensive Cardiac Hypertrophy

Hiromichi Sekiguchi, Hideyuki Fujikawa, Hideaki Mito, Hisao Watabiki, Toshio Yaginuma, and Takashi Natsume[1]

This study was designed to evaluate left coronary vascular bed reserve in hypertensive cardiac hypertrophy. The subjects were eight patients with hypertensive cardiac hypertrophy (HT) and nine patients with normal coronary flow (N). Coronary sinus flow (CSF) was measured by a continuous thermodilution method, and the ascorbic acid-platinum reaction was employed for the mean transit time measurement (T_0-T_2) of left coronary flow. T_0 is ascorbic acid injection into the left coronary orifice and T_2 is the maximal point of detecting ascorbic acid volume in the coronary sinus. Volumes for the coronary vascular bed (CVB) were obtained by multiplying CSF and T_0-T_2. CVB value in the HT patients was more than that in the control group at control heart rate, but was less than in N at a high pacing rate. CSF in the HT patients was more than that in the N group at any heart rate, and the CVB values per unit of myocardial mass were less in the HT patients than in the N group. The pressure rate product in the HT patients was more than that in the normal coronary flow group. The transit time of left coronary flow in the HT patients was less than that in the N group. The CVB in the HT group was more than that in the normal coronary group at control heart rate and less than that in the normal group at a high pacing rate. Cardiac work in the HT patient was thus shown to be greater than that in the normal group. These findings therefore suggest that the coronary vascular bed reserve in patients with hypertensive cardiac hypertrophy is lower than that in normal human subjects, particularly at high heart rates.

HR (beat/min)		Control	80	100	120
N	CSF	79.1	96.9	111.0	116.5
	CVB/W	0.92	1.06	1.08	1.08
	PRP	8640	10 754	13 500	13 525
HT	CSF	90.8	102.4	107.0	114.1
	CVB/W	0.71	0.72**	0.66**	0.68**
	PRP	11 122††	12 480†	16 813†	18 840†

PRP, Pressure rate product; *N*, wall thickness; $*P < 0.02$ *vs* N; $†P < 0.01$ *vs* N; $††P < 0.02$ *vs* N

---

[1] Department of Cardiology, Jichi Medical School, Tochigi, Japan

Reversal of Coronary Circulation Abnormalities After Relief of Pressure Overload in the Rat

Shogen Isoyama, Fumitoshi Sato, and Tamotsu Takishima[1]

We tested the hypothesis that impaired coronary autoregulation and decreased coronary flow reserve in hypertrophied hearts may regress after relief of pressure overload. We banded the ascending aortas of Wistar rats for 4 weeks. Debanding was then performed on some of these banded rats. Four weeks after banding or debanding, the chest was opened and in vivo left ventricular pressure was measured. The heart was isolated and perfused with Tyrode's solution containing bovine red blood cells (Ht, 30%) and serum albumin (15 g/l). The perfusate was oxygenated with a gas mixture containing 20% O_2, 3% CO_2, and 77% N_2. The left ventricular cavity was kept empty. Coronary perfusion pressure-flow relationships were obtained over a range of coronary perfusion pressures between 25 and 175 mmHg. Reactive hyperemic response was estimated after 40-s ischemia at 50, 100, and 150 mmHg perfusion pressure. In vivo peak systolic left ventricular pressure increased to 178 ± 8 mmHg in the hearts of the banded group (103 ± 6 in the hearts of sham-operated rats). The increased pressure produced significant myocardial hypertrophy (3.51 ± 0.16 *vs* 2.41 ± 0.05 mg/g heart weight/body weight ratio). After debanding, peak systolic left ventricular pressure decreased to the control level, and myocardial hypertrophy regressed to 15% above the control value. Coronary autoregulation was observed in control hearts, but was impaired in the hearts of the banded group over the whole range of perfusion pressures. Flow reserve (peak reactive hyperemic flow minus resting flow) and the duration of reactive hyperemia were also reduced in the hearts of the banded group. After debanding was carried out, the impaired autoregulation and the decreased flow reserve were completely normalized. These results indicate that abnormalities of the coronary circulation in pressure-overloaded cardiac hypertrophy are highly reversible after this pressure load is relieved.

---

[1] First Department of Internal Medicine, Tohoku University School of Medicine, Sendai, Japan

Paradoxical Response of Epicardial Coronary Arteries and Small Resistance Vessels to Intracoronary Acetylcholine in Vasospastic Angina

Akira Hirosaka, Yasuhiro Sakauchi, Yukihiko Abe, Kayano Kido, Masahiko Sato, Naohiko Watanabe, Naoyuki Awano, and Yukio Maruyama

This study was carried out to examine the response of epicardial arteries (A-epi) and small resistance vessels (A-res) to intracoronary (IC) acetylcholine (Ach) in patients with and without vasospastic angina. Injection of Ach (20, 50 μg) into the left coronary artery (LCA) was performed in ten patients whose LCAs were angiographically normal. Ach-induced spasm of the left anterior descending artery (LAD) occurred in five patients (group A), and did not occur in the other five (group B). We injected nitroglycerin (NTG; 0.2 mg) into the LCA to remove A-epi tone, followed by the same amount of Ach to evaluate the response of A-res. Thereafter, papaverine (10 mg) was injected into the LCA to evaluate the maximum vasodilation capacity. Flow in the great cardiac vein, measured by Webster's method, and coronary angiography were performed following the IC injection of each drug. We calculated the coronary vascular resistance ratio (R) to maximum dilator response and measured LAD diameter after each procedure. Percent changes in LAD diameter after IC Ach following IC NTG did not differ in the two groups. The R(NTG)-R(Ach) value was larger in group A than in group B (0.35 ± 0.1 *vs* 0.12 ± 0.1; $P < 0.01$). These findings suggest that IC injection of Ach produces significant vasodilation in A-res in vasospastic angina, in contrast to the constriction produced in A-epi by this treatment.

Fukushima Medical College, Fukushima, Japan

Keyword Index